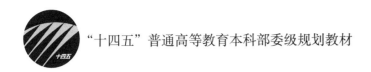

"十四五"普通高等教育本科部委级规划教材

服饰文化汉英翻译教程

肖海燕 编著

中国纺织出版社有限公司

内 容 提 要

中华服饰文化博大精深，在传承和弘扬中华优秀传统文化，增强国际传播能力的战略背景下，要让中国传统服饰文化和设计理念走向世界，离不开服饰领域的对外翻译与传播。本教程将翻译理论、翻译知识、翻译技能教学置于中国传统服饰文化语境，从词、句、段、篇多层面系统训练学生的服饰汉英翻译能力。教程包含丰富的传统服饰文化语料，从深衣到旗袍，从头饰到肚兜，全方位呈现中国传统服饰文化精髓，辅以系统的翻译知识与技能教学，旨在为服饰行业打造具有国际视野、一专多能的复合型人才，为中华优秀服饰文化对外传播提供教学资源。

本书可供服装服饰专业学生、服装类院校的英语专业生等用作教材，也可以供服饰行业的从业人员以及对服饰文化译介感兴趣的科研工作者参考。

图书在版编目（CIP）数据

服饰文化汉英翻译教程 / 肖海燕编著． -- 北京：中国纺织出版社有限公司，2022.9

"十四五"普通高等教育本科部委级规划教材

ISBN 978-7-5180-2132-1

Ⅰ．①服…　Ⅱ．①肖…　Ⅲ．①服饰文化 – 英语 – 翻译 – 高等学校 – 教材　Ⅳ．①TS941.12

中国版本图书馆 CIP 数据核字（2022）第 117074 号

责任编辑：宗　静　　特约编辑：韩翠娟
责任校对：寇晨晨　　责任印制：王艳丽

中国纺织出版社有限公司出版发行
地址：北京市朝阳区百子湾东里A407号楼　邮政编码：100124
销售电话：010—67004422　传真：010—87155801
http：//www.c-textilep.com
E-mail：faxing@c-textilep.com
中国纺织出版社天猫旗舰店
官方微博 http：//weibo.com/2119887771
三河市宏盛印务有限公司印刷　各地新华书店经销
2022年9月第1版第1次印刷
开本：787×1092　1/16　印张：15.5　彩插：8
字数：357千字　定价：68.00元

教学内容及课时安排

章/课时	课程性质/课时	节	课程内容
第一章 （4课时）	基础知识/4	●	服饰汉英翻译概论
		一	翻译的定义
		二	翻译的标准
		三	翻译的过程
		四	中国传统服饰文化译介
第二章 （36课时）	讲练结合 I /36	●	传统服装汉译英
		一	深衣
		二	胡服
		三	龙袍
		四	唐装
		五	传统裙装
		六	官服
		七	戎装
		八	旗袍
		九	中山装
第三章 （6课时）	讲练结合 II /16	●	传统服饰品汉译英
		一	头饰
		二	肚兜
第四章 （2课时）		●	丝绸汉译英
			丝绸汉译英
第五章 （4课时）		●	刺绣汉译英
			刺绣汉译英
第六章 （4课时）		●	新时代中国服饰汉译英
			新时代中国服饰汉译英

注　实际教学过程中，任课教师可以根据教学实际情况灵活调整教学内容和时间安排。

前言

　　服饰是记录人类跨向文明大门的具象符号，被誉为人类形体美的外延，可谓"人的第二皮肤"，承载着丰富的物质文化和精神文化内涵。中国是一个服饰大国，中华服饰文化博大精深，是中华民族几千年来智慧与技能的结晶，凝聚着中华民族独有的文化内涵。在加强国际传播能力建设和中国文化走出去的大背景下，要打造融通中外的服饰话语，让中国传统服饰文化和设计理念走向世界，离不开服饰领域的对外翻译与传播。

　　语言是人类文化的载体，也是人类交流思想的媒介。翻译作为最重要的语言活动之一，是文化跨语言、跨时空传播的桥梁。在中国文化走向世界的过程中，要让异域读者全面真实地了解和接受中国服饰文化，翻译发挥着至关重要的作用。中国服饰文化通过语言符号对外传播是我们主动与国际服饰话语体系接轨的重要途径之一。中国服饰文化对外译介既有助于中国传统服饰文化走出去，又有助于我国服饰行业真正做到与国际接轨，提高整个行业的国际化水平。因此，培养既有深厚的服饰专业知识与技能，又具有良好的翻译素养和跨文化交际能力的复合型人才，对于中国服饰文化的对外传播以及中国服饰行业的国际化发展有着举足轻重的作用。

　　服饰文化翻译教学和实践表明，翻译的重点在于语言的理解与表达。笔者针对服饰文化翻译难点，根据自己多年积累的教学材料与实例，结合长期从事服饰文化翻译教学与研究的心得，笔耕数载，终于成书。本教程把翻译理论讲授和翻译技能培养置于中国传统服饰文化语境中，将翻译理论、翻译知识、翻译技能的教学融入服饰文化汉英翻译实践，系统培养学生的服饰文化汉英翻译能力，打造与国际接轨、一专多能的复合型人才，为中国服饰行业国际化和中国服饰文化对外传播提供教学资源，贡献绵薄之力。

　　本教程作为服饰文化领域为数不多的汉英翻译教程之一，具有以下三个特点。

　　一、贴近行业。在国际化和中国文化走出去的大背景下，培养学生的翻译技能是培养"应用型"人才的重要内容，尤其对于与国际紧密接轨的服饰类人才而言，培养他们熟练使用外语进行服饰文化翻译的能力显得尤为重要。本教程尽可能从服饰类相关专业学生的实际出发来构思，语料以正式出版的服饰行业里较为权威的著作和期刊中的真实语料为主，全部语篇和段落语料都来源于中国传统服饰文化文本及其英译文，同时所有例句翻译语料也以服饰文化为主，内容涵盖深衣、龙袍等代表性的中国传统服饰。教材紧密贴合服饰行业，为学生构建起较为全面系统的中国传统服饰文化翻译语境。

　　二、授人以渔。本教程注重对学生实际翻译能力的培养，在语篇对译讲解的基础上，每章循序渐进翻译技能训练知识点，从句子到语篇、从技巧到策略，多层次地训练学生的翻译能力，内容包括：增译与减译、合译与分译、替换与重复、转换法、重构法、汉语流水句的翻译、汉语无主句的翻译、被动句的使用与翻译、形合与意合、异化与归化、习语及文化负载词的翻译、语篇翻译中的增删与重组、语篇翻译中的照应、连接与省略等。本教程密切结合各章服饰文化语料内容进行分析与讲解，让学生熟练掌握如何使用翻译技能在实践中进行服饰文化翻译。同时，教程在每章开篇精心设计并挑选与服饰文化翻译相关的"经典译言"，目的在于进一步启迪或激发学生对翻译的兴趣与思考。

三、重视实践。学习翻译要经历从无意识到有意识再到下意识的过程，学生看懂了书中的翻译知识与译例分析并不意味着就能做好翻译了，因为理解并不等于习得，关键在于实践。"纸上得来终觉浅，绝知此事要躬行"，翻译能力的提高还需要不断地在翻译实践中运用翻译知识，磨炼翻译技能。换句话说，要使学生通过翻译学习真正提升翻译技能与水平，必须突出翻译学习的实践性，使其在学习过程中做到举一反三。本教程在讲解各种翻译原理与技巧的同时，注意结合行业实践，提供精当的服饰文化翻译练习题及其答案。习题以服饰文化翻译为主要内容，包含句子翻译、段落翻译、语篇翻译三种题型，使学生能够从不同层面即学即练，及时巩固，在扎实的学练结合中逐步掌握翻译知识与技能，切实提高服饰文化翻译水平。

本书共分为六章。第一章介绍服饰文化汉译英的相关翻译理论与概念，奠定本书的理论基础。第二章至六章为中国服饰文化专题的翻译，内容涵盖深衣、胡服、龙袍、唐装、传统裙装、官服、戎装、旗袍、中山装、头饰、肚兜、丝绸、刺绣及新时代中国服饰等。每章内容包含经典译言、汉语原文、英语译文、词汇对译、译文注释、翻译知识、译文赏析、翻译练习、译笔自测、知识拓展十个部分的内容。每章节介绍一个与服饰文化相关的翻译知识，并辅以翻译技能，使翻译知识与技能训练形成完善的训练体系，并由浅入深系统培养学生的服饰汉英翻译技能。教程配句、段、篇翻译练习，每章节附一个汉英双语中国传统服饰文化点睛。此外，书末还以附录形式列出了较为重要的汉英服饰词汇对译。六章内容结合起来构成中国服饰文化英译系统，拆开来也可各自独立成为教学篇章，教师可以根据教学对象或教学条件进行选择、取舍或重组。

本书从选题论证到付梓出版历时三载，其间得到了诸多支持与帮助。首先，本书的出版要感谢北京市教委社科计划项目（项目编号：SM201910012004）和北京服装学院高水平教师队伍建设专项资金（项目编号：BIFTXZ202005）的资助。其次，为贴近服饰行业翻译实践，保证翻译语料的质量，本书在翻译讲解与训练中使用的语料基本选自国内服饰领域有代表性的权威著作及期刊。在此，向这些服饰著作及期刊的作者表示衷心的感谢！此外，本书得以顺利出版离不开北京服装学院文理学院的领导和同事们的大力支持；北京服装学院文理学院中外服饰文化研究中心的研究生王静雯、聂楠、苏敏和叶海燕参与了本书不同章节的语料收集与整理工作；中国纺织出版社的宗静女士为本书的出版付出了无数辛勤的汗水。在此，一并表示诚挚的感谢！

由于笔者学识和水平有限，书中难免有舛误和不足之处，恳请广大读者朋友批评指正。

肖海燕
2022年1月于北京

目录

第一章 服饰汉英翻译概论

学习目标：

1. 掌握翻译的定义及基本概念
2. 理解翻译的标准及基本要求
3. 了解翻译的过程及基本步骤
4. 了解中国传统服饰文化译介的现状及重要性

经典译言

译事三难：信、达、雅。求其信，已大难矣！顾信矣，不达，虽译犹不译也，则达尚焉。

——严复

翻译是人类社会普遍存在的一种语言活动和交流手段。自人类产生语言交流以来，翻译就伴随着人类历史的绵延而不断发展。谭载喜指出："在整个人类历史上，语言的翻译几乎同语言本身一样古老。两个原始部落间的关系，从势不两立到相互友善，无不有赖于语言和思想的交流，有赖于相互理解，有赖于翻译。"（1991：3）翻译不仅使人类各种语言与文化之间的沟通成为可能，还推动着人类文明向更高阶段发展。

据《周礼·秋官》记载："象胥，掌蛮夷闽戎狄之国使，掌传王之言而谕说焉，以和亲之。若以时入宾，则协其礼与其言辞传之。"（马祖毅，1998：2）可见，我国早在周代就已经设有翻译官职。季羡林曾指出，中华文化之所以能够常葆青春，很大程度上得益于历史上两次外来文化对中华文化的补充，一次是从印度来的佛经文化，另一次是从西方来传来的科学技术与文化。这两股洪流能够汇入中华文化的大河，翻译活动在其中起到了重要的媒介与促进作用。我国从东汉至唐宋的佛经翻译、明清之际的科技翻译、清末民初的西学翻译、"五四"以来的社会科学和文学翻译，一直到中华人民共和国成立以来各领域的翻译，在中国几千年的历史发展进程中，翻译一直发挥着重要作用。

21世纪的中国，翻译仍然发挥着重要作用。一方面，中国与世界其他国家和民族之间政治、经济、科学文化的交流与合作进一步加强，对国外先进技术和文化的引进是任何一个历史时期都无法比拟的。另一方面，改革开放以来，中国国力显著增强，中国在国际上的地位和影响日益提高。为了让世界更好地了解中国，需要将我们自身的优秀文化和科学技术介绍到国外去。从历史的角度来看，无论是文化输入还是输出，都离不开翻译这一重要的桥梁和纽带。

第一节　翻译的定义

英国学者诗人理查兹（Ivor Richards）曾将翻译称作"整个宇宙演进中最为复杂的活动之一"。（Richards, 1953: 250）一般来讲，我们所理解的翻译是指两种不同语言之间的翻译，即用一种语言文字解释另一种语言文字所传达的意义。从更广泛的意义来看，翻译还涉及更多的类型。英国著名语言学家雅各布森将翻译划分为三种类型：语际翻译（interlingual translation）、语内翻译（intralingual translation）、符际翻译（inter-semiotic translation）。其中，语际翻译就是上面提及的两种不同语言之间的翻译；语内翻译是指同一语言中用一些语言符号解释另一些语言符号，如将古代汉语译为现代汉语；符际翻译是指用一种符号去解释另一种符号。本书中翻译指两种不同语言之间的翻译，即从汉语翻译为英语。

关于什么是翻译，翻译学界不同学者有不同的定义。"翻译"一词既可以指翻译活动的主体，即译者；也可以指翻译的行为和过程；还可以指翻译活动的结果，即译文。《新牛津英语词典》（the New Oxford Dictionary of English）将"翻译"定义为："express the sense of（words or text）in another language"，即用另一种语言表达（词语或文本）的意义。《汉语大词典》则将翻译定义为：把一种语言文字的意义用另一种语言文字表达出来。可见，翻译是用另一种语言文字表达一种语言已经表达的意义。英国翻译理论家彼得·纽马克这样定义翻译："it is rendering the meaning of a text into another language in the way the author intended the text"，即翻译就是把一个文本的意义按原作者所意想的方式移入另一种文字。（Peter Newmark, 2001）美国翻译理论家尤金·奈达认为："Translation consists in reproducing in receptor language the closest natural equivalence of the source language, first in terms of meaning and secondly in terms of style."即翻译是指在译语中用最切近而又自然的对等语再现原语的信息，首先在语义上，其次是在文体上。（Eugene A. Nida, 1982: 12）苏联翻译理论家巴尔胡达罗夫（Barkhudarov）认为，翻译是把一种语言的言语产物，在保持意义不变的情况下，改编为另一种语言的言语产物的过程。（巴尔胡达罗夫，1985）

我国翻译研究著名学者王克非教授这样定义翻译："翻译一词兼有动词、名词意义。作动词时，它指的是翻译行为，即把一种语言的信息用另一种语言表达出来的行为；作名词用时，它可以抽象地指称翻译行为的过程，也可以指翻译行为的产物即译品。"（王克非，1997）冯庆华认为："翻译是许多语言活动中的一种，它是用一种语言形式把另一种语言形式里的内容重新表现出来的语言实践活动。翻译是一门艺术，是语言形式的再创造。"（冯庆华，2002）张今认为翻译是两种文化沟通、互渗的过程，他从文学翻译的角度将翻译定义为："翻译是两个语言社会（language-community）之间的交际过程和交际工具，它的目的是促进本语言社会的政治、经济和（或）文化进步，它的任务是要把原作中包含的现实世界的逻辑映像或艺术映像，完好无损地从一种语言移注到另一种语言中去。"（张今，1987：8）

总之，翻译的再创作过程极其复杂，涉及原作者和译者双方的语言、文化、社会诸因素，迄今为止，似乎还没有人给翻译下一个令所有人都接受的定义。不管采用哪种定义，翻译作为一项跨语言跨文化的交流活动，承担着维护文化多元性、语言多元性，促进各民族文化良性互动的使命。

第二节　翻译的标准

　　翻译标准是衡量译文的尺度，是译者在翻译过程中要努力达到的目标。英国翻译家亚历山大·泰特勒早在18世纪就提出了著名翻译三原则，即：一、译文应该完全复写出原作的思想；二、译作的风格和手法应与原作属于同一性质；三、译作应该具备原创作品的通顺。（Lefevere, 1992：128）在中国，严复在1898年提出了"译事三难：信、达、雅"，其中"信"就是忠实于原文的本意，"达"就是通顺流畅，"雅"就是文字古雅。我国翻译界对这个"雅"字一直存有异议，20世纪30年代初，林语堂提出了"忠实、通顺和美"的三原则。后来，刘重德将严复的翻译三原则改为"信、达、切"，以符合时代的发展。其中，"切"指切合原文的风格。1954年，茅盾在全国文学翻译工作会议上指出，"对于一般翻译的最低限度的要求，至少应该是用最明白畅达的译文，忠实地传达原作的内容。"翌年，陈允福在讨论翻译标准的文章中提出将"忠实"和"通顺"作为翻译的总标准，这一提法受到了翻译教学界的支持，张培基编写的《英汉翻译教程》和孙致礼编写的《新编英汉翻译教程》都将"忠实"和"通顺"作为翻译标准。

　　事实上，对初学翻译的人来说，较为合适的翻译标准主要为"忠实"和"通顺"。"忠实"主要指内容，翻译是在理解别人用某种语言表达的意思之后，把同样的意思用另一种语言表达出来。即译者的任务是表达别人的思想，而不是自己进行创作。因此，译者要力求准确表达原作者的思想。在翻译过程中，由于语言文化的差异，往往难以百分百地传达原文的意义或信息，但译者不可因此降低"忠实"这一标准，而应该尽最大努力，全面传达出原文的意义或信息。"通顺"是指语言。如果原文是通顺易懂的，那么译文也应该做到通顺、易懂。真正做到这两点并不容易。尤其是汉译英，要把中国的事物用英语写出来，让外国人看懂，做到用词正确，句子平稳，合乎英语用法，不是一件轻而易举的事。一般来说，译文应该使用明白晓畅的现代语言，不可逐字逐句地硬译，否则难以达到"通顺"的标准。

　　在翻译过程中，"忠实"标准和"通顺"标准两者相辅相成，不可分割。如果只注重"忠实"，读者则可能看不懂，或看不下去，这样"忠实"也就失去了意义；如果只注重"通顺"，则有可能歪曲了原文的信息，读者能读下去，却没法了解真实的原文，这样"通顺"也就失去了意义。对初学者而言，要把握好分寸，做到忠实、通顺兼顾，才可以说基本达到了翻译标准的要求。下面对汉英翻译中的"忠实"和"通顺"原则分别举例说明。

　　例1　原文：旗袍作为一种光滑、合身的服饰，展现了穿着者内在的优雅，体现了中国女性的身份认同与文化传承。

　　译文：The *qipao*, a sleek and fitting dress, brings out the inner elegance of the wearer and embodies the identity and cultural heritage of Chinese women.

　　【分析】原文描述旗袍这一中国传统服饰的特点，先说明面料和形制特点，然后说明其对穿着者气质、身份、文化方面的体现。译文忠实再现了原文的信息内容，主要体现在两方面。其一,原文主语"旗袍"是中国传统服饰的代表之一,译文采用音译法,将其译为"*qipao*",原汁原味地呈现给英语读者,忠实再现了这一具有中国服饰特色的文化意象。其二,在信息表述顺序方面,译文完全依照原文信息块的出现顺序,先译出旗袍的面料和形制特点,然后再译旗袍对穿着者气质、身份、文化方面的体现,忠实再现了原文的所有信息,充分体现了汉译英的忠实性原则。

例2 原文:初兴的式样是一种蓝布旗袍,袍身宽松,廓型平直,袍长及踝,领、襟、摆等处不施镶滚,袖口微喇,看上去严冷方正。

译文: The early prevailing style was a kind of blue cotton *qipao* with loose body, straight and flat outline and bell-mouthed sleeves. The dress, which looked very serious and formal, was ankle-long with no edgings or lacework in collar, front garment piece and hem parts.

【分析】原文描述蓝布旗袍的形制特征,详细说明了蓝布旗袍的松量、廓型、长度、装饰、袖口以及整体观感。全由七个流水分句构成,流泻铺排,形散神聚,一气呵成,体现了汉语意合语言的特点。英语是形合语言,英语语法要求句子焦点突出,各成分之间层次分明,逻辑清楚,这样才称得上通顺。例句译文为顺应翻译的"通顺"性要求,对原文句子结构、信息表述顺序等方面进行了调整。译文首先进行了分译,将原文的一句拆为两句,先说明蓝布旗袍的松量、廓型、袖口方面的细节。然后,另起一句说明蓝布旗袍的长度、装饰以及整体观感。分译使英译文表意清楚,减轻了每个句子结构容量方面的负担。译文第一句介绍蓝布旗袍细节时使用了"with..."这一介宾结构,作后置定语,使句子表意明确,逻辑清晰;第二句则使用"which"引导非限定性定语从句修饰先行词"the dress",使句子焦点突出,层次分明。可见,译文两句在通顺原则的指导下,对原文进行了适当调整和灵活处理,在充分传递原文信息的同时,做到符合英语语言表达规范,形成了通顺易懂的译文。

第三节 翻译的过程

翻译需要遵循一定的科学程序。如果拿到原文便急于求成,还没看懂原文就提笔开始翻译,则很难译出好的译文。对于翻译过程,译界不同学者有不同的划分。比如,尤金·奈达从语法和语义角度将翻译过程分为分析(analysis)、传译(transfer)、重组(restructuring)、检验(testing)四步。乔治·斯坦纳从解释学角度将翻译过程分为信赖(trust)、侵入(aggression)、吸收(incorporation)、补偿(restitution)四步。国内针对初学翻译者,一般将翻译过程分为理解、表达、检验三步,下面将对此三步骤分别予以讲述。

一、理解

理解是表达的前提,没有准确透彻的理解,表达就无从谈起。对原文本理解的好坏直接影响到翻译的质量,翻译中的大多数失误都是由于理解这一步没有把握好。理解可分为广义理解和狭义理解,其中广义理解指对原文内容、原文作者、原文产生的时代背景、原文读者等方面的理解;狭义理解仅指对原文文本的理解,包括语法分析、语义分析、语体分析和语篇分析。翻译过程的起点就是对原文现象的理解。在翻译实践中，译者首先必须联系上下文，从原文的语言现象入手，理解词汇含义、句法结构和惯用法，其次还要注意语言的文化背景、逻辑关系和具体语境。下面举个例子说明由于理解不到位而导致的译文失误。

例3 原文：Grading is the method used to increase or decrease the sample-size production pattern to make up a complete size range.

译文：为了做出一个完整的尺寸范围，人们采用尺寸划分，就能增加或减少生产纸样的样本尺寸。

【分析】从原文的结构中可以看出，这句话主要是给Grading下定义，显然译文没有精确译出定义，其根源在于译者对"grading"一词的理解有误。查阅服饰类专业词典，可以明确该词意为放大或缩小（裁剪样本）、推档、放码等意思，由此可以清楚地看出本句就是介绍推档的定义。译者显然对服饰行业不了解，同时没有注重理解句中的关键词汇，没有结合上下文语境和行业语境，借助有效的工具加深对原文的精确理解，从而导致译文表达的意思与原文相距甚远。

二、表达

翻译过程中，理解不是目的，而是翻译过程的开始。译者在透彻理解的基础上，还必须将自己的理解传达给读者。表达是译者把自己所理解的内容，忠实而通顺地传达给译文读者，是理解的深化和体现。在翻译过程中，译者应以忠实、通顺为己任，在表达中力求做到：一是对原文意义或信息的传达要尽可能准确、充分，二是译文要尽可能自然、通顺。

翻译过程中，表达可以很简单，如杨绛所说，"就三件事：一是选字，二是造句，三是成章"（杨绛，1986）；表达也可以很复杂，涉及所有的翻译技巧。表达的核心问题是结构重组，即组织译文中的词汇、句法和语篇特征，使读者能够最大限度地理解和领会译文。这一过程中，译者要恰到好处地再现原文的思想内容和语体色彩，使译文既忠实于原文又符合译入语的语法和表达习惯。因此，译者要在选词用字、组词成句、组句成篇上下功夫，灵活应用翻译技巧，将自己对原文的理解忠实而通顺地传递给译文读者。比如：

例4 原文：男耕女织成了这一时期的重要经济特征，种植桑麻，从事纺织是一种典型的社会经济图景。

译文：One important feature of the economy at that time was the division of labor between men and women, who were engaged separately in farming and weaving. Planting of mulberry trees and weaving of textiles were typical scenes of economic activity of that time.

【分析】原文例句由三个流水分句构成，呈现汉语形散神聚的特点。译文在表达方面根据英文表意的需要进行了分译，将原文拆分为两个句子，分别对当时的经济特征和经济图景进行说明，使译文结构清晰，表意明确。此外，为使译文读者更好地了解"男耕女织"的含义，译者没有硬译为"men plough and the women weave"，而是采用解释性表达，并且以who作为关系代词引导非限制性定语从句，对男女之间的分工劳动作进一步说明，符合英语语言的表达习惯。封建社会的小农经济，男人耕田、女人织布是普遍的社会分工，译文将原文隐含之意展现在读者面前，具有明晰化的效果。

三、检验

检验是确保译文质量的最后一步，其目的在于检查译文是否精确、自然、简练。一般来说，译文需要至少检验两遍：第一遍对照原文校对，检查有没有疏漏、误译之处；第二遍抛开原文，检查有没有生硬拗口之处。翻译过程中，译者由于疏忽、疲劳等原因会造成漏译，在检验时要对照原文，逐词、逐句、逐段地检查译文，将漏译的地方补译出来。翻译实践中，译者难免偶尔出现差错，检验步骤就是为了发现误译并纠正误译。此外，译者还要特别注意检查译文是否精练，在不影响译文表达的前提下，尽量删去可有可无的词语，选用最精练的表达方式，使译文晓畅自如。如因表达不当使得译文生涩，译者需要进一步揣摩原文，在彻底理解原文的基础上重新表达，尽量做到文从字顺，通达自如，消除翻译腔。总之，检验是翻译过程的一个重要环节，是确保译文质量的重要关口。

例5 原文：从此，传统的满装旗袍又被注入了时代的血液，赋予了青春的活力。

译文：Since then, the traditional full-style *qipao* has rejuvenated the blood of the times and been given youthful vitality.

【分析】对翻译过程而言，在理解原文的基础上，用译文将原文内容表达出来，貌似就已经完成了翻译任务。但对于高质量的翻译而言，检验是必不可少的一步。例5译文貌似从内容上忠实传递了原文的信息，从语言形式上也最大限度地再现了原文的结构特点。但倘若略掉检验这一步，则会出现较大问题。因为，对照例5原文仔细检查，会发现译文中出现了一个明显的错误。原文中的"满装"指的是起源于20世纪20年代的中国满族服饰，又称"满服""旗装""旗服"，译文以"full-style"来译属望文生义，是误译。"满装"的正确译法应该是"Manchu clothing"。没有检验这一步，再好的译者都可能会在译文中留下错误。检验不仅可以发现和修正译文中的错误，有时还可以通过检验对译文进行进一步打磨与推敲，形成更高质量的译文。

第四节　中国传统服饰文化译介

中华民族历来享有"衣冠古国"的盛誉。中华服饰文化历史悠久，是近千年来中华民族智慧与技能的结晶，凝聚着中华民族独有的思想内涵。随着中国综合国力提升与国际地位日益提高，中国"软实力"逐步增强，文化"输出"已经成为国家战略。在世界文化格局全球化的大背景下，各个区域拥有着各具特色的思维方式，各民族的传统文化内涵与使用语言的方式存在较大的差异。语言差异在某种程度上造就了不同文化间根深蒂固的文化冲突。翻译作为化解文化冲突的手段，是解决文化间差异的主要途径。古往今来，翻译是各国各民族交流活动的桥梁，不可或缺，可以说凡有表述，必有翻译。在中华服饰走向世界的过程中，要让中华服饰文化更真实更全面地被了解和接受，服饰文化翻译发挥着无可替代、至关重要的作用。

中国是个服装大国，产量和销量均居世界前列。纺织服装行业的迅猛发展和国际化脚步的加快，使对高级服饰翻译类人才的需求日益增长，既懂服饰又懂翻译，能够从事服饰外贸和翻译工作的人才尤为急需。然而，懂翻译的人对服饰知识知之甚少，懂服饰的人往往不一定能熟练地进行翻译，因而培养具有服饰翻译能力的人才变得十分重要。此外，当前我国在世界服饰行业中依然缺乏足够的话语权。要吸取借鉴世界服饰业的先进理念和行业经验，同时让中华民族的服饰文化和设计理念走向世界，都离不开服饰翻译。另外，服饰翻译不同于文学翻译，它是一种专门用途语言翻译，广义上属于非文学翻译范畴。它既遵循翻译的一般规律，又有其自身的特点，融合了面料、款式、制作工艺、服饰设计等专业知识，具有很强的行业翻译特点。随着全球经济的发展和各国之间交流的增多，翻译活动分工更细，越来越专业化和领域化。

服饰是无声的语言，展现着人类身体的美，表达着人类的思想。服饰翻译是跨语言跨文化甚至跨时空的交际行为，受语言、文化、社会等多层次语境的限制，是一个复杂的选择顺应过程。本书将翻译理论的讲授和翻译技能的培养置于中国传统服饰文化语境中，运用服饰文化语料详细讲述翻译技巧，将翻译理论、翻译知识、翻译技能的教学融入服饰文化汉英翻译实践，系统培养学生的服饰文化汉英翻译能力，为服饰行业的国际化打造一专多能的复合型人才，为中国传统服饰文化走出去提供教学资源。

本书共六章，第一章介绍服饰文化汉译英的相关翻译理论与概念，奠定本书的理论基础；第二章为九个不同类别中国传统服装的专题翻译，包含深衣、胡服、龙袍、唐装、传统裙装、官服、戏装、旗袍以及中山装；第三章为中国传统服饰品的专题翻译，包含头饰和肚兜；第四章为具有中国传统特色的面料——丝绸的专题翻译；第五章为具有中国传统特色的工艺——刺绣的专题翻译；第六章为新时代中国服饰专题翻译。章节内容包含经典译言、服饰专题语篇汉语原文、英语译文、词汇对译、译文注释、翻译知识、精彩段落译文赏析、翻译练习、译笔自测、知识拓展——中国服饰文化点睛（双语）十个部分的内容。介绍与服饰文化汉英翻译相关的翻译知识，并辅以翻译技能训练，使六章的翻译知识与技能训练形成完善的翻译知识体系，并由浅入深系统训练学生的服饰汉英翻译技能。教程有配套的句、段、篇翻译练习及其答案，每章节附一个中国传统服饰文化点睛，拓展学生对该服饰专题理解的深度和广度。此外，本书末以附录形式列出常用的汉英服饰词汇对译表。

第二章 传统服装汉译英

第一节 深衣

学习目标

1. 了解深衣的历史及文化内涵
2. 熟悉深衣的基本语汇及其汉英表达方式
3. 掌握增译法和减译法的相关知识与技能
4. 通过实践训练提高深衣文化汉英翻译能力

一、经典译言

以效果而论，翻译应当像临画一样，所求的不在形似而在神似。

——傅雷

二、汉语原文

深衣

古代中国人在相当长的时间里都采用上衣下裳的着装形制，认为这种服饰结构象征着天地秩序，郑重场合时穿用的礼服大多如此。但与此同时，也向来不乏上下连属的服式，从战国时期的深衣、始于汉代的袍服、魏晋的大袖长衫，一直到近代的旗袍，都是属于长衣样式。中国服装也因此呈现出两种基本形制。

深衣由上衣下裳连接而成，裁剪制作自有特点，与其他衣服不同。《礼记》中专门设了一章，题目就叫"深衣"，详细介绍深衣的制作。主要内容大致如下：战国时期，深衣的样式是符合礼仪制度的，它的造型既合乎规矩，有圆有方，又对应均衡；尺寸上也有一定要求，短不能露肤，长不能拖地，前襟加长，呈一个大三角形。腰上为衣料的直幅，腰下取衣料的斜幅，以便于举步；衣袖的腋部要能够适于肘的活动，袖的长短大约是从手部再折叠回来时恰到肘部。深衣既可以文人穿，也可以武士穿；可以做侯相时穿，也可以行军打仗时穿。深衣属于礼服中的第二等，功能完备且不浪费资财、风格上也朴实无华。这一时期着深衣的形象，可以从一些出土于古墓的帛画上看到，同一时期的陶俑、木俑也有不少这类人物形象，不仅款式清楚，而且花纹历历可见（图2-1）。

深衣的材料多为白色麻布，祭祀时则用黑色的绸，也有加彩色边缘的，还有的在边缘上绣花或绘上花纹。穿深衣时，将加长的呈三角形的衣襟向右裹去，然后用丝带系在腰胯之间。这种丝带被称为"大带"或"绅带"，带子上根据需要可插笏板，笏板并非仅供大臣上朝时使用，还相当于记事用的便携笔记本。后来随着游牧民族服饰对中原人的影响，革带出现在了中原地区的服饰中。革带再配上带钩，用作系结。带钩做工精致，已成为战国时期新兴的工艺美术品种之一。长的带钩可以达到30厘米左右，短的也有3厘米。石、骨、木、金、玉、铜、铁等质料应有尽有，奢华的带钩镶金饰银，或雕镂花纹，或嵌上玉块和琉璃珠。❷

前

　　译前提示：深衣作为汉族文化的代表，历史悠久，雍容典雅。作为一种服饰制度，深衣在礼法慎重的中国传统社会产生了广泛而深远的影响。本文概括了中国的几种传统的上衣下裳服饰样式，着重介绍了深衣的裁剪、长短、功能、穿着对象等，对深衣进行具体详尽的描述。深衣的材料以及装饰也是其重要特征，作者的精彩描绘为我们呈现了一幅栩栩如生的深衣画卷。

后

深衣前后片示意图（臧迎春提供）

图2-1　深衣前后片示意图❶

三、英语译文

In ancient China, for quite a long time, the dress style of upper and lower garments was adopted, because people believed in its symbolism of the greater order of heaven and earth and wore it on important occasions. In the meantime, one-piece style co-existed, starting from the *shenyi* of the Warring States Period, the robe of the Han Dynasty, the large-sleeved *changshan* of the Wei and Jin Period, down to the *qipao* of the contemporary times, all in the form of the one-piece long robe. Therefore, Chinese garments took the above-mentioned two basic forms.

The *shenyi* is made up of the upper and lower garments, tailored and made in a unique way. There is a special chapter in the *Book of Rites* detailing the make of the *shenyi*. It is mainly explained in the book that in the Warring States Period, the look of the *shenyi* conformed to the rites and rituals. Its style fit for the rules with the proper square and round shapes and the perfect balance. It has to be long enough not to expose the skin, but short enough not to drag on the floor. The forepart is elongated into a large triangle. The upper part and the lower part of *shenyi* are cut separately, with the part above the waist in straight cut and the part below the waist bias cut for ease of movement. The underarm section is made for flexible movement of the elbow, therefore the general length of sleeves reaches the elbow when folded from the fingertips. Moderately formal, the *shenyi* is fit for both men of letters and warriors, and both ministers in court and soldiers in war as well. It ranks

---

❶ 图片来源：华梅. 中国服饰［M］. 北京：五洲传播出版社，2004：12.
❷ 语料参考：华梅. 中国服饰［M］. 北京：五洲传播出版社，2004：9–11.

second in ceremonial wear, functional, not wasteful and simple in style. *Shenyi* of this period can be seen in silk paintings unearthed from ancient tombs, as well as on clay and wooden figurines found in the same period, with clear indications of the style and often even the patterns.

Material used for making *shenyi* is mostly white linen, except for the black silk employed in the garments for sacrificial ceremonies. Sometimes a colorful decorative band is added to the edges, which are sometimes embellished with embroidered or painted patterns. When *shenyi* is put on, the elongated triangular hem is rolled to the right and then tied right below the waist with a silk ribbon. This ribbon was called *dadai* or *shendai*, on which is attached a *huban* that is not only used for the ministers to go to court, but also for them to take down notes. Later on, leather belt appeared in the garment of the central regions as an influence of nomadic tribes. A belt buckle is normally attached to the leather belt for fastening. Belt buckles are often intricately made, becoming an emerging craft in the Warring States Period. Large belt buckles can be as long as 30 centimeters, whereas the short ones are about 3 centimeters in length. Materials can be stone, bone, wood, gold, jade, copper or iron, with the extravagant ones decorated with gold and silver, carved in patterns or embellished with jade or glaze beads. ❶

四、词汇对译

上衣下裳	upper and lower garments	腋部	underarm section
直幅	straight cut	指尖	fingertips
斜幅	bias cut	挖掘	unearth
文人	men of letters	帛画	silk paintings
武士	warriors	木俑	wooden figurines
游牧民族	nomadic tribes	陶俑	clay figurines
天地秩序	order of heaven and earth	祭祀	sacrificial ceremonies
大袖长衫	the large-sleeved *changshan*	革带	leather belt
上下连属	one-piece style	战国时期	the Warring States Period
近代	contemporary times	琉璃珠	glaze beads
《礼记》	the *Book of Rites*	带钩	belt buckle
汉代袍服	the Han Dynasty robe		

五、译文注释

1. 但与此同时，也向来不乏上下连属的服式，从战国时期的深衣、始于汉代的袍服、魏晋的大袖长衫，一直到近代的旗袍，都是属于长衣样式。

In the meantime, one-piece style co-existed starting from the *shenyi* of the Warring States Period, the robe of Han Dynasty, the large-sleeved *changshan* of the Wei and Jin Period, down to the

❶ 语料参考: Hua Mei. *Chinese Clothing*［M］. 北京: 五洲传播出版社, 2004: 9–11.

qipao of the contemporary times, all in the form of a long robe in one piece.

【注释】本例句介绍了中国历代上下连属服饰。上衣下裳是我国古人最早创制的服装样式。春秋战国之交，开始出现了上下连属样式的服装，即"深衣"。《礼记·深衣》郑注："深衣，连衣裳而纯之以采者。"钱玄《二礼名物通释·衣服·衣裳》也称："古时衣与裳有分者，有连者。男子之礼服，衣与裳分；燕居得服衣裳连者，谓之深衣。"例句原文用"从……到……"的结构来体现深衣的历史，对应的英文句式为"from...to..."。译文将其增译为"starting from...down to..."，对上下连属服饰的发展进行更为生动的描述，使译文的逻辑更加严谨，从而能够让译文读者对上下连属服饰的发展脉络有较清晰的理解。同时，译文省略了对原文中"也向来不乏"的翻译，将其含义融入下文关于上下连属服饰由远而近的发展时间顺序中，使译文显得更加简洁明了，表意清楚。

2. 深衣由上衣下裳连接而成，裁剪制作自有特点，与其他衣服不同。

The *shenyi* is made up of the upper and lower garments, tailored and made in a unique way.

【注释】该句表达的核心内容是深衣独特的剪裁方式，译者翻译此句的过程中采用了"减译法"。例如，译文中减译了"连接而成"一词，直接使用"is made up of"对深衣上衣下裳的结构进行描述；在表达其裁剪方式时，原文既指出深衣独具特色，又将其与其他服装进行比较，实际上突出的都是深衣裁剪制作的独特性。在译文中，作者直接使用"unique"一词对其裁剪进行概括，既能够体现其裁剪特别，又包含深衣区别于其他服饰的含义。译者减译了原文中的"与其他衣服不同"，使译文精炼而不显累赘，高度凝练地表达了原文内容。

3. 衣袖的腋部要能够适于肘的活动，袖的长短大约是从手部再折叠回来时恰到肘部。

The underarm section is made for flexible movement of the elbow, therefore the general length of sleeves reaches the elbow when folded from the fingertips.

【注释】该句介绍了深衣衣袖的功能与长度。译者用介词"for"表示目的，加强了衣袖腋部的功能表述，突出了深衣衣袖要适应肘部活动的特点；原文在衣袖功能与长度之间并未使用衔接词体现两者的逻辑关系，而译文增添了副词"therefore"来体现前后两个分句间的因果关系。翻译过程中，译者通过恰当增译词语将衣袖功能与长短联系起来，使译文语句逻辑通顺、流畅自然。

4. 深衣既可以文人穿，也可以武士穿，可以做侯相时穿，也可以行军打仗时穿。

Moderately formal, the *shenyi* is fit for both men of letters and warriors, and both ministers in court and soldiers in war as well.

【注释】该句介绍深衣的穿着对象，说明了深衣的两类穿着人群和两种使用场景。为帮助读者更好地理解深衣样式，译文增译了状语"moderately formal"，为介绍深衣的适用对象做铺垫。此外，译文运用两个"both...and..."结构使译文结构紧凑，且与增译的"moderately formal"形成照应，表意清楚明确。在翻译过程中，我们在准确传达原文语义时，要对译文进行恰当调整，按照不同情况进行增译或减译，使译文做到通顺流畅。

5. 这种<u>丝带</u>被称为"大带"或"绅带"，带子上根据需要可插笏板。

This ribbon was called *dadai* or *shendai*, on which is attached a *huban* that is not only used for the ministers to go to court, but also for them to take down notes.

【注释】该句对穿着深衣时所用的丝带进行了介绍。将"大带"音译为"*dadai*","绅带"译为"*shendai*","笏板"译为"*huban*"是典型的异化翻译。译者为了保留中华文化的异域色彩,将中国传统服饰用音译表达,能够让英美读者更好地体验译名所蕴含的异国风情。这些译名的文化含义在上下文中有清楚的解释,其中"*dadai*"和"*shendai*"是丝带的别称,"*huban*"则是中国传统文化中一个重要的文化意象,又称手板、玉板或朝板,是古代臣下上殿面君时的工具。考虑到英美读者可能对笏板并不熟知,译文使用后置定语从句将原文对于笏板的解释完整呈现出来。这种附带释义的异化译法,使译文读者能够在轻松理解译文的同时,感受中国服饰的异域风采。

6. 石、骨、木、金、玉、铜、铁等质料应有尽有,奢华的带钩镶金饰银,或雕镂花纹,或嵌上玉块和琉璃珠。

Materials can be stone, bone, wood, gold, jade, copper or iron, with the extravagant ones decorated with gold and silver, carved in patterns or embellished with jade or glaze beads.

【注释】由于中国人特有的语言习惯,汉语以意统形,是散点句法,多用流水分句。如例6的原文为四个形式上较为独立的分句,共同说明带钩的材质与装绘。虽然形式上看,有的分句成分残缺,结构并不完整,但在汉语语境下表意仍然一气呵成,并无任何不妥。英语是焦点句法,句子有完整的主谓,句中各成分之间都要保持语法上的一致性。译文以带钩材料"materials"作主语建构起完整的主谓结构,然后以"with+名词"的独立结构来描写带钩上的豪华装饰,使译文结构焦点突出,主次分明,体现出英语语言的形合特点。

六、翻译知识——增译与减译

在翻译过程中,增译法和减译法是最常用的两种翻译方法,它们看似互相排斥实则却又相辅相成、互为补充。由于英汉语言在词法、句法与修辞上存在较大差异,在翻译时如果照原文词句直接搬至译文中,在很大程度上会影响译文通顺度与可读性。翻译时,译者应根据具体情况对信息进行增减,确保语言交际有效进行。

(一)增译

增译法指根据英汉两种语言不同的思维方式、语言习惯和表达方式,在翻译时增添一些词、短句或句子,以更准确地表达出原文所包含的意义。

增译法多在汉译英过程中使用,这是由汉英两种语言之间的差异决定的。就语法而言,汉英翻译中需要增译的原因很多,比如汉语有较多无主句,而英语句子要求结构完整,一般都有主语,因而在汉语无主句译为英文时,往往需要根据语境补出主语。另外,英语句子的逻辑关系一般用连词来表示,而汉语往往通过上下文和语序来表示这种关系。因此,在汉译英时要适当增补连词。再者,英汉两种语言在词汇的使用方法上也存在很大差别。比如,英语中代词使用频率较高,在说明某物归某人所有或与某人有关时,必须在该事物前面加上物主代词,因而在汉译英过程中有时需要增补物主代词。英语句子大量使用介词和冠词,汉译英时需要适当增补这两类词语。最后,增译有时还是上下文的需要,在汉译英时有时需要增补一些原文中没有明言的词语或一些概括性、注释性的话语,以确保译文意思完整,更充分地传递原文信息。

总而言之，使用增译方法的好处首先在于提高译文语法结构的完整性，其次还可以增强译文表意的明确性。下面举例进行说明。

例1　原文："深衣"，从字面上看，就是用衣服把身体深深地遮蔽起来。

译文：*Shenyi*, or deep garment, literally means wrapping the body deep within the clothes.

【分析】本句介绍深衣的中文字面意义。将深衣音译为"*shenyi*"是典型的异化翻译，保留了中国传统服饰的独特魅力。但在译文中，译者增译"deep garment"一词作为插入语，阐释深衣在英文中对应的字面意义，属于归化翻译。这一译法让译文读者从英文词义的角度理解深衣内涵，有利于英语读者迅速领会原文意义，对其学习下文中关于深衣的背景知识有较大帮助。

例2　原文：深衣是连体式的长衣服，它最大的特点是把左边的衣襟缝接出一片三角形。

译文：*Shenyi*, a one-piece long gown with a piece of triangle cloth sewn to its left front, was popular during the pre-Qin Dynasty Period.

【分析】该句对深衣形制进行介绍。在译文中，译者并未按照原文句式翻译而是采用了增译的方式在译文中增加了对深衣流行时代背景的介绍。译者将深衣的特点作为插入语置于主谓之间，将其形制与构成组合为一句，并且根据上下文语境，在句末增加了对深衣流行时期的补充说明。这样的增译方法有助于译文读者加深对深衣形制特征背景的认识，让对中国历史与文化不熟悉的英语读者更好地理解"深衣"的总体特色，体现了以读者为中心的翻译策略。

例3　原文：春秋战国时期的衣着，上层人物的宽博、下层社会的窄小，已趋迥然。

译文：As far as the clothing of the Spring and Autumn-Warring States Period is concerned, there is an obvious difference between the loose and comfortable dress of upper circles and the narrow and fit dress of the lower class.

【分析】该句原文对春秋战国时期不同阶层衣着差异进行了介绍。中文原文的语序比较自由、分散，如果译者不改变原文语序，对其直译照搬，译文读者极有可能一头雾水，无法理解核心内容。但例句中译者抓住了本句的核心内容"差异"，采取了增译法。译者增译"there be"结构，将"obvious difference"作为句子主语，突出"上下阶层的着装差异"，使内容焦点一目了然。此外，译者在句首增译了"as far as…is concerned"结构作为状语，引出话题背景，使得译文焦点突出，层次分明，增加句子结构的紧凑性和译文表述的严谨性。

例4　原文：在形式上，值得注意的一是深衣，二是胡服。

译文：In the dress form, there are two styles worth mentioning: one is *shenyi* (a large and loose garment), the other is *hu* dress (dress developed by *hu*, an ethnic minority group of ancient China).

【分析】该句是对春秋战国时期深衣与胡服的介绍。中文原文的主语并不明晰，是一个典型的无主句。同样，译者采用了"there be"句型引出主语，突出"two styles"的形式不同。中文原文只列出了两个服装名称，并未对深衣与胡服多作补充说明。译文中，译者以括号加注的方式增加了介绍二者主要特点的内容，让译文读者对两种服装名称背后的形制特点和穿着背景有一定了解，避免因为对中国文化的陌生而导致文化误读，充分考虑了译文读者的文化背景因素。

例5 原文：胡服特征是衣长齐膝，腰束郭洛带，用带钩，穿靴，便于骑射活动。

译文：The typical *hu* dress includes a dress that is not too long (just reaches one's knees), a metal-decorated belt at the waist with belt hooks and a pair of boots. The *hu* dress is known for its convenience in shooting on horse.

【分析】该句介绍胡服的特征以及穿搭方式。原文使用五个流水分句，字数不多，但内容较复杂。译者对原文句式进行了重组，以一个包含三个并列宾语的主谓结构介绍胡服的外形特征与穿戴方式，结构紧凑，一气呵成。同时，译者还增译评价性话语 "that is not too long" 来解释胡服长度，有利于更清楚地阐释胡服的形制与穿搭特点。然后，译者另起一句说明胡服的功能，这种分译法避免了译文语句烦琐冗杂的问题。译者还在胡服的功能介绍中增译 "be known for" 这一短语，既能突出胡服广受欢迎的便利性，也使得英文表达显得地道、流畅。

（二）减译

减译，指从全文出发根据逻辑、句法、修辞的需要在译文中删减一些不必要的语言表达的翻译方法。减译法是在不改变原文意思的基础上，省略原文中部分语句或文字，使译文更加简洁明了。具体而言，减译法是删减一些可有可无的，或者违背译文表达习惯的一些成分，但不可以漏译原文重要内容，不能对原文词句进行随意删减。使用减译法可以化繁为简，使译文更加简洁明了。

例6 原文：由于其形制简便，穿着适体，后来用途越来越广，最后无论男女，不分尊卑，皆可穿着。

译文：Owing to its simple design for comfortable wear, it was widely spread among common people.

【分析】该句对深衣的着装群体进行简要介绍——几乎涵盖了社会大部分人群。原文描述其用途广泛并且适宜男女、各种阶层人士穿着。译者将"男女"与"尊卑"二者进行综合，减译为 "common people"，即普罗大众皆可穿，简化了对深衣穿着对象的各种分类，使表达简洁、直观。

例7 原文：采用圆领方袖，以示规矩，寓意做事要合乎准则。

译文：Round sleeves and square collar meant faying in with a rule.

【分析】该句介绍深衣圆领方袖的寓意。"方、圆"是中国传统文化中常见的一对概念，内涵多样。"方"具有部分、规则、原则等含义；"圆"则意味着整体、灵活与圆满。方与圆相辅相成，是中国传统文化的精髓之一。深衣的圆领方袖也凝聚着中国古人对人格与处事准则的追求。原文为无主句，主语不明晰，故在译文中，译者将"采用"一词删去，以 "Round sleeves and square collar" 作为译文主语，并且将"以示规矩"一词删去，因为该词所表达的意义与下文"合乎准则"相互交叠。采用减译法之后，译文句子结构完整、简练，表达不显拖沓，清晰易懂。

例8 原文：久而久之，宽衫大袍成为不事劳作的有闲阶级的典型服饰，也是汉民族的一种传统服饰形象。

译文：Time went by, and the *changpao* became a typical garment for those with leisure, as well as a traditional garment of the Han people.

【分析】该句解释汉代的广袖深衣、唐代的圆领襕袍、明代的直身长袍渐渐成为有闲阶级的专属服装。译文中，"宽衫大袍"在上下文中指中国传统服装"长袍"，译者将其减译为"*changpao*"，属于异化翻译，体现了中国传统服饰的风采，同时也足够简洁，因语篇前文已有对"*changpao*"的描述，减译不会对读者造成理解障碍。此外，译者将"不事劳作"一词省去不译，因为该词的词义与"有闲"有重合，如果直译则显得多余。此句翻译中，译者运用减译法的同时，精准地保留了原文内涵，化繁为简，使读者对重点内容把握得更好。

例9 原文：曲裾深衣是深衣的一种，前襟被接长一段，穿着时须将其绕至背后。

译文：*Quju shenyi* was one kind of *shenyi*. Elongated front piece had to be wound to the back.

【分析】该句原文介绍深衣的一种类型——曲裾深衣，作者对其形制与穿着方法进行了简要说明。原文对其"前襟"形态进行描写，但汉语流水分句的特点使句子主干显得不太清晰，直译为英文容易使译文读者分不清句中各部分的逻辑关系。在译文中，译者将曲裾深衣的属性和其穿着方式分两句来介绍，并且省略"前襟被接长一段"中的"一段"，直接将其译为"its elongated front piece"作为第二句主语。不仅如此，译者还将"穿着时"进行减译，使译文更为简练，既能突出曲裾深衣的属性，又能简练说明其穿着方式，一举两得。

例10 原文：日益完善的印花技术和设备正在进入服装工业领域，用来降低生产成本，缩短生产时间，并且减少对劳动力的需求。

译文：Increasingly sophisticated printing technology and equipment are entering apparel industry to cut down production costs, time and labor requirement.

【分析】例句主要论述科学技术对服装行业的影响，具体影响方面使用了三个动词来体现，分别是"降低""缩短"和"减少"，语言流畅，富有节奏感。译者在翻译例句时巧妙采用了减译法，将原来的三个动词译为一个"reduce"，使三个宾语共享同一个谓语动词。这样译不仅使译文更简洁，还使句子结构更紧凑，表达更精练。

尽管增译法与减译法两种翻译方法产生的效果各有不同，但二者都有助于使译文更加通俗易懂，在翻译过程中对二者的运用相辅相成。英语与汉语这两种语言在思维方式和表达习惯方面存在极大差异，虽然增译法和减译法是较常见的翻译方法，但翻译时需要权衡，不能过度运用增译或减译，否则会降低译文的翻译效果。此外，在使用增译法和减译法的过程中，始终要忠于原文，尽量做到"信""达""雅"。

七、译文赏析
段落1
曲裾深衣是深衣的一种，前襟被接长一段，穿着时须将其绕至背后。汉代，曲裾深衣是女子服装中非常流行的款式，因为它通身紧窄，衣长曳地，下摆呈喇叭状，行不露足，最能体现女子的婀娜多姿。绕襟深衣是在曲裾深衣基础上的变体形式，西汉时期的女子深衣多为此样式，即将衣襟接得极长，穿时在身上缠绕数道，用带子匹结固定，每道花边则

显露在外。❶

Quju shenyi was one kind of *shenyi*. Its elongated front piece had to be wound to the back. In Han Dynasty, *quju shenyi* was the most popular style among women's clothes. It was a long and tight gown with the trumpet-shape hem. Women looked graceful with their feet veiled. Front piece winding *shenyi* is the variant of *quju shenyi*. Women in Western Han Dynasty (206B.C.-25A.D.) mostly wore this style. They made the front piece of the gown extremely long to wind several times around their body and fixed by sashes revealing every beautiful border.❶

【赏析】段落原文对曲裾深衣和绕襟深衣进行了详尽的介绍，将两种深衣的形制以及典型特征做了重点描述，使完整的深衣形态跃然纸上。译文中，译者不仅准确传达了原文语义，还运用了减译法，使得译文简练流畅，中心突出。首先，译者为了突出曲裾深衣的外形特征，采用了减译法对原文进行处理。由于原文对汉代曲裾深衣的介绍为体量较大的长句，在进行英译时，译者减去了"因为"一词的翻译。这样将原句拆分为几个简练的短句，一方面点明曲裾深衣出现的年代，另一方面突出曲裾深衣的特点，同时避免了句式冗长，给予译文读者良好的阅读体验。在对绕襟深衣的英译处理中，译者同样采用了减译手法，将原文的长句拆解成数个短句。先对绕襟深衣与曲裾深衣的关系进行介绍"Front piece winding *shenyi* is the variant of *quju shenyi*"，然后对其形制特点展开描述，重点明确。其次，在对"曲裾深衣"与"绕襟深衣"的英译中，译者分别采用了异化和归化的翻译手法。将"曲裾深衣"译为"*quju shenyi*"，将"绕襟深衣"译为"front piece winding *shenyi*"突出的是两种不同的深衣形制。对曲裾深衣采用异化译法，是遵循该词的传统译法，保持与前人译法的一致性；而对绕襟深衣采用归化译法，旨在突出其形制特征，以更好地传达中国传统服饰的内在含义。

段落2

深衣和袍服有同有异，都是上下连属的长衣，但深衣没有延续下来，袍服倒是一直穿用到近代——即使是21世纪的中国人，也还能想起它的模样——宽大笔直的袍身，斜在右腋下的大襟，朴素简洁的款式配着一些细腻精致的织绣花纹。汉代的广袖深衣、唐代的圆领襕袍、明代的直身长袍都是典型的宽身长袍，穿着者多为文人及统治阶层，久而久之，宽衫大袍成为不事劳作的有闲阶级的典型服饰，也是汉民族的一种传统服饰形象。❷

There are both similarities and differences between *shenyi* and *paofu*. They are both one-piece gowns but *shenyi* died out while *paofu* survived up until the present day. Even today in the 21st century, the mere mention of the *changpao* will bring up an image of a straight gown with side opening under the right arm, its simplicity in style enhanced by the elaboration of weaving and embroidery. The Han Dynasty *shenyi* with wide sleeves, the Tang Dynasty round collar gown and the Ming Dynasty straight gown are all typical wide *changpaos*, mainly preferred by the intelligentsia and the ruling class. Time went by, and the *changpao* became a typical garment for those with

---

❶ 语料参考：沈周. 中国红：古代服饰［M］. 合肥：黄山书社，2012：59.
❷ 语料参考：华梅. 中国服饰［M］. 北京：五洲传播出版社，2004：14.

leisure, as well as a traditional garment of the Han people.❶

【赏析】段落原文对深衣与袍服从结构与历史方面进行了比较，并将二者与各朝代的典型袍服进行类比，为读者呈现了中国传统袍服的精美画卷。在译文中，译者同时采用了增译法和减译法，灵活处理文章结构，使得深衣与袍服的特征更加直观。首先，译者灵活地将原文开头长句进行拆分，用 "There are both similarities and differences between *shenyi* and *paofu*"，统领本段，直指段落大意——深衣与袍服的不同，引领了全文，给译文读者后面的阅读增加了铺垫。其次，译者采用增译法，描述了现代中国人脑海中对于袍服的印象，在译文中巧妙地增加了 "mere mention" 一词。"mere mention" 意为 "只要一提到"，该词的使用突出了袍服 "宽大笔直" 的典型特征所给人留下的深刻印象，有利于加深译文读者对袍服外形的理解。最后，在对宽衫大袍着装阶层的英译处理中，译者采用了减译法。原文将宽衫大袍的着装阶层描述为 "不事劳作的有闲阶层"，译者省略 "不事劳作"，将其简译为 "those with leisure"，突出穿着人群，同时避免了语义重复，清晰直观地传达了宽衫大袍为有闲阶层所用这一信息。

八、翻译练习

（一）请将下列汉语句子译成英文

1. 根据儒家《礼记》所说，长袍必须足够长、足够宽，以完全覆盖身体，这是中国的传统伦理。

2. 随着中国经济的自由化，对民族服饰的兴趣促使人们开始寻找流行的、真实的和本质上的中国风格。

3. 深衣成为宋、明两代士大夫的一种正式服装。

4. 士大夫所穿的深衣在高句丽时期传入韩国并为儒家信徒所穿着。

5. 战国到前汉时期，深衣越来越流行，其形状与早期描述的差异更大。

6. 西汉前期与后期的深衣，形式上虽然有渐进的变化但是区别却很明显。

7. 到了明代，随着洪武帝试图取代元朝蒙古人所使用的所有洋装，以及中国精英阶层的支持，深衣开始流行起来。

8. 衣服的下半部分由十二片布缝在一起组成，代表一年十二个月。

9. 绕襟深衣是在曲裾深衣基础上的变体形式，西汉时期的女子深衣多为此样式。

10. 禅衣是一种单衣，即没有衬里的单层外衣，其外形与深衣相似，有直裾和曲裾两种款式，不论男女均可穿用。

（二）请将下列汉语段落译成英文

春秋战国时期周王室衰微，五霸七雄等诸侯国各自为政，一方面竞相发展生产，注重商品流通，另一方面兼并弱小，掠夺土地和财富。特别是对大量技术工匠的掳掠占有和铁工具的推广应用，促进了各种手工业的交流提高。各方面竞争的成就对纺织材料、服装剪裁工艺和装饰艺术，无不发生重大影响，服饰材料日益精细，品种名目日见繁多，形成了百花齐放

---

❶ 语料参考：Hua Mei. *Chinese Clothing*［M］．北京：五洲传播出版社，2004：14.

的服饰局面和推陈出新的深衣服饰。同时，春秋战国学术界"百家争鸣"的空气对当时文化学术发展有极大的推动作用，也促进了精美服饰的流行。❶

九、译笔自测

深衣及其发展

深衣是连体式的长衣服，它最大的特点是把左边的衣襟缝接出一片三角形，穿的时候，将它缠绕至背后，再用带子系住，这样做的目的是不显露出肌肤和尚不完善的内衣。深衣一般用柔软的材料制成，再用挺括的锦缎材料缘边，使其既具有强烈的装饰效果，又持久耐穿。

深衣的形制，每一部分都有其独特的寓意。比如在制作中，先将上衣下裳分裁，然后在腰部缝合，这是为了尊祖承古；采用圆袖方领，以示规矩，寓意做事要合乎准则；在后背垂直如绳的背线，寓意做人要正直；水平的下摆线，寓意处事要公正、公平。

"深衣"，从字面上看，就是用衣服把身体深深地遮蔽起来，这与中国传统的伦理道德相关。古代中国社会的主流思想强调男女有别，两性间不可太亲近，不能随便往来，即便是夫妇，也不能共用一个浴室、共用一个衣箱，甚至于晾晒衣服的衣架都要分开；婚后的妇女回到娘家，自己的兄弟也不能与她在一个桌子上吃饭，女子出门必须遮蔽得很严。儒家经典著作《礼记》等书中详细记述了这些着装上礼仪规定。

深衣最早出现于春秋战国时期，最初只是士大夫阶层居家的便服，由于其形制简便，穿着适体，后来用途越来越广，最后无论男女，不分尊卑，皆可穿着。魏晋以后，深衣被袍衫所取代，才逐渐退出历史舞台。但后世的裤褶、襦裙等服装都以深衣为原型。

在深衣出现之前，古人的衣服都是由上衣和下裳组成的，而深衣却打破了这一风俗，上下连为一体，呈直筒式。不过，深衣的形制特征仍是上下分裁而合制，上下保持着一分为二的界限。

到了汉代，深衣变形为曲裾袍——一种有三角形前襟与圆弧形下摆的长衣。同时还时兴直据袍，即直襟的长袍，也称"襜褕"。汉代的广袖深衣、唐代的圆领澜袍、明代的直身长袍都是典型的宽身长袍，穿着者多为文人及统治阶层，久而久之，宽衫大袍成为不事劳作的有闲阶级的典型服饰，也是汉民族的一种传统服饰形象。❷❸❹

十、知识拓展

春秋战国时期的衣着，上层人物的宽博、下层社会的窄小，已趋迥然。在形式上，值得注意的一是深衣，二是胡服。深衣有将身体深藏之意，是士大夫阶层居家的便服，又是庶人百姓的礼服，男女通用，可能形成于春秋战国之交。从马山楚墓出土实物观察，深衣是把以前各自独立的上衣、下裳合二为一，却又保持一分为二的界线，故上下不通缝、不通幅。据

❶ 语料参考：《中国文明史话》编委会. 服饰史话［M］. 北京：中国大百科全书出版社，2009：11.
❷ 语料参考：华梅. 中国服饰［M］. 北京：五洲传播出版社，2004：9-11.
❸ 沈周. 中国红：古代服饰［M］. 合肥：黄山书社，2012：59.
❹ 臧迎春. 中国传统服饰［M］. 北京：五洲传播出版社，2003：16.

记载，深衣有4种不同名称：深衣、长衣、麻衣、中衣。从出土文物看，春秋战国时衣裳连属的服装较多，用处也广，有些可以看作深衣的变式（图2-2）。❶

图2-2 汉代女服❷

As far as the clothing of the Spring and Autumn-Warring States Period is concerned, there is an obvious difference between the loose and comfortable dress of upper circles and the narrow and fit dress of the lower class. In the dress form, there are two styles worth mentioning: one is *shenyi* (a large and loose garment), the other is *hu* dress (dress developed by *hu*, an ethnic minority group of ancient China). *Shenyi* means to hide the body inside, which serves as either the informal dress of scholar-officials when staying at home or the full dress of the common people as well. There is no difference between men's dress and women's dress. The style might be formed at the turn of the Spring and Autumn Period. Through observing the objects unearthed at the site of a Chu State tomb in Mashan, we notice that *shenyi* is a combination of previously separate upper and lower garments, but remains a line between the two parts, so the upper and lower garments are not sewn together. According to records, *shenyi* has four different names: *Shenyi*, *changyi*,*mayi*, and *zhongyi*. According to unearthed artifacts of the Spring and Autumn-Warring States Period, the dress with the upper and lower parts combined together accounts for a large portion and can be worn for many purposes; some may be regarded as an altered style of *shenyi*.

第二节　胡服

学习目标
1. 了解胡服的历史及文化内涵
2. 熟悉胡服的基本语汇及其汉英表达方式
3. 掌握翻译中替代法与重复法相关的知识与技能
4. 通过实践训练提高胡服文化汉英翻译能力

一、经典译言

凡是翻译，必须兼顾着两面：一当然力求其易解，一则保存着原作的丰姿。

——鲁迅

---

❶ 内容参考：《中国文明史话》编委会. 服饰史话［M］. 北京：中国大百科全书出版社，2009：92.
❷ 图片来源：华梅. 中国服饰［M］. 北京：五洲传播出版社，2004：9.

二、汉语原文

胡服

早在战国时期，赵国的第六位国君赵武灵王意识到赵国军队的武器虽然比胡人优良，但大多数是步兵和兵车混合编制的队伍，加上官兵都身穿长袍，甲胄笨重，动辄就是几万、几十万，而灵活多变能够迅速出击的骑兵却很少，于是力排众议倡导本国的军队效法西北游牧民族的胡服骑射，也就是穿短袍、长裤操练骑马射击。结果，赵国的军队很快就强盛起来。

不仅如此，这种当初屡遭排斥才被汉族人认可的服装式样，到了魏晋南北朝已经由军服变为中原地区的日常服装。当然，这里还有一个很重要的原因，就是这一历史阶段战乱频仍，南北民众因躲避战争而大规模迁徙，客观上为服饰文化的交流提供了便利。

图2-3 胡服❶

裤褶和裲裆就是所谓的"胡服"，从形象上不难看出，这两种服式都便于骑马，而且适宜气候寒冷的地带（图2-3）。

所谓"裤褶"，是一种上衣下裤的服式。裤褶是一种套装，上衣为齐膝袍服。中原人民为了能够让其便于骑行与捕猎，改变了原来的窄袖和西北人左掩的习惯。因此，当时的中原人也将西北人称为"左衽之人"。此时的袍，实际上就相当于一个短上衣，上衣式样虽大同小异，却也多种多样。我们从资料中看，魏晋南北朝时期的裤褶上装，既有左衽，也有右衽，还有相当多的对襟，甚至有对襟相掩，下摆正前方两个衣角错开呈燕尾状，服装结构因此丰富了许多。下身裤褶穿起来特别精干，在南朝墓内画像砖和陶俑中很常见。

裤褶的下身是合裆裤。这种裤装最初是很合身的，细细的，穿起来相当利落，一副健步如飞的样子。传到中原以后，尤其是当某些文官大臣也穿着裤褶装上朝时，引起了一些保守派的质疑，认为这样两条细裤管立在那简直不合体统，与古来的礼服上衣下裳的样式实在是相距甚远。在这种情况下，有些人想出一个折中的办法，将裤管加肥，显得与裙裳无异，

❶ 图片来源：扬子晚报网和汉程网。
扬子晚报网网址：https://m.yzwb.net/wap/news/1168063.html
汉程网网址：http://minsu.httpcn.com/info/html/2019727/PWCQTBAZMECQ.shtml

待抬腿走路时，仍是便利的裤子。为了避免陷入荆棘或在泥沼中，有人想出个好办法，将裤管轻轻提起用两条丝带系在膝下。这种被称作"缚裤"的形象在南朝画像砖和陶俑上也屡见不鲜。20世纪80年代初，青年人流行穿喇叭裤，有人就将魏晋时的缚裤指认为喇叭口裤的源头，其实不是，只是缚裤呈现出的廓形很像是从膝部以下扩展的喇叭口裤。

裲裆是这一时期的另一种代表服式，也是由西北引入中原的服饰风格。从当时的随葬品陶俑和墓葬壁画、画像砖上的服饰形象看，裲裆的样式多是前后两片，肩上和腋下以襻扣住，也有可以穿在里面的，有皮的、棉的、单的、夹的，尺寸也可大可小。裲裆的说法虽然很久都听不到了，但这种风格一直沿用至今。

裤褶与裲裆，这两种服装在当年风行一时，男女皆穿。裲裆下配裤装的穿法未脱开上衣下裳的中华民族服饰原形，但也体现了服饰文化的交流与融合。❶

译前提示：胡服是古代诸夏汉人对西方和北方各族胡人所穿的服装的总称，与当时中原地区宽大博带式的汉族服饰，有较大差异。胡服的一大特征是短衣齐膝，《史记·赵世家》曰：自战国赵武灵王，胡服骑射，裤褶传入中国，历代皆以为戎服，或用其冠，或用其履，或用其衣服及带，或三者全用，晋代民间，始偶用之。胡服之制，冠则惠文，带则贝带，履则靴，裤则上褶下裤。文章从历史背景出发，对裤褶和裲裆的形制与功能进行介绍，揭示了服饰背后的文化现象。

三、英语译文

Hu Dress

In as early as the Warring States Period, the sixth emperor of Zhao already realized that although the Zhao army, which was mainly composed of infantry and chariots, had better weapons, the long robes worn by generals and warriors were too cumbersome for an army, especially when they had to drag their armors and supplies around. They had tens of thousands of soldiers, but few riders were flexible to make a quick attack. He went against all objections and advocated for changes towards the *hu* or western minority clothing style of the nomadic riders. The Zhao soldiers wore shorter robes and trousers to practice riding and shooting and soon became a better army.

Moreover, this style that was once frowned upon and rejected became the daily wear of the common folks by Wei, Jin and Southern and Northern Dynasties in the central plains. One reason for this change, unfortunately, was the frequent migration of the people both from north and south to run away from the incessant wars and chaos. This process also helped the exchange of garment culture.

Kuzhe and *liangdang* were the typical "*hu*" or minority wear of that time. It is not hard to see that both styles were fit for riding and for life in the cold climate.

The so-called *kuzhe* was a style with upper garment and lower trousers. It was a suit with the upper garment being a short robe above knees with wide sleeves, a central China adaptation to the

❶ 语料参考：华梅. 中国服饰［M］. 北京：五洲传播出版社，2004：19-22.

original narrow sleeves fit for riding and hunting. What also changed was the closure of the robe, which moved from left to right. Interestingly, people of central China called the northwesterners "people with left closure." The robes at this time were shortened significantly, and varied in style. Historical materials show a number of styles of these upper garments in Wei, Jin, Southern and Northern Dynasties, which had left, right and middle closure, or even swallowtails at the front hem. Thus, the costume structure was greatly enriched. A set of these garments made the wearer sharp and agile, as is frequently seen in clay burial figurines in the Southern Dynasty.

The lower garment of the *kuzhe* was a pair of trousers with closed crotch. Initially these trousers were close fitting, showing off slender legs that could freely move around. When this style appeared in central China, especially when some officials wore them in court, the conservatives questioned the appropriateness of the two thin legs that cried out rebellion against the loose fitting traditional ceremonial wear. Widening the legs was a compromise, so that the pants still appeared similar to the traditional robe. When the wearers walked about, these pants were more flexible and convenient than the robe. To avoid being caught in thorns or dragged in mud, someone came up with a brilliant idea of lifting the trouser legs and tying them up just below knee-level. This kind of pants called "*fuku*" can be frequently seen in the Southern Dynasty's burial figurines and brick paintings. They were considered by someone as the source of the bell-bottomed pants popular among the young people in the early 1980s, but in fact, it was not the case. The only similarity between the two lied in the fact that the silhouette of *fuku* was like the bell-bottomed pants extending from below the knee.

Liangdang or double-layered suit was another style typical of this period, and it came from the northwest into central China. It was no more than a vest, which could be seen in many burial pieces of that time. Judging from day figurines and wall paintings in tombs, *liangdang* was in two separate pieces fastened on the shoulders and under the arms. There were also *liangdangs* worn inside in materials of leather or cotton, lined or unlined, close or loose fitting. The name changed over the years but the style remained.

Kuzhe and *liangdang* were all the rage at that time for both women and men. *Liangdang* with pants did not deviate from the separate piece style that has always been the prototype of the Chinese clothing, but the modifications were made due to the exchange and fusion of different garment cultures.[1]

四、词汇对译

胡服骑射　wearing *hu* dress and shooting on horse

左衽之人　people with left closure

合裆裤　trousers with closed crotch

日常服装　daily wear

传统礼服　traditional ceremonial wear

中华民族服饰原型　the prototype style of the Chinese clothing

服饰文化的交流与融合　exchange and fusion of different garment cultures

上衣下裤　upper garment and lower trousers

❶ 语料参考：Hua Mei. *Chinese Clothing* [M]. 北京：五洲传播出版社，2004：19-22.

喇叭裤　bell-bottomed pants

墓葬俑　burial figurines

画像砖　brick paintings

屡见不鲜　be frequently seen

裲裆　*liangdang*

墓葬壁画　wall paintings in tombs

对襟　middle closure

保守派　the conservatives

捆绑　tying up

轮廓　silhouette

从……判断　judging from

五、译文注释

1. 当然，这里还有一个很重要的原因，就是这一历史阶段战乱频仍，南北民众因躲避战争而大规模迁徙。

One reason for this change, unfortunately, was the frequent migration of the people both from north and south to run away from the incessant wars and chaos.

【注释】例句原文的前一句者描述了魏晋南北朝人民对胡服穿着场合的转变，即胡服由军服转变为人们的日常服装，本句对前句胡服地位变化的原因进行进一步解释。原文并未出现"变化"一词，如果译者直接按照中文原文进行直译，很容易让译文读者摸不着头脑，找不到"reason"要解释的对象。译者结合上下文，巧妙地增译"this change"来概括上文描写的胡服穿着场合的变化，既起到对前句进行总结的作用，又完美地过渡到下文描述的原因之中，衔接连贯自然。此外，原文中有"战乱频繁""战争"二词，如果直译出来，译文可能会显得冗余，影响其可读性。译者将战乱频繁导致民众迁徙巧用"frequent migration"替代，既没有偏离原文句意，又精简了译文。而为了突出"战乱频繁"，译者在"wars"一词之后又增加"chaos"，强调了战争严重给人们带来的不便。本例句中，译者还从语篇整体的角度进行了巧妙的增译，增加了"unfortunately"一词作为插入语，对人民所受战争之苦深感同情，结合语篇上下文，突出了本句的感情色彩。

2. 这种裤装最初是很合身的，细细的，穿起来相当利落，一副健步如飞的样子。

Initially these trousers were close fitting, showing off slender legs that could freely move around.

【注释】例句原文是对裤褶的下身——合裆裤进行描写，意在突出合裆裤纤细、便于行走的特点。"showing off"原有炫耀之意，译者将其巧妙地用在译文中，使着装者对于轻便利落的合裆裤所给予的摇曳生姿的得意之情跃然纸上，更加突出了合裆裤的特点。原文中，走路利落与健步如飞实际上是一个意思，在进行汉译英时，译者为避免重复，使用指示代词"that"替代"slender legs"，使健步如飞的动作活灵活现，同时不显得累赘。

3. 裤褶是一种套装，上衣为齐膝袍服。中原人民为了能够让其便于骑行与捕猎，改变了原来的窄袖和西北人左掩的习惯。

Kuzhe was a suit with the upper part being a short robe above knees with wide sleeves, a central China adaptation to the original narrow sleeves fit for riding and hunting. What also changed was the closure of the robe, which moved from left to right.

【注释】例句对裤褶上装的样式进行了介绍，突出西北民族与中原汉族之间着装习惯的差异——西北民族习惯左掩，而汉族人习惯右掩。但如果按照中文原文中的语序直译为英文可能会显得冗余。原文中"西北人左掩的习惯"，实际上就是袍服左掩，中原人将其改变为右掩的样式。译者首先对原文句式进行拆分，将中原人民对西北袍服袖宽的修改和左掩样式的修改分别进行说明，各自单独成句，突出重点，从而避免句式过长造成的阅读不便。在英译过程中，还可以采用增译策略来更深入地说明相关内容。比如，例句译文用关系代词"which"引导非限定性定语从句，对左掩习惯的改变进行了更为明确的阐释，句子层次清晰，表意充分，使译者更容易理解相关的中国服饰文化。

4. 我们从资料中看，魏晋南北朝时期的裤褶上装，既有左衽，也有右衽，还有相当多的对襟，甚至有对襟相掩，下摆正前方两个衣角错开呈燕尾状，服装结构因此丰富了许多。

Historical materials show a number of styles of these upper garments in Wei, Jin, Southern and Northern Dynasties, which had left, right and middle closure, or even swallowtails at the front hem. Thus, the costume structure was greatly enriched.

【注释】例句原文是对魏晋南北朝时期裤褶上装样式的介绍，包括左右衽和对襟。原文第一个分句主语是人称代词"我们"，译文用指物的"historical materials"代替原文的主语，这源于汉英两种语言不同的使用规范：汉语多倾向于以人作主语；英语则多倾向于以物作主语。此外，原文中的左衽、右衽和对襟，译者直接译成了"left, right and middle closure"。文中"衽"和"襟"都是"closure"的意思，译文采用省译法，简洁说明了裤褶上装的三种样式，清晰、直观。译者这样处理的原因在于，汉语原文反复使用相同语义的词能赋予语言一种节奏感，有强调效果，但在英文中，重复并没有加强语气的功能，也不属于对偶排比的形式，属于不必要的重复，在汉译英的过程中要避免重复。

5. 有些人想出一个折中的办法，将裤管加肥，显得与裙裳无异。

Widening the legs was a compromise, so that the pants still appeared similar to the traditional robe when they are worn in the court hall.

【注释】本句说明由于合裆裤裤管太细，不符合传统礼节，人们于是对其进行加肥改造。原文是典型的流水分句，分句间的逻辑结构隐含在字里行间，如果不采用适当的翻译策略进行调整，译成英文可能会出现歧义。例句译者并未完全按照原文结构进行直译，而是更换了描述视角，将原文的人称主语"有些人"替换成了物称主语"Widening the legs"，然后根据主语调整谓语表达，体现了汉译英过程中的思维方式的转换。译文还增译了连接词"so that"来明示译文的目的状语，然后反话正说用"similar to"替换原文中的"无异"。多种翻译方法灵活使用，使译文充分传递原文信息的同时，做到主次分明，逻辑清晰，表达流畅。

6. 20世纪80年代初，青年人流行穿喇叭裤，有人就将魏晋时的缚裤指认为喇叭口裤的源头，其实不是，只是缚裤呈现出的廓形很像是从膝部以下扩展的喇叭口裤。

They were considered by someone as the source of the bell-bottomed pants popular among the young people in the early 1980s, but in fact, it was not the case. The only similarity between the two lied in the fact that the silhouette of *fuku* was like the bell-bottomed pants extending from below the knee.

【注释】例句原文将缚裤与现代喇叭裤样式进行比较，有人认为缚裤是喇叭裤的源头，但作者否认了这一说法，认为喇叭裤只是廓形与缚裤相似而已。译者对原文进行了拆分与调整。首先将原文拆分为两句，将喇叭裤与缚裤的相似之处另起一句，单独说明。译者选用缚裤作主语，统领第一句，以被动语态构建句子，自然引出喇叭裤之源的说法，结构紧凑，层次分明。为进一步阐明原文的观点，译者增译了"the only similarity between the two lied in the fact that..."，帮助读者廓清前后两句之间的逻辑关系，以明晰化的手段充分传递原文信息，体现出对读者认知的关照。

六、翻译知识——替代与重复

汉英两种语言分属不同的语言系统，在表达方式上存在显著差异。总的来说，汉语比较习惯于重复，其重复倾向表现在词语搭配、句式安排、篇章结构等各方面。究其原因，汉语的重复倾向与其语音文字的特点有密切关系。汉语讲究均衡美，往往造成用词造句方面的重复倾向。英语则比较忌讳重复，常用替代、省略等表达方法。英语中的重复无论在使用范围、出现频率和表现形式的多样性方面都远远低于汉语。讲英语的人对于重复相同的音节、词语或句式往往感到不习惯。在汉英翻译实践中，除非有意强调或出于修辞的需要，一般来讲在能够明确表达意思的前提下，英译文中可以采用替代或省略等方法来避免无用的重复。这样不仅能使行文简洁、有力，而且比较符合英语读者的语言习惯。下面将分别对替代和省略两种翻译方法进行说明。

（一）替代

汉英翻译中的替代指用某种语言形式来代替句中或上文已出现过的词语或内容，它是英语语言表达中的一个常用方法。一般来讲，英语多用代称，以避免重复；汉语少用代称，多用实称，因而较常重复。

例1 原文：服装配套极为繁复，不仅穿起来费时，活动起来也极为不便；而胡人服装则是短衣长裤，衣身瘦窄，腰束带，穿靴，非常便于骑射活动。

译文：The matching of clothes were so complicated that it's not only time-consuming to wear but also inconvenient to dress, while the clothes worn by the northwestern barbarian tribes were short tops and pants in narrow shape tied with the waistband and boots. Such costumes were suitable to ride and shoot on horseback.

【分析】例句是对中原地区服装与胡服进行比较，突出胡服轻便的特色。在对胡服特色与作用进行介绍时，译者首先描述胡服特征，然后另起一句单独说明胡服便于骑射的特点，使用"such costumes"来替代前文对胡服特点详细描述，这种替代方法的使用不仅使译文简洁明了，表意清晰，还增强了与前句的连贯性。另外，译文还以"so...that"句型结构来突出繁复的服装给生活带来的不便，使译文结构更紧凑，逻辑更清楚。

例2 原文：赵武灵王下令士兵们也穿上这种靴子，并在靴面和靴筒表面还装饰着几十、甚至上百个青铜泡，更显威武夺目。

译文：King Wuling ordered his soldiers to wear this style of boots, and have dozens even hundreds of dazzling bronze bubbles decorated on their upper surfaces to show off his mighty army.

【分析】例句描述的是赵武灵王在军队中引入胡人皮靴的情况。原文有"靴子""靴面" "靴筒" 等词汇，其中 "靴" 出现次数较多。汉译英时如果把所有这些 "靴" 全部译出，会显得重复累赘。译者在翻译时，使用 "their" 来指代前面出现过的 "boots'"，避免了不必要的重复，还增强了前后文的连贯性。

例3　原文：公元前307年，赵国的武灵王颁胡服令，推行 "胡服骑射"，进行了服饰史上的一次重要变革。

译文：In 307 B.C., King Wuling of Zhao State promulgated the decree of *hu* dress to implement "wearing *hu* dress and shooting on horse", which was an important reform in the history of costume.

【分析】英文中定语从句是指由关系词引导的语法功能为定语的从句，一般用来修饰名词或代词，被修饰的名词或代词叫先行词，引导定语从句关系代词有who，whom，whose，which，that等和关系副词where，when，why等。从语法结构来分析，引导词在定语从句中替代被修饰的先行词，并同时起到衔接先行词和定语从句的作用，是英语句子中典型的替代表现。本例句对 "胡服骑射" 的推行做简要介绍。"胡服骑射" 指赵武灵王为了使国家强盛，推行胡服，教练骑射。胡服与中原地区的服装相比，更为轻便、利于活动，对骑兵来说也更合适。例句原文中 "胡服令" "推行'胡服骑射'" "重要变革" 属于并列关系，汉译英时，译者对原文句内各部分之间的关系进行调整，以 "which" 作为引导词引出定语从句，将 "服饰史上的一次重要变革" 变为 "胡服骑射" 的修饰语，从句中的 "which" 替代的是先行词 "胡服骑射"。这样译使译文结构紧凑，主次分明，突出了服饰变革的重要意义。

例4　原文：胡人的帽子不同于当时中原人的冠，是用于暖额和防风沙的，由动物皮革制成。

译文：Hu's hat was made of the leather or hide to prevent against sand and to keep forehead warm which was different from the hat of the Central Plains.

【分析】例句介绍胡人帽子的功能与材质，并将其与中原人的冠进行对比。原文将胡人帽子与中原人的冠之间的对比置于句首，随后描述其功能与材质。汉译英时，译者根据英文表达的需要对原文语序进行调整，先描述其材质与功能，再以 "which" 引导非限定性定语从句，将其与中原人的冠进行对比。定语从句中，"which" 替代的是前面的整个分句所描述的内容，使译文句子重点更突出，层次更分明。

例5　原文：裲裆下配裤装的穿法未脱开上衣下裳的中华民族服饰原形，但也体现了服饰文化的交流与融合。

译文：*Liangdang* with pants did not deviate from the separate piece style that has always been the prototype of the Chinese clothing, but the modifications were made due to the exchange and fusion of different garment cultures.

【分析】例句原文分析了 "裲裆下配裤装的穿法" 的服饰文化背景，强调这种穿法没有背离中国传统。译者在翻译例句时运用替代的方法，对于前文提及的 "裲裆下配裤装的穿法"，译者在第一个分句中使用 "*liangdang* with pants"，使译文在语篇中紧密衔接上一句。在第二个分句中，译者为清楚说明 "裲裆下配裤装" 这种穿法所体现的文化意义，根据语篇上下文增补了主语，但并没有重复使用前一分句里相同的名词短语，而是用 "the modifications"来替代，突出了这一穿法的创新性，简洁流畅，减少了不必要的重复。此外，例句还采用"due to"来说明这种穿法形成的原因，虽然例句原文并未体现任何因果关系，但译者使用这一短语使译文逻辑更清晰，句子重点也更为突出。

（二）省略

这里的省略是指汉英翻译过程中，省略汉语原文中反复出现的相同表达方式，以使译文简洁流畅，避免重复累赘，更符合英语语言的表达规范。

例6 原文：服装风格是一种着装特点，是一种选择态度，是穿着者个性的表现。

译文：Style of clothing is a kind of dress feature, attitude of choice and indication of the wearer's personality.

【分析】例句原文是对服装风格的说明，从三个方面进行定义，重复使用了三个"是"字。在汉语中，这样的重复可以产生一种节奏感，起到一定的强调作用。但在英文中，如果直译出重复使用的"是"字，会使译文结构显得繁杂，不自然。译者在翻译例句时省略了后两个"是"字，将三种定义并列共同作为"is"的表语，结构清楚，语言简洁流畅。

例7 原文：我们谈到背心裙，谈到斜裙，谈到鱼尾裙，谈到西服裙——谈到一切，只是不谈旗袍，我怕她思乡。

译文：We talked of the jumper skirt, of the bias skirt, of the fish tail skirt, of the tailored skirt—of everything but *qipao* because I cared of her homesick feelings.

【分析】例句中人们谈及各种裙装，唯一不提旗袍，生怕有人思乡。原文使用了六个"谈"字，充分体现了汉语这种声调语言特有的节奏感，同时，前五个重复使用的"谈到"词语和最后一个"不谈"造成强烈的对比反差，将谈话者的细心体贴表现得淋漓尽致。但英文表达忌讳过多的重复，因此译者在翻译这句话时，采用了省译的方法，只保留了第一个"谈（talk）"字，后面五个"谈"字全部省去，使英译文读起来焦点突出，层次清楚，表达流畅。

例8 原文：服装设计要顺应设计规律，不要违反设计规律。

译文：Fashion design has to work with, and not against design principles.

【分析】例句说明了服装设计与设计原则的关系。原文"设计规律"这个词组出现了两次，这种重复使用同一个词组的现象在汉语中很普遍。译成英文时需要根据英文表达规范，尽量避免重复，因此需要省略其中重复使用的词组。译者将"顺应"和"不要违反"合译为"work with, not against"，两个介词共用同一个宾语，使译文避免了重复，简洁流畅。

例9 原文：对于面料的质地，设计师不能主观臆断，其颜色必须目见，声音必须耳闻，手感必须亲自去触摸。

译文：The designer should not judge the texture of a fabric by subjective assumptions. The color must be seen, the sound heard, and the hand touched by himself.

【分析】例句说明了设计师必须通过亲身实践来把握面料的质地。为强调亲身实践的重要性，原文使用了三个"必须"，汉译英时如果全部译出，译文会显得冗余。译者遵循英文重简洁，尽量避免重复的特点，英译时只保留了第一个"必须"，而将后两个"必须"直接省略掉了，使译文更符合英语语言的表达规范。

以上分析了汉英翻译中针对汉语中普遍存在的重复现象的处理方式,替代和省略是两个基本的策略。在具体的汉英翻译实践中,译者要注意做到替代明确、省略合理,遵循译入语的表达规范,充分传递原文信息为主,做到清晰和准确。

七、译文赏析
段落1
穿胡服的唐女性形象与唐朝宫廷艺术中以男性视角描绘女性的传统精英场景形成鲜明对比。如唐代早期至晚期两位著名的男艺术家阎立本和周昉所描绘的宫廷艺术作品,展示了男性艺术家对图像的常态化处理和对女性的视角。作为宫廷画家,周昉与皇帝保持着密切的关系,像他这样的画家描绘了精英们的威望生活。❶

The Tang women's persona in *hu* dress contrasts with the traditional elite setting found in Tang court art that depicted women in a male perspective; such as those portrayed by two flourishing male artists, Yan Liben and Zhou Fang, from early to late period of this dynasty. The court art demonstrates the normalcy of male artists staging the images and their perspectives on women. As a court artist, Zhou Fang remained in a close relationship with the emperor , and artists like him painted the prestige lives of the elites.

【赏析】该段落由着胡服的女性形象出发,介绍了唐代著名画家对于画作中的女性形象的处理。在译文中,译者在准确传达原文信息的同时,运用了适当的翻译策略,使得译文中心明确,语义清晰。首先,在段落开头,作者想向读者阐明穿胡服的唐女形象与唐代宫廷绘画中唐女形象的不同。译者在进行英译时,运用"contrast with"这一短语,将原句中的两个主语分别置于其前和其后,凸显该句对于二者的对比,使得译文逻辑清晰。其次,译者运用了回指的方法,用"those"一词对原文中"阎立本与周昉所描绘的宫廷艺术作品"中"作品"一词进行替代。这是因为在前文中,译者已经提到了"the traditional elite setting found in Tang court art that depicted women in a male perspective",回指策略的使用可以减少不必要的重复,使译文句意更加直观,有助于增强句子的连贯性,提高读者的阅读效率。原文第二句中的"唐代"英译为"this dynasty"也同样是采用了回指策略。再次,原文将两位艺术家的作品与男性艺术家对于画作中女性形象的处理置于同一语句中,但译者在进行英译时将二者分译为两个各自独立句子,如此一来,减少了译文句子的结构容量,从而减轻了译文读者阅读负担,使得段落大意更加简洁明了。

段落2
胡服与唐装的古典服饰风格不同,这也说明了唐装风格选择广泛。胡服第一次在中国成为流行可以追溯到战国时期,胡服被定义为非中国人口所穿的服装。在唐代,胡服是通过丝绸之路从西方传入的,胡语能够让人联想至中国的西部和西北部。然而,胡服的主要文化渊源是复杂的,因为这种风格结合了不同的外国服饰风格。在关于唐代的学术研究中,胡服的

---

❶ 语料参考:Berman G. *Tang Elite Women and Hufu Clothing: Persian Garments and the Artistic Rendering of Power*. [C]. Milwaukee: the University of Wisconsin-Milwaukee, 2020: 4.

服饰风格源于唐人对粟特人、萨珊波斯人等西亚外国人服饰风格的模仿。❶

The *hu* dress differed in style from the Tangs' classical dress but illustrates the broad choice of styles in Tang fashion. The history of the first time *hu* dress styles became fashion in China can be dated to the Warring States Period, with *hu* dress defined as clothing worn by the non-Chinese population. In the Tang Dynasty, the *hu* dress came from the West through the Silk Road with the *hu* terminology implying a link towards the foreigners from the West and Northwest of China. However, the main cultural origin of *hu* dress is complicated, since the style combined different possible foreign dress styles. In the Tang scholarship, the *hu* dress style derives from the Tangs' imitation of the dress styles they observed from West Asian foreigners such as the Sogdians and Sasanian Persians.

【赏析】该段落详细描述了胡服的历史，从胡服的传入时间、传入地点到胡服的含义以及风格渊源等方面进行了细致的介绍，向读者展现了中华传统服饰——胡服的文化内涵，以弘扬中国传统服饰文化。译文中，译者运用了多种连词对原文短句进行处理，使译文逻辑更加严谨。首先，译者将段落开头的两个分句用"but"连接成为一句话，使得文中胡服与唐装风格多样之间产生了联系，逻辑显得更为严密。其次，译者在翻译时注意选词的准确性。比如，译者将"唐装风格"译为"styles in Tang fashion"，而不是"styles in Tang costume"。"fashion"一词比起"costume"或者"clothing"的风格特征更加强烈，能更为贴切地对应原文的唐装风格。原文作者想要突出的是其服装特征而不是种类，故用使用"fashion"一词显得更为灵活生动，充分体现了译者在翻译中的主观能动性。另外，译者基于目标语表达规范，对部分语句的语序进行了调整，如"胡服是通过丝绸之路从西方传入的"，译者将其调整为"the *hu* dress came from the West through the Silk Road"。原文中状语"通过丝绸之路"位于句子中间位置，译者遵循英语句法的表达需要，进行灵活调整，将其置于译文句末，避免产生歧义，同时使译文表达自然流畅。

八、翻译练习
（一）请将下列汉语句子译成英文

1. "丝绸之路"被认为是联结亚欧大陆的古代东西方文明的交汇之路，而丝绸则是最具代表性的货物。

2. 东汉时期的一些汉服开始受到胡人服饰的影响，圆领长袍开始出现。

3. 军服改革吸收了胡服样式，将传统宽衣博带改成窄袖短衣，以便于射箭；将套裤改成有裆的、裤管连为一体的裤子，并用带系缚，这便是裈。

4. 唐装是中国的一种服饰，为汉族服饰系统中一种款式，特征是交领、右衽、系带。

5. 到了中晚唐时期，服装中加强了华夏的传统审美观念，开始复古，从以显出女子身材为主逐步恢复到秦汉那种宽衣大袖，飘逸如仙的风格。

6. 鲜卑统治者继续穿着自己独特的鲜卑服装，以保持自己的民族身份，避免与中国多数人融合。

---

❶ 语料参考：Berman G. *Tang Elite Women and Hufu Clothing: Persian Garments and the Artistic Rendering of Power*. [C]. Milwackee: the University of Wisconsin–Milwaukee，2020: 6.

7．战国时期，中国人穿的短上衣，用丝织带束腰，开口向右，也影响了胡服；它和裤子一起穿，活动起来更方便。

8．唐装是以清代对襟马褂为雏形，加入立领和西式立体裁剪所设计的带有传统元素的现代服饰。

9．徐显秀墓壁画上的一些女仆人，由于使用了窄袖，看起来更像鲜卑式服装，而不是中式服装。

10．唐朝首都长安是丝绸之路的东端，是外国人从西方卖东西、从中国买东西回国的地方。

（二）请将下列汉语段落译成英文

衣冠服饰是社会风尚的表现，各种思想意识与生活习俗都会不同程度地在服饰文化上反映出来。魏晋南北朝时期玄学的兴起与道教、佛教的传播，冲破了儒学一统天下的局面，也推动了服饰文化的多样化发展。当时的文人都喜欢穿着宽大的服饰，一是因为时代风貌的改变，人们不再崇尚武力而是变得贪图享乐；二是文人的心态发生了很大的变化，开始不再关心世事，不再重视礼教。❶

九、译笔自测

胡服骑射

公元前307年，赵国的武灵王颁胡服令，推行"胡服骑射"，进行了服饰史上的一次重要变革。

胡服，是指当时"胡人"（西北方的少数民族）的服饰。秦汉以前，中原地区的服装都是宽衣博带，且服装配套极为繁复，不仅穿起来费时，活动起来也极为不便；而胡人服装则是短衣长裤，衣身瘦窄，腰束带，穿靴，非常便于骑射活动。

当时，位于中国西北方的赵国，经常与胡人交战。由于那里的地形多为崎岖的山谷，这使得擅长战车作战的赵国人不占优势。于是，赵国的君王武灵王决定进行军事改革，训练骑兵。既然要发展骑兵，首先就需要进行服装的改革，以适应马上作战。

赵武灵王吸收了胡人的军服样式，将传统宽衣博带改成窄袖短衣，以便于射箭；将套裤改成有裆的、裤管连为一体的裤子，并用带系缚，这便是裈。裈能保护大腿和臀部肌肉皮肤在骑马时减少摩擦，且裤外不必再加裳。

在鞋帽方面，赵武灵王也进行了改革。胡人的帽子不同于当时中原人的冠，是用于暖额和防风沙的，由动物皮革制成。于是，赵武灵王下令士兵用黑色的绫绢围戴在头上，后来逐渐发展成为帽。至于鞋，胡人穿的是皮靴，软硬适中，便于跑动；鞋帮很高，一直到膝盖下面，用以护腿，也适于骑马。赵武灵王下令士兵们也穿上这种靴子，并在靴面和靴筒表面还装饰着几十、甚至上百个青铜泡，更显威武夺目。

通过赵武灵王的服饰改革，中国最早的正规军装出现了，以后逐渐演变改进为后来的盔甲装备。胡服的推广，不仅使赵国屡次赢得胜利，还开创了中国骑兵史上的新纪元。同时，这项服饰改革也弱化了传统服饰的身份等级标示功能，强化了其实用性，对此后的服饰发展

---

❶ 该段落语料参考：《中化文明史话》编委会．服饰史话［M］．北京：中国大百科全书出版社，2009：22．

产生了重要的影响。

从此以后，"习胡服，求便利"便成了中国服饰变化的总体倾向，汉族人不断吸取少数民族的服饰精华来丰富自己的服饰文化。❶

十、知识拓展

唐代的胡服，实际上是指西域地区的少数民族服饰和印度、波斯等外国服饰。在武则天时代，"胡服胡帽"就已经成为流行。开元、天宝时期已经有了新的变化，很难发现所谓的"胡服"特征。

比较常见的胡服形式是翻领窄袖袍、条纹小口裤、透空软锦靴和锦绣浑脱帽，有的还佩有蹀躞带。唐代前期胡服和唐代流行的柘枝舞、胡旋舞不可分。唐代诗人咏柘枝舞、胡旋舞的，形容多和画刻所见胡服相通（图2-4）。❷❸

In fact, *hu* dress in the Tang Dynasty refers to the ethnic costumes of minority nationalities in the western region and foreign costumes from India, Persia and other countries. During Emperess Wuzetian's reign in the Tang Dynasty, "*hu* dress and *hu* hats" had become popular. New changes have taken place in Kaiyuan-Tianbao Period, and it is difficult to find the so-called *hu* dress characteristics.

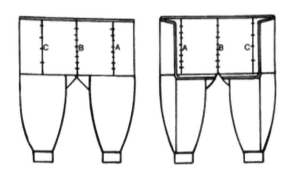

图2-4 战国时期男裤结构示意图❹

A typical *hu* dress comprises a robe with a turnover collar and tight cuffs, a pair of trousers with narrow leg bottoms, a pair of brocade boots and an embroidered hat. From time to time, *hu* dress wearers also had a walking belt. *Hu* dress in the early Tang Dynasty can not be separated from *zhezhi* dance and *huxaun* dance which were popular in the Tang Dynasty. Poets of Tang Dynasty chanted the dance of *zhezhi* and *huxuan*, which is often described in the same way as *hu* dress seen in paintings.

第三节 龙袍

学习目标

1. 了解龙袍的历史及文化内涵
2. 熟悉龙袍的基本语汇及其汉英表达方式

❶ 语料参考：沈周. 中国红：古代服饰［M］. 合肥：黄山书社，2012：144.
❷ 知识拓展内容参考：臧迎春. 中国传统服饰［M］. 北京：五洲传播出版社，2003：81，139.
❸ 沈从文. 中国古代服饰研究［M］. 上海：世纪出版集团，2005：308.
❹ 华梅. 中国服饰［M］. 北京：五洲传播出版社，2004：22.

3. 掌握合译法与分译法相关的知识与技能

4. 通过实践训练提高龙袍文化汉英翻译能力

一、经典译言

翻译是一种创造性的工作，好的翻译等于创作，甚至还可能超过创作。

——郭沫若

二、汉语原文

龙袍

冕服和龙袍是中国古代皇帝的典型服饰。它们体现了中国人独特的审美观、宇宙观和宗教观。在古代中国社会，哪一个等级的人在哪一个场合穿哪一种衣服，都是有严格规定的。

古代的大臣们每天上朝必须身穿朝服，这一点皇帝也不例外。皇帝的朝服就是我们所说的龙袍。早在先秦时期，各国君主就已经拥有了象征一国最高统治的独一无二的袍服。一开始中国古代帝王的龙袍并不是黄色，历朝皇帝龙袍的色彩也不尽相同。

中国的阴阳五行说认为，金、木、水、火、土相生相克。白色代表金，青（绿）色代表木，黑色代表水，赤（红）色代表火，黄色代表土。周代以红色为高级服色，秦时以黑为最高地位的服装颜色，秦始皇以黑色作为龙袍的基本色调，百官都穿黑色衣服。汉灭秦后逐渐以黄色为最高级的服装颜色，皇帝穿黄色龙袍。晋代"以赤色为贵"，龙袍大都采用红色。隋唐开始重新采用黄色龙袍，唐代时宫廷下令，除皇帝以外，官员一律不许穿黄衣服。宋朝和明朝也流行红色龙袍。到了清朝，龙袍以明黄色为主，也可用金黄、杏黄等色（图2-5）。

图2-5 黄纱绣彩云金龙单龙袍❶

在清朝执政的满族精英正式确立了严格的宫廷服装标准。清朝皇帝在采用汉代风俗的同时，也渴望建立自己的文化身份，并通过对着装规范施加严格的指导，从而能够在控制社会的同时施加秩序和和谐感。1759年，乾隆皇帝对皇室服饰法规进行了编纂，并严格概述了谁

---

❶ 图片来源：故宫博物院官网。

可以穿什么衣服，什么时候穿。一个人在朝廷中的地位可以通过其服装的颜色、工艺质量、材料和装饰来确定。

清朝时，鲜黄色是供皇帝、皇太后、皇后和第一任妃子使用的服装颜色。在特定的庆典场合，人们可以穿其他颜色的衣服，但是明亮的黄色是为这些最高级别的宫廷人物保留的。较不重要的官员根据职级和职位分配颜色。此外，清代只有皇帝才能穿十二章龙袍，龙袍是圆领、大襟、右衽、窄袖加综袖，马蹄袖端长袍，明黄色，用刺绣作金龙九条，再装饰十二章纹样，前后饰正龙各一条，左右及交襟处饰正龙各一条，马蹄袖端饰正龙各一条。领和袖均用石青色镶织金缎边饰。随季节变换棉、纱、裘等材料。由于历史原因，目前存世的龙袍，以清代龙袍居多。❶❷

译前提示：龙袍是中国古代皇帝的朝服，绣有龙形图案。文章对龙袍的图案、形制、材质以及颜色等特征做了详尽的介绍。作为中国传统服饰的典型代表，龙袍不仅拥有自身独一无二的文化内涵，还凝聚着中国封建社会的礼仪规范、穿衣习俗等方面的文化。龙袍，称得上是中国服饰史上一颗璀璨的明珠，散发着无穷的魅力。

三、英语译文

The Dragon Robe

The *mianfu* and the Dragon Robe were typical garments for ancient Chinese emperors. They served as a micro cosmos that exemplify the unique Chinese aesthetic, cosmological and religious views. In ancient Chinese society, it was all strictly specified which class should wear what on what occasions.

In ancient times, ministers had to wear court clothes when going to the court every day, and the emperor was no exception. The emperor's court robe was the so-called Dragon Robe. As early as the pre-Qin period, the monarchs of all countries had unique robes symbolizing the supreme rule of a country. At first, the Dragon Robes of ancient Chinese emperors were not yellow, and the colors of the Dragon Robes in previous dynasties were also different.

The Chinese theories of the Yin and Yang and the Five Elements all try to explain the interdependence and mutual rejection of gold, wood, water, fire and earth. White represents gold; green represents wood; black represents water; red represents fire, and yellow represents earth. In Zhou Dynasty, red was regarded as the superior color for garments, so the Dragon Robe of the dynasty was in red. In Qin Dynasty, black was ranked highest among all garment colors, so Emperor Qin Shihuang used black as the basic color for his Dragon Robe, and all officials followed suit and wore black as often as they could. When Han Dynasty replaced Qin, yellow was promoted to the highest place, so the Dragon Robe of that dynasty was in yellow. The Jin Dynasty ranked red as a noble color, and Dragon Robes of that period took red as the main color. The Sui and Tang Dynasties

❶ 语料参考：臧迎春. 中国传统服饰［M］. 北京：五洲传播出版社，2003：142.
❷ 华梅. 中国服饰［M］. 北京：五洲传播出版社，2004：15–16.

began to adopt yellow Dragon Robes again. In the Tang Dynasty, the court made it official that no one, except the emperor, had the right to wear yellow clothes. In turn, red dragon robes were popular in the Song and Ming Dynasties. In the Qing Dynasty, the Dragon robes were mainly bright yellow, but also golden and apricot yellow.

During the Qing Dynasty, the ruling Manchu elite formalized a strict standard of court dress. While adopting some of the Han Dynasty customs, the Qing emperors were keen to establish their own cultural identity, and by imposing strict guidelines on the code of dress, they were able to institute control over society while imposing a sense of order and harmony. In 1759 the imperial dress regulations were codified under the direction of the emperor Qianlong, outlining in strict details who could wear what and when. A person's place in the court could be established by the colour, quality of workmanship, materials and embellishments of his clothes.

In the Qing Dynasty, bright yellow was reserved for the clothing of the emperor, empress dowager, empress, and the first concubine. Other colours were worn on specific ceremonial occasions, but bright yellow was reserved for these court figures of highest level. Less important officials were assigned colours according to their ranks and positions. In addition, only emperors had the privilege of wearing the Dragon Robe decorated with twelve symbolic emblems, whose design included circular collar, big front, right lapel, narrow and comprehensive sleeves, horse-hoof-shaped cuffs. The robe of bright yellow color was embroidered with nine golden dragons and embellished with twelve symbolic emblems. The front and the back of the robe were embroidered with one front dragon each. There was a front dragon on the left side, the right side and the joining of the left and right borders each. There is also a front dragon at the end of each horse-hoof sleeve. The collar and sleeves were brimmed with azurite gold satins. The materials could be changed according to seasons, for instance, cotton, gauge, or fur, etc. Due to historical reasons, most of the existing Dragon Robes are those of the Qing Dynasty.❶❷

四、词汇对译

十二章龙袍	Dragon Robes decorated with twelve symbolic emblems	最高级的颜色	superior color
大襟	big front	审美观	aesthetic view
右衽	right lapel	宇宙观	cosmological view
综袖	comprehensive sleeves	宗教观	religious view
马蹄袖	horse-hoof sleeves	严格规定	strictly specified
中国宫廷服饰	Chinese court dress	制定严格标准	formalize a strict standard
庆典场合	ceremonial occasions	文化身份	cultural identity
		着装规范	a code of dress

❶ 语料参考：臧迎春. 中国传统服饰［M］. 北京：五洲传播出版社，2003：142.
❷ 语料参考：Hua Mei. *Chinese Clothing*［M］. 于红，张蕾，译. 北京：五洲传播出版社，2003：15-16.

中国的阴阳五行说　The Chinese theories of
　　Yin and Yang and the Five Elements
相生相克　interdependence and mutual rejection

秩序感与和谐感　a sense of order and
　　harmony
装饰　embellishment

五、译文注释

1. 它们体现了中国人独特的审美观、宇宙观和宗教观。

They served as a micro cosmos that exemplify the unique Chinese aesthetic, cosmological and religious views.

【注释】该句前文是对冕服与龙袍的简要介绍，本句概括二者的价值所在——代表中国人的价值观念。该句原文中的"体现"一词被作者巧妙地增译为"serve as a micro cosmos that exemplify"，以形象化和具体化的表述，集中表达了冕服与龙袍所凝聚的多重重要意义与价值。这样的译法比直接译为"they exemplify..."更为生动且更能增强表达效果，充分体现了译者的主体性。

2. 在古代中国社会，哪一个等级的人在哪一个场合穿哪一种衣服，都是有严格规定的。

In ancient Chinese society, it was all strictly specified which class should wear what on what occasions.

【注释】该句对中国古代社会的穿衣规定进行了概述。由于汉语重意合，与英语重形合的语言表达习惯不同，因此在汉译英的过程中，译文会与原语有所差异。该句原文为一个复合句，由几个小分句组成，主要说明中国古代穿衣准则的严格。在译文中，译者对原文语序进行了调整，以"it"作为形式主语，采用主语从句的形式来传达原文语义，这样的调整策略，使译文结构紧凑，主次分明，符合英语读者的阅读习惯，能更充分地体现作者的写作意图。

3. 中国的阴阳五行说认为，金、木、水、火、土相生相克。

The Chinese theories of Yin and Yang and the Five Elements all try to explain the interdependence and mutual rejection of gold, wood, water, fire and earth.

【注释】该句对中国的阴阳五行说进行了简要介绍。五行学说以日常生活的五种物质：金、木、水、火、土元素作为构成宇宙万物及自然现象变化的基础。其源于古代人民对星辰的自然崇拜，是中国道家的一种哲学思想。译者将"土"译为"earth"而不是"soil"，这是因为在整个五行说之中强调的是宏观的宇宙世界，而"soil"只能指微观的土壤，强调的是土壤本身的性质特点。"earth"指的是与天空相对的大地，词义更为广泛，也与五行说所表达的内涵更为贴切。

4. 周代以红色为高级服色，秦时以黑为最高地位的服装颜色，百官都穿黑色衣服，汉灭秦后逐渐以黄色为最高级的服装颜色，皇帝穿黄色衣服。

In Zhou Dynasty, red was regarded as the superior color for garments, but by Qin Dynasty black ranked highest among all garment colors. All officials followed suit and wore black as often as they could. When Han Dynasty replaced Qin, yellow was promoted to the highest place, and favored by the emperors of the time.

【注释】该句对中国周、秦、汉三个朝代的服饰着装色彩规范进行了介绍。原文为由五个流水分句组成的长句，以时间为顺序，表意连贯，富含大量信息。译者在翻译这句话时采用了分译法，将原文拆分为三个独立的句子，以此来避免译文冗杂、表意不清。译文第二句看似简单，译者却对其进行了精心的处理。译者在这一句中增加了"followed suit"这一词组，意在体现官员们对于指令的遵从，以及句尾的"as much as they could"也体现了百官对于皇帝指令的听命与服从，侧面反映了中国封建社会严格的等级秩序，呼应了语篇上下文对服饰色彩变迁背后的权力这一因素的强调。这样的译法一方面体现了中国封建社会的观念与氛围；另一方面能够帮助译文读者更准确理解原文。此外，译文第三句充分体现了英语形合语言的特点，译者先用"when"引导状语从句点明时间背景，然后以"yellow"做主语，谓语部分则采用被动结构"was promoted…and favored by…"，以体现黄色在当时的崇高地位，即使译文层次分明，逻辑清楚，又避免了重复与多余，可谓一举两得。

5．至唐代时，宫廷下令，除皇帝以外，一律不许穿黄衣服。

By Tang Dynasty the court made it official that no one, except the emperor, had the right to wear yellow.

【注释】该句说明黄色服装具有至高无上的地位，只能为帝王所有。译者在汉译英过程中，将"官员一律不许穿黄衣服"译为"no one had the right to wear yellow"，将原文对谓语动词的否定"不许穿"替换为对主语的否定"no one"，且明示"right"一词，突出体现了皇权的威严与崇高，侧面反映出黄色在唐朝至高无上的地位。

六、翻译知识——分译与合译

分译与合译是两种常见的翻译技巧。由于英汉两种语言在词汇、结构上的差异，两者不可能在句式上完全对应，所以在汉译英时必须从词汇搭配、句子结构等方面进行灵活的拆分或融合，即所谓的分译或合译。下面对分译与合译这两种翻译技巧分别进行说明。

（一）分译

分译法指把原文中的一个长而复杂的句子译成两个或两个以上的句子。汉语为意合语言，只要逻辑相关，语意连贯，可以由许多流水分句组成长句。英语为形合语言，句子结构必须遵循一定的英语语法规范，句内的逻辑关系须通过语言标记予以明示，因此在表达上往往有严格的语法限制。在翻译汉语长句时，如果原文内容超过了英文句子结构的容量，无法在既有语法的框架下完全表达，则需要进行分译，即将原来的汉语长句拆分为两个或两个以上的句子。

例1　原文：草鞋是用芒草为料编成的鞋子，通常为贫困者所用。

译文：Straw sandals are woven by Chinese silver grass. They are usually worn by the poor.

【分析】例句原文对草鞋进行了介绍，既描述了草鞋的制作材料，又介绍了草鞋的使用人群。译者在翻译例句时采用翻译中常用的分译法将原文拆分为两个句子，将草鞋的制作材料与使用人群分开描述，各自单独成为一句，使得译文更加简洁、直观。

例2　原文：新娘子在结婚当天要以红色盖头蒙面，象征着童贞、年轻、纯洁，婚礼结束后由新郎亲手揭开。

译文：The bride wore the red veil that symbolized virginity, juvenscence and purity on the day of marriage. After the wedding, it was unveiled by the groom.

【分析】该句是对清代婚服中的红盖头进行介绍。原文是由三个分句组合而成的长句，既介绍红盖头的使用场景，又表达其象征意义。译者在翻译时采用了分译的方法，将原文分译为两个句子，其中第一个句子说明红盖头的穿着场合和象征意义，第二个句子说明婚礼后由新郎揭红盖头的仪式步骤。按时间先后断句分译，逻辑清楚。同时，分译减轻了译文句子结构容量方面的压力，使译文中的两个句子简洁流畅，表意清楚。

例3　原文：中国军服的历史，最早可以追溯到战国时期的"胡服骑射"，从那时起，中国才有了正式的军服。

译文：The history of Chinese martial attire could be traced to "wearing *hu* dress and shooting on horse" in Warring States Period. From then on, China had its formal martial attire.

【分析】该句介绍了中国军服的起源时间。原文由四个分句组合成一个长句，既指出中国军服最早出现的时间，又点明中国军服正式形成的时间。如果直接将原文译为一个句子的话，译文会显得过于拖沓、冗杂，且不容易直接明了传递原文信息。译者在翻译例句时，采用了分译法，将原文一分为二，译为两句。第一句译文先说明战国时期"胡服骑射"是中国最早的军服，第二句译文说明"胡服骑射"的产生代表中国有了正式的军服，两个独立的句子简洁流畅，各有侧重，充分传递了原文信息。

例4　原文：梁，指的是位于鞋头的装饰，通常用皮革为料，裁制成直条，有一梁、二梁、三梁多种。

译文：*Liang* refers to the ornament on the toe cap. It's made of leather cut into straight strips. There are many types, such as Single *liang*, Double *liang* or Triple *liang*.

【分析】该句对双梁鞋的形制与材料进行了介绍，既包含了对"梁"的定义，还包括对其材质、形制、类型的介绍，信息量较大，属于典型的汉语流水句的平行铺排结构。在进行英译时，应先分析原句结构，找出逻辑语义重心，然后根据表达需要进行分译。译者将原文分译为三个句子：先介绍了什么是"梁"；其次介绍其用料和形制；最后介绍"梁"的类型。译文以三个各自独立的短句展开表述，层层递进，逻辑清晰，使译文读者能够对"梁"有多角度的深入理解。

例5　原文：云头履是一种高头鞋履，用布帛为料，鞋首用棕草，高高翘起，形状似翻卷的云朵，故名。

译文：Cloud head shoes are named after their raised toe caps, which are very high and look like rolling clouds. They are made of cloth and silk, with brown grass at the head part of the shoes.

【分析】该句对云头履的材质、形制、名称来源进行了介绍。原文是由六个流水分句构成的长句，信息量十分丰富。但由于每个分句都很简短，原文现得语言流畅，一气呵成，有助于读者迅速掌握云头履的特征。由于汉英语言结构方面的差异，如果翻译时照搬原文的流水句结构，会使译文表达混乱，层次不清。译者在翻译例句时，采用了分译加调整的方法，将原文的一句话分译为两句，并且将云头鞋的命名由来提前与定义相结合。译文第一句说明云头鞋的命名由来，并以"which"引导非限定性定语从句详细说明云头鞋的形制。第二句描述云头鞋的材质。这种分译加调整的方法更大程度上体现了译者的主动性，使译文表达符合目标语的语言规范，同时也使译文逻辑层次更清楚，更具有可读性。

（二）合译

合译指在认真分析源语多个句子间逻辑内涵的基础上，找出各句在逻辑语义上的主次关系，然后在译文中按逻辑主次进行调整与组合，构建符合译入语规范的译文句子。下面举例进行说明。

例6 原文：秦汉以前，平民百姓多以布裹头。这些布的颜色多以青、黑为主。

译文：Ahead of Qin and Han Dynasties, common people wrapped their heads mainly with black and cyan cloth.

【分析】例句为两个独立的句子，分别介绍秦汉以前百姓头巾样式和头巾颜色。翻译例句时，译者将两句合译为一句，将原文第二句的内容译为定语，跟在介词"with"之后修饰"cloth"，说明头布的颜色。这样合译使译文精准地表达了原文的信息，简洁且恰当。

例7 原文：短袖衣的衣袖为长袖衣之半，故称为"半袖"。半袖是一种很特殊的背心。

译文：Sleeves of this garment are half length of that of long sleeve garment, hence its name of "half sleeves", which is a special type of vest.

【分析】这两句对中国古代半袖进行介绍。前一句介绍半袖的名称由来，后一句介绍其从属的服装类型。在译文中，译者采用合译的翻译策略将其合并为一个复合句，用"which"引导非限定性定语从句，将第二句译作"半袖"的定语。这样的合译方法能使译文结构主次分明、逻辑清晰，能将原句信息更为精炼地体现出来，有助于读者更好地了解半袖的名称由来与种类。

例8 原文：笼裙是一种桶形的裙子。它用轻薄沙罗为料制作，呈桶状。

译文：Barrel-shaped dress is made of light and thin gauze which shapes a barrel.

【分析】这两句对笼裙的外形以及材料进行介绍。笼裙最初常见于西南少数民族地区，隋唐时期传入中原，穿着时从头部套入。译者将第一句和第二句后半部分关于裙形的介绍相融合，译为定语，用"which"引导定语从句修饰"笼裙"，使得译文既简洁又准确，将笼裙的特点完美地展现在译文读者眼中。

例9 原文：马褂的式样为圆领或立领，大襟或对襟。马褂下摆开衩，袖口部分大多齐平。

译文：There were additional styles of "ridding jacket"—stand collar or round collar, garment with full front or opposite front pieces, a slit on the lower hem and even cuffs.

【分析】这两句对马褂的样式以及长度进行介绍。褂在明代时指罩在外面的长衣，清代时则指罩在袍服之外的外服，男女皆可穿着。原文用两句分别说明马褂的式样与形制，译者在进行英译时，用合译法将二者合二为一，借助具有解释作用的破折号，清楚列举马褂的式样与形制，细致又不显冗余。另外，结合语篇上下文，使用"there be"句型呼应前文，使译文更为连贯。

例10 原文：她们都梳着高发髻，上穿襦，下穿长裙。服装的颜色都比较清新淡雅。

译文： With their hair in high buns, they wear *ru* on the upper part of the body and long skirt on the lower part, both in plain colors.

【分析】这两句是对穿着襦裙的仕女进行描绘。前一句主要介绍妇女的穿着以及发式，后一句介绍妇女服装的颜色。在汉译英过程中，译者采用了合译的方法，先准确说明妇女们的发型与穿着，然后用介宾短语"both in plain colors"作状语，简明扼要地突出服装色彩特点，避免句式多余，同时又完整再现了原文对妇女服装颜色特征的描绘。

七、译文赏析

段落1

龙袍，从字面上理解就是绣有龙形花纹的袍服，在中国古代服饰文化史上占有重要地位。龙纹最早出现在周朝，当时出现画有龙纹的服装，只是深衣的一种款式，没有形成真正意义上的龙袍。秦汉时期，上下相连的袍服出现，但袍服上仍没有龙形的花纹，所以也没有严格意义上的龙袍。直到隋唐时期，冕服上才开始出现龙纹，但冕服为"上衣下裳"的形制，而帝王平时穿用的常服上并没有龙纹。到了元明时期，帝王袍服上绣有龙纹的现象才越来越多，皇帝也被称为"真龙天子"帝王和龙才为融为一体，自此也就出现了真正意义上的龙袍。❶

The Dragon Robe literally means a robe that is embroidered with images of the Chinese dragon, which played an important role in the history of Chinese ancient dress culture. The design of Chinese dragon appeared as early as in the Zhou Dynasty. At that time, the dress with the design of Chinese dragon was just one pattern of *shenyi* and the real dragon robe had not yet existed. By the Qin and Han Dynasties, the robe with upper and lower parts as a whole was designed, but still no pattern of Chinese dragon in the robe. By the Sui and Tang Dynasties, the crown clothing that consisted of both upper clothes and lower skirt began to have the pattern of Chinese dragon. However, the ordinary dress worn by the emperor wasn't embroidered with the pattern of Chinese dragon. Only by the Yuan and Ming Dynasties, the imperial robe embroidered with the pattern of the Chinese dragon began to become more common, and the emperor was also called "Son of Heaven". In this way, the emperor and Chinese dragon became associated together, and the real Dragon Robe came into being.

【赏析】该段落对龙袍龙纹的产生和发展进行了细致的介绍，对中国传统中龙纹与帝王的关系进行说明，并深入分析了龙袍与龙纹的文化意蕴。英译时，译者运用了不同的翻译策略对原文的细节进行调整。在完整呈现原文大意的基础上，译者灵活的翻译技巧给译文增色不少，尤其是译者对于分译法与合译法的巧妙运用。例如该段原文的第二句是较长的复合句，由几个分句组成，既介绍龙纹出现的时期又介绍龙纹服装与深衣的关系，包含较多的信息。译者在进行英译时，将该句进行拆分，把龙纹出现的时期与后文独立开来，各自成句。这样在保留该句句意的同时又能够精准呈现重点信息，有利于帮助读者增强对龙纹的理解。

---

❶ 语料参考：《中华文明史话》编委会. 服饰史话［M］. 北京：中国大百科全书出版社，2009：60.

此外，对于隋唐时期冕服上龙纹的介绍，译者也采用了分译法对该句进行英译处理，将原文中的转折部分用"however"引导独立成句。而在前半句中冕服上龙纹的出现，译者又采用了合译策略将其统整为一个句子，以关系代词"that"引导定语从句，将上衣下裳的形制译为冕服的后置定语，使译文焦点突出，更符合英文表达规范。针对不同的句意与句式，译者在该段落的翻译中采用了不同的翻译策略，灵活而又巧妙，可见在进行翻译时我们应该"见机行事""顺势而为"。

段落2

中国宋代历史上有"黄袍加身"的说法，意思就是做皇帝，做国家的统治者。但是为什么"黄袍"就代表着皇帝呢？根据中国的阴阳五行学说，世界上所有的事物都是由金、木、水、火、土五种元素构成，这五种元素相生相克，白色代表金，青（绿）色代表木，黑色代表水，赤色代表火，而黄色则代表土。中国社会发展到汉代以后，黄色开始被所有人接受，并被认为是最尊贵的颜色。唐代以后，服饰制度规定只准皇室穿用黄色服饰，其他人一概不允许。从唐代开始一直到中国封建社会的最后一个朝代——清朝，黄色一直都是帝王专用的颜色，是尊贵和权力的象征。❶

There was a saying in the Song Dynasty, "draping an imperial yellow robe over somebody's shoulder" which means to acclaim somebody as the emperor or the ruler of the country. But, why is it a "yellow robe" that stands for the emperor? According to the Chinese thought of Ying, Yang and the five elements (metal, wood, water, fire and earth), all things in the universe consist of the five elements. And there is a phenomenon of mutual promotion and restraint between the five elements. The white stands for metal, the blue for wood, the black for water, the red for fire, and the yellow for earth. Since the Han Dynasty in Chinese history, the yellow had been accepted by all people and was regarded as the most distinguished color. By the Tang Dynasty, the dress system stipulated that except the royal family all others were prohibited from wearing yellow dress. From the Tang Dynasty to the Qing Dynasty, the last one in Chinese feudal society, yellow had been the special color for the royal family, and a symbol of dignity and power.

【赏析】中国传统服饰中，对于颜色的运用十分严谨也十分考究，每种服装颜色可使用的人群与文化内涵都大不相同。本段以"黄袍加身"为例，对中国古代服装中的黄色内涵进行剖析，同时也分析了服装中其他颜色的意义。在译文中，译者不仅采用了分译法，还在某些句子之间增添连词，加强了译文逻辑关系。首先，在该段句首，译者将"黄袍加身"出现的时间与意义分隔开来。译者先是点明该词出现的时间——宋代，然后用"which"对该词的内涵进行分析，使得译句主次分明。其次，原文中对于中国传统中五色内涵的介绍较长，如果将其译为英文的长句可能不便于外国读者理解中国文化。考虑到这一点，译者将其分译为几句话，先是交代这五种元素来自"阴阳五行"，为下文颜色内涵的描述埋下伏笔，接着交代五种颜色的关系，然后对每种颜色的内在意义进行介绍，层层递进。在翻译时，我们应该考虑到译者对原语文化背景的接受程度，然后适当地对原文进行处理，以译者为中心，这样

❶ 语料参考：《中华文明史话：编委会》. 服饰史话 [M]. 北京：中国大百科全书出版社，2009：60–61.

才更有利于原语文化的传播与发扬。

八、翻译练习

（一）请将下列汉语句子译成英文

1. 坐龙纹一般位于袍胸正中，躯干弯曲七弯，尾动而稳，进一步凸显统治者不可动摇的至高无上地位。

2. 中国皇服最突出的特征是绣龙，明清时有袍绣九龙的定制——在龙袍前后两肩、两袖等处绣成对称的八龙，然后再绣一龙在大襟的里面，以象征君权神授，彰显九鼎之尊的皇家威仪。

3. 在清朝，黄色的使用比以往任何时候都更加严格，并且建立了黄色的层次结构以区分帝国氏族中的等级。

4. 长袍是用大量的丝绸制成的，有时分几层或用丝绸填充以增加保暖性。

5. 朝靴是古代帝王和官吏上朝时所穿着的靴子，用乌皮、黑绸缎等材料制作，始于唐代。

6. 龙的形象以丝鳞、丝金的手法演绎出来，金丝的运用在整体图案中更显出皇室的威严与富贵。

7. 马蹄袖身窄小，紧裹手臂，袖口裁为弧线，袖口上半部可以覆盖在手背上。因为便于射箭，也称"箭袖"。

8. 清军入关后，将箭袖用于礼服之上，因外形看起来像马蹄，故名。

9. 夏披领是清代帝后朝服使用的种领饰，一般以绸缎为料，裁剪成菱形，上绣龙蟒等图纹，并加以缘饰。

10. 披领常缝缀在衣上，也有的与衣身分开，使用时罩在肩上，在颈部扣结。

（二）请将下列汉语段落译成英文

到了清朝，统治者延续了中国几千年的封建传统，清朝皇帝的龙袍样式是圆领、右衽、箭袖，颜色以明黄为主，也可以用金黄或杏黄色进行辅助，袍子的上面绣着五爪金龙和五色祥云，在祥云中间还分布着"十二章"图纹。根据中国古代文献记载，皇帝的龙袍上共绣有九条龙，但从实物来看，前后加起来只有八条。这样做的目的是让其他的人认为，皇帝本身就是那第九条龙。其实这第九条龙只是被绣在衣襟里面，一般情况下看不到。龙袍作为皇帝的服饰代表，能更好地表现皇室对龙的崇拜，显示天子的威严。清代统治者还规定，龙纹只能用于皇室，其他任何人都不得使用。❶

九、译笔自测

龙袍及其由来

龙袍，又称"黄袍"，是中国古代帝王的袍服。黄袍作为帝王的专用服装源于唐代。唐代以前，黄色服饰在中国一直都较为流行，普通民众也可穿着。唐代以后，黄色变为皇族的专用色彩，黄袍也成为皇族的专用服饰。

---

❶ 该段落语料参考：《中华文明史话》编委会. 服饰史话［M］. 北京：中国大百科全书出版社，2009：61-63.

黄袍之所以又被称为"龙袍",是因为上面绣有龙形图案。龙是古代中国人创造的一种虚拟动物,乃万兽之首。传说龙身长若蛇,有鳞似鱼,有角仿鹿,有爪似鹰,能翻江倒海,吞风吐雾,兴云降雨。在中国古代社会,龙是帝王的象征,凡是皇帝君王的器物,大都以龙为纹样。

龙袍的基本形式很简单。它是一件长袍,伸到脚踝,长袖,脖子上有一个圆形开口。但是,龙袍的织物和装饰却极具复杂性和丰富性。龙袍上的关键元素是龙。大多数龙袍在服装中央有一条大龙,袖子上有较小的龙,下摆较低。巨龙在繁复的海洋图案中游泳,一些几何图案、自然风光、波浪以及其他色彩鲜艳的人物装饰在衣服的下半部分和袖子上。长袍由大量的丝绸制成,有时分几层或用丝绸填充以增加保暖性。有时,长袍会在颈部固定处或袖口处绣上刺绣。

中国历史上第一个身穿黄袍的皇帝为唐高祖李渊。他不仅以黄袍为常服,还下令禁止庶民(普通百姓)穿黄色的服饰。到唐高宗时,更是重申"一切不许着黄"。李渊之所以选择穿黄色的袍服,与中国传统文化有很大关系。在中国古代,开国皇帝登基时,为了给自己建立的王朝寻找统治根据,总要遵循"终始五德说"。"终始五德说"是战国时期阴阳家邹衍主张的一种观点。邹认为,历史上的每一个朝代都以土、木、金、火、水这五德的顺序交替统治,周而复始。而唐朝按"终始五德说"属"土",土的代表颜色为黄色,因此李渊选择着黄袍。

从唐朝皇帝首穿黄袍开始,黄色正式成为皇权的象征。宋太祖赵匡胤"黄袍加身"后,更是禁止任何人穿黄袍,否则便以谋反论处。宋仁宗时又规定:百姓着装不许以黄袍为底或以黄色配制纹样。从此,黄袍为皇室所独有,黄色亦为皇室所专用。这种制度为此后中国历代皇朝所沿袭,直至清朝灭亡。❶

图2-6　清代皇帝画像❷

十、知识拓展

清朝是由满族建立的政权,因其长期处于游牧生活和征战状态,所以紧身、简洁、便于骑射是其服饰文化的主要特征,这与传统的汉族服饰文化差异较大。清朝统治者一直对自己的民族服饰有着独特的理解,他们不仅认为民族服饰是祖先的传统,而且认为这是他们屡战不败的重要因素,所以对民族服饰的继承和发展极其重视。清朝服饰也是中国历代服饰中最为庞杂和繁缛的,对于近世纪的中国服饰影响较大(图2-6)。

The Qing Dynasty that was found by the Manchu people. Because Manchu was a nomadic and martial tribe, the costume culture in the Qing Dynasty were characteristic of close fitting, simple garments that were convenient for horse-back shooting.

❶ 语料参考:沈周. 中国红:古代服饰 [M]. 合肥:黄山书社,2012:21-24.
❷ 图片来源:华梅. 中国服饰 [M]. 北京:五洲传播出版社,2004:17.

The style of the Qing costume differed greatly from the traditional costume culture of the Han people. The Manchu emperors had distinctive understanding of their own national costume. Special attention was paid to inheritance and development of their national dress, because they were of the opinion that their national costume was not only their ancestor's tradition, but also an important factor to establish full control over the conquered territory. The Chinese costume in the Qing Dynasty that had great influence to modern Chinese costume in past century was the most cumbersome and over- elaborated in comparison with proceeding dynasties.❶

第四节　唐装

学习目标

1．了解唐装的历史及文化内涵
2．熟悉唐装的基本语汇及其汉英表达方式
3．掌握翻译中转换法的相关知识与技能
4．通过实践训练提高唐装文化汉英翻译能力

一、经典译言

翻译是画画，不是照相；是念台词，不是背书。

——王宗炎

二、汉语原文

隋唐服饰

隋唐时最流行的女子服饰是襦裙——短上衣加长裙，裙腰用绸带系得很高，几乎到腋窝的下面。这种服饰在汉代就已经出现，魏晋时期裙腰变得更高，上衣越来越短，衣袖越来越窄；后来又走向另外一个极端，衣袖被加宽到二三尺。隋朝统一全国以后，社会上又时兴小袖，追逐流行的妇女往往里面穿着大袖的衣服，外面再披一件小袖衣，又叫披袄子，并讲究用金缕蹙绣，听任小袖下垂以为美，竟成一时风尚。唐代女子长期穿用小袖短襦和曳地长裙，但盛唐以后，贵族妇女衣着又转向阔大拖沓，衣袖竟大过四尺，裙子也比一般的长很多，后因行走不便被朝廷颁布法令加以限制。一般妇女穿青碧缬（印花或染花织物），戴平头小花草帽。

隋唐女子爱好打扮。"半臂"在宫廷中出现并流传开来，它有对襟、套头、翻领或无领等式样，袖长齐肘，身长及腰。因领口宽大，穿时上胸裸露。这种服装样式存在了很长时间，后来男子也开始穿着。唐代女子的发饰多种多样，而且都有自己专用的名字。早期女子的头发都是高高耸起的，后期流行用假发做成的发髻，显得头发蓬松。女鞋一般是花鞋，多用锦绣织物、彩帛、皮革做成。线鞋则用彩线或麻线编结而成，这些鞋子有许多不

❶ 内容参考：臧迎春. 中国传统服饰［M］. 北京：五洲传播出版社，2003：139.

图2-7　唐代妇女胡服展示图（高春明绘）❶

同的样式。

唐朝是中原文化与少数民族文化及外来文化交流、融合较多的朝代，唐人非常善于吸收和融合西北少数民族的文化以及天竺、波斯等外来民族的文化，在服饰尤其是妇女服装上有明显的反映。627—741年，社会上非常流行胡服等新式服饰，比较普遍的穿着是：头戴金锦浑脱帽，上身穿翻领小袖、齐膝长袄或男式圆领衫子，下身穿条纹间道锦小口裤子，腰里系着有金花装饰带，脚上穿靴（图2-7）。还有的人发髻向上高高地耸立，似展翅欲飞的大鸟，脸上有黄色星的装饰，还有的画着两个月牙，也有的在嘴角酒窝之间涂上胭脂。

唐代初年，女子骑马出行一定要用一种大纱帽遮住全身，叫作幂，后来发展成了帷帽。帷帽的形状像斗笠，帽子周围垂下来的网帘用丝做成，一直垂到脖子的下面。盛唐以后，帷帽被朝廷明令废除，但这种帽子款式还是被人们保存了下来，稍作改动仍然非常流行，对之后的很多朝代都有较大的影响，如今中国南方地区的农村妇女仍然存有类似的遮阳帽。

唐代女子在服饰上比较追求阳刚之美，再加上当时朝廷统治政策的松懈，一些女性开始公开穿着男子的服饰出现。女子喜爱的男装，主要是男子常穿的戎装，戎装的样式是头上戴软脚幞头，身上穿翻领或圆领的缺胯袍，腰系整蹀带，下穿小口裤，脚上穿黑红皮靴或锦履，这种穿着被叫作"丈夫服",是唐代妇女服饰的又一大特色。❷

译前提示：本文主要是对隋唐女子服饰进行描写，如襦裙、半臂、各种鞋履以及发饰等，介绍了其形制特色以及产生的文化背景。隋唐时期是中国封建社会发展史上的一个巅峰，相对富足的物质生活条件和较为宽松的社会文化环境，促使隋唐文化呈现出空前的发展动向。隋唐服饰以其宽松随意、流畅自然、洒脱不羁的风格在中国服装史上画上了浓墨重彩的一笔。

三、英语译文

Costume in Sui and Tang Dynasties

The most popular women's wear during the Sui and Tang Periods was *ruqun*—short upper clothes plus long skirts, and the waist of the skirt was tied as high as nearly to the armpit. The dress appeared as early as in the Han Dynasty. By the Wei and Jin Dynasties, the waist of the skirt was much higher, the upper clothes became shorter and shorter, and the sleeves became narrower and

❶ 图片来源：华梅. 中国服饰［M］. 北京：五洲传播出版社，2004：32.
❷ 语料参考：《中华文明史话》编委会. 服饰史话［M］. 北京：中国大百科全书出版社，2009：32-34.

narrower. Afterwards, it went to the other extreme. The sleeves were widened by two to three *chi* (c. 1/3 meter). After the nation was unified by the Sui Dynasty, small sleeves became a fashion. Women pursuing for fashion began to wear shirt with loose sleeves within and a dress with tight sleeves outside, which was also known as *pi' aozi*. They preferred golden thread for embroidery, and the random hanging of the tight sleeves was considered to be beautiful and became a vogue. In the Tang Dynasty, women wore short coat and full-length skirts, but after the prosperous period of the Tang Dynasty, women of upper class began to wear loose dress with wider sleeves of four *chi* (c 1/3 meter) and long skirt that was much longer than the ordinary ones. Later, it was prohibited by court laws for it was very inconvenient for walking. The ordinary women wore bluish green dyed or printed fabrics and small flat straw hats with flower patterns.

Women in the Sui and Tang dynasties paid attention to dressing up. In that period, "half-length sleeves" emerged and became popular in the court, with various styles, such as front-opening, pullover, turn-down collar, collarless styles. The sleeves were long to the elbow and the jackets were to the waist. As the collar was large, the part above the bosom was exposed. This dress style lasted for quite a while which was also worn by men later. In the Tang Dynasty, the women's hairdos were also various, and for each style there was a special name. In the beginning, the hairdos were upward, and the bob made of wig became popular later, which made the hair bulky. The women's shoes were embroidered with flowers, mostly made from brocade fabrics, colorful silk and leathers. The string shoes were knitted with colorful thread or twines in different styles.

In the Tang Dynasty, the culture of the Central Plains had made its greatest achievements in the exchanges and mix with the cultures from other minority groups and nations. The Tang people were quite good at absorbing and merging with minority cultures in the northwestern regions and foreign cultures from ancient India, Persia and other places, which was clearly shown in the women's wear. From 627 to 741, some new dress, such as the *hu* dress, were very popular. Dress of common people usually included a golden-thread brocade hat, upper dress of either knee-length jackets with turn-down collars, small-sleeves coat, or dress shaped like the male round-collar shirt and lower dress of brocade trousers with the decoration of stripes, a waistband decorated with the golden flowers, and a pair of boots. Some people had their buns stood high, just like a bird on the wing, some had yellow star or moon-shaped decorations on the faces, and some others would apply kermes at the dimples in their cheeks.

In the early Tang Dynasty, while riding on a horse, woman had to cover her whole body with a kind of large black gauze cap, which was also called *mi* (covering cloth). Later, it evolved into veiled cap shaped like a bamboo hat, the hanging part of the veiled cap was made of silk and reached to the shoulders. After the prosperous period of the Tang Dynasty, the veiled cap was abolished by a court decree, but the style was still preserved. It became popular after some alterations, and exerted influence on the dynasties afterwards. At present, many rural women in the southern China still wear sun shading cap of the similar style.

The Tang women tended to pursue a kind of manly beauty in dress. Besides, due to the slackening of the ruling policies at that time, some women began to show up in public in men's dress. The men's dress favored by women mainly referred to martial attire, that is, a soft gauze scarf, a turn-down or round-neck collar robe, a *zhengxie* band (a special waistband), narrow-end trousers, ruby leather boots or brocade shoes. This kind of dress is called "*zhangfu fu*" (manly dress), which is another feature of the women's wear in the Tang Dynasty.❶

四、词汇对译

小袖　small sleeves

小袖外套　small-sleeves coat

曳地长裙　full-length skirt

圆领衫　round-collar shirt

青碧缬　bluish green dyed or printed fabrics

条纹裤　trousers with the decoration of stripes

披帛　silk shawl

胭脂　kermes

线鞋　string shoes

酒窝　dimples

中原文化　the culture of the Central Plains

大纱帽　large black gauze cap

天竺　ancient India

帷帽　veiled cap

金锦浑脱帽　a golden-thread brocade hat

遮阳帽　sun shading cap

齐膝长袄　knee-length jacket

统治政策　the ruling policies

翻领　turn-down collars

戎装　martial attire

五、译文注释

1. 隋朝统一全国以后，社会上又时兴小袖。

After the nation was unified by the Sui Dynasty, small sleeves became a fashion.

【注释】该句对隋朝时期的流行服装进行了介绍。原文中"社会上又时兴小袖"中"时兴"一词为谓语动词，而"小袖"为名词，在该句中做宾语。在翻译过程中，译者运用了转换法，将"small sleeves"作为译文主语，同时将动词"时兴"转换为名词"fashion"，做动宾短语"became"的宾语，使得译文主谓结构清晰、句式严谨。

2.（她们）讲究用金缕蹙绣，听任小袖下垂以为美，竟成一时风尚。

They preferred golden thread for embroidery, and the random hanging of the tight sleeves was considered to be beautiful and became a vogue.

【注释】该句前文是对当时女子衣袖着装特点的主要介绍。蹙金绣，最早出现在隋代文献中，到唐代成为宫廷刺绣的主要针法：以捻紧的金线作为刺绣线材，使用跨线将金线横向钉于绣地上，形成一种类似皱纹状的线型刺绣结构工艺。译文中，译者为减轻读者的阅读负担，没有对"金缕蹙绣"工艺进行详细而烦琐的介绍，而是将其减译为"golden thread for embroidery"，表达了这一刺绣工艺的主要特色，使译文简洁易懂。原文是三个短句组合而成的一个复合句，由于汉语重意合，在句子成分上显得松散；而英文重形合，句式较为严

---

❶ 语料参考：《中华文明史话》编委会. 服饰史话［M］. 北京：中国大百科全书出版社，2009：122–126.

谨。在译文中，译者将原文后两个分句组合在一起，并用"and"一词将其与首句连接起来，使得译文句式结构严谨、逻辑清楚。

3. "半臂"在宫廷中出现并流传开来，它有对襟、套头、翻领或无领等式样。

In that period, "half-length sleeves" emerged and became popular in the court, with various styles, such as front-opening, pullover, turn-down collar, collarless styles.

【注释】该句对"半臂"的样式进行了介绍。原文是由两个单句组成的复合句，每个单句都有自己的主语，前一句的主语是"半臂"，后一句的主语是"它"。虽然两者表示的都是"半臂"，但这种将两个完整的主谓结构并列且无连词衔接的句型结构，显然与英文的习惯表达方法不同，不能直接搬到英译文中。译者在翻译过程中对句子结构进行了灵活调整，选用"half-length sleeves"为译文主语直接译出原文中的第一个分句，然后"with+名词"的结构来介绍样式，从而将原文中的第二个分句译作状语，使得句子重点突出。这样的翻译策略能够突出原文的中心内容"半臂"，让译文读者对其样式有深入的认识。

4. 这种服装样式存在了很长时间，后来男子也开始穿着。

This dress style lasted for quite a while which was worn by men later.

【注释】该句前文是对"半臂"服装样式的介绍。"半臂"原来流行于女性，后来男性也开始穿着。原文是由两个分句组合而成的一个汉语复合句，但在英译的过程中，作者对其句式进行了转换，将原文的第二个并列分句转换为了译文中的定语从句。译者在翻译时将"This dress style"作为译文主语，用which引导的定语从句修饰主语，补充描写这种样式的新的穿着人群——男性。译文句式简练，同时突出主语"半臂"及其穿着人群广的特点。

5. 女鞋一般是花鞋，多用锦绣织物、彩帛、皮革做成。

The women's shoes were embroidered with flowers, mostly made from brocade fabrics, colorful silk and leathers.

【注释】该句对唐代女鞋进行了介绍，包括样式与面料。原文中"花鞋"一词为名词，而在译文中，译者转变其词性，将其用动词词组来表达，译为被动结构"were embroidered with flowers"，突出花鞋的装饰特点。这样的词性转换译法能够帮助读者更准确地理解花鞋的文化内涵与形制特点。

6. 如今中国南方地区的农村妇女仍然存有类似的遮阳帽。

At present, many rural women in the southern China still wear sun shading cap of the similar style.

【注释】例句前文是对唐代帷帽的介绍。该句主要表达的内容是帷帽在当代仍然存在于我国南方农村。原文的谓语动词是"存有"，在翻译过程中，译者使用"wear"一词来表达。"存有"于"戴着"看似意义不同，实则表达的都是帷帽依然存在的现状。译者使用"wear"能够更好地与主语"rural women"相匹配，使译文更符合英文的表达规范。

六、翻译知识——转换

根据汉英语言差异，在翻译时，我们常常要应用各种翻译技巧，改变表达方式，使译文在忠实于原文的基础上通顺自然。转换法是诸多翻译技巧中比较常用的一种。所谓转换，即对原语做出词性、句式等方面的调整，使其更符合译入语的表达习惯，避免译文语言受原文影响而过于异化。转换的形式多种多样，既可以是词性的转换，又可以是句式等的转换。下面对汉译英过程中的词类转换与句式转换进行详细介绍。

（一）词类转换

词类转换指在翻译过程中，按照目的语的规范，把原文中某些词类的词转换成另一种词类的词。英语和汉语的词类大部分是重合的。但汉语中，一个词可充当的句子成分比英语多。例如，英语中充当主语的只有代词、名词或相当于名词的动名词或不定式，充当谓语的只有动词；而汉语中，名词、动词、形容词都可以做主语、谓语、宾语和表语。因此，翻译时不要拘泥于原文中所用的词性，而要根据具体语境和译入语的特点进行灵活转换。

例1 原文：我们一贯认为荷包是古人用来盛放碎银子、票据、印章、手帕、针线等零碎物品的小型佩囊。

译文：It has been our long-held view that pouch was a small bag for carrying pocket money, receipts, seals, handkerchiefs, needle and thread, and other odds and ends in ancient times.

【分析】在该句中文原文中，"认为"一词是动词，而在译文中被译为名词"view"。该句译文采用了主语从句，it为形式主语，that引导的从句作为整个句子真正主语。这样的结构使译文句子成分与原文有所不同，但同样准确描述了荷包的主要作用。一般而言，汉语中动词的使用频率很高，译为英语时顺应英语中的名词化倾向，将原文中的动词转换为名词，使译文更符合目标语的表达习惯。

例2 原文：他在面料上不加选择。

译文：He is indiscriminate in fabric.

【分析】该句原文中，"不加选择"一词为动词词组。译者在翻译过程中进行了词性转换，在忠于原文信息的基础上，灵活处理，将其译为"is indiscriminate"，用形容词替代了原文中动词所表达的词义。这样的转换策略在向译文读者充分传递原文信息的同时，也使译文表达地道、流畅，符合目标语的表达规范。

例3 原文：清代身披云肩的女子给观众的印象很深。

译文：Women wearing cappa in Qing Dynasty impressed the audience deeply.

【分析】在汉译英的过程中，汉语的动词可以转换成英语的名词，而有时汉语的名词也可以转换成英语的动词。比如例句原文中"印象"是名词，译文将其译为动词"impress"，用词较精练，表意直接。当然，根据译者的偏好，例句的翻译中也可保留"印象"的名词形式，将原句译为"Women wearing cappa in Qing Dynasty left a deep impression on the audience."

例4 原文：他们全体赞成他的建议：官吏按照品级高低，穿用不同材质的腰带。

译文：They were all in favor of his suggestion that officials, based on their ranks, wore waistband in different materials.

【分析】该句原文"赞成"一词为动词，而在译文中被译为以名词"favor"为中心词的词组"in favor of"，虽然词性由动词转换为了名词，但随着译文表达形式的调整，两者表达的都是赞同的意思。此外，该句的"赞成"一词也可译为"be for"的形式，与之相对的"反对"可译为"be against"。

例5　原文：唐代官服的皮革腰带的主要特点是工艺精湛，经久耐用。

译文：Leather belt on official costume of Tang Dynasty is chiefly characterized by its fine workmanship and durability.

【分析】该句是对唐代官服上的皮革腰带特点进行介绍。原文中"特点"一词为名词词性，而译文将其译为动词词组"be characterized by"，这是一个英文中表达事物特征的高频词组。类似的还有中文表示"目的在于"用英文词组"aim at"表达，也是将原文中的名词转换为译文中的动词形式。在翻译过程中，我们可以关注一些中英文中发生词性转换的对应表达，以此提高翻译速度，使译文更加地道。

（二）句式转换

因为原文与译文的句子结构具有各自的特点与规则，汉译英过程中的句式转换是一个比较复杂的问题。原文和译文句子之间不是单一的对应关系，两者之间的转换并不是一成不变的，翻译时应该根据表达需要进行灵活处理。一般来说，原文和译文之间的句式转换主要是出于句法结构、惯用法、行文表达方式以及语感和情态等方面的需要。

例6　原文：这位设计师有一种习惯叫人受不了，意见反复不定，一会儿一个变化。

译文：The designer had a disconcerting habit of expressing contradictory ideas in rapid succession.

【分析】例句原文是由三个流水分句构成的长句，体现了意合语言的特点。译文为一句焦点突出、层次清楚、结构紧凑的英文句子，体现了形合语言的特征。汉英翻译过程中，这种句式上的转换源于汉英两种语言系统的差异。汉语的句法结构一般使用意合的方法，即句子中各个成分靠内在的逻辑关系贯穿起来，没有一定的连接词，因此，句子的结构形式较为松散，而词句较为简洁。而英语的句法结构往往重形合，即句子中各个成分（词、词组、短句）的结合，都有适当的连接词表达其相互的关系，所以结构形式上比较严谨，在该句中就有较典型的体现。如若按照中文语序进行英译会使句子整体较为冗长，因为英语句法较为刻板，缺少弹性，而汉语句法灵活，富于弹性。因此，在进行汉译英时我们要将汉语中的意合转为英语中的形合，这样才能使译文通畅，符合英文表达习惯。

例7　原文：这个牌子的衣服高雅、简洁，这点早已远近闻名了。

译文：The brand was already spreading a fame for its elegance and simplicity.

【分析】该句汉语原文句子由两个单句构成，而英文译文句子只含一个单句，这属于句式转换中复合句与简单句的转换。由于英汉两种语言的表达方式不同：英语中常借助词形变化和连接词用一个单词或短语表达一层意思，一个简单句里可以借助词汇和语法手段表达多层意思；而汉语则常用一个包含谓语动词的短句来表达一层意思，一句话里可以用几个短句来表达多层意思。所以在进行翻译时，我们先要理解原文句意，然后根据译文表达习惯采用适当的翻译策略，使得译文简洁、自然。

例8　原文：不信你就自己来看看这个设计。

译文：If you don't believe, you can come over to see the design.

【分析】汉语注重"意合"，有些语法形式上的词语可以简化或省略；英语注重"形合"，形式上追求结构完整。例句文是一个汉语单句，但在英译时，译成了一个复句。原文中"不信"一词被译为了条件状语"If you don't believe"，这样的翻译方法才符合英文的句法逻辑，主次分明，同时句子结构完整。

例9 原文：我们召开这次关于时尚的会议不是来讨论这个问题的。

译文： We haven't held the fashion conference to discuss this issue.

【分析】该例句体现了汉英翻译中句式结构方面的否定转移现象。例句原文是对"讨论"一词进行否定，而在译文中，却将否定词"haven't"置于"held"前面。这种现象称为否定转移，该句是对不定式的否定。一般来说，英语中的否定词总是紧紧出现在被否定部分的前面，但有时否定词所否定的并不是紧随其后的部分，而是后面的某个部分，这就是否定的转移。下面将再举一例进行说明。

例10 原文：这位年轻的设计师并不是因为薪水高而接受那份工作。

译文： The young designer didn't accept the job because it was well paid.

【分析】该句是"not...because（of）..."结构的否定转移。原文中"不是"直接位于"因为"之前，否定设计师接受那份工作的原因。译文中的"didn't"位于动词"accept"之前，但其否定对象却并非动词本身，而是后面由"because"引导的原因，属于英语句式结构中的否定转移现象。在英语否定句结构中，如果not否定的是主句中的主语或谓语，则不属于否定转移；如果像例句一样否定的是because分句，则属于否定转移。英文中这类否定转移句式主句与分句之间不需要用逗号隔开。

七、译文赏析

段落1

就封建社会的文化和经济发展状况而言，中国的唐代无疑是人类文明发展史上的一个巅峰。唐朝政府不仅对外国实行开放政策，允许外国人到中国经商、吸引外国留学生，甚至允许外国人参加选拔官员的科举考试和出任官职，对外来的文化、艺术、宗教采取欣赏和包容的态度，使当时的首都长安成为中外文化交流中心。特别值得一提的是，唐朝妇女不必恪守传统规范，她们可以穿袒露胸臂的宽领服装或吸收其他国家服饰风格穿出异国情调，还可以穿胡服男装骑射，并享有选择配偶和离婚的自由。❶

In terms of cultural and economic development of the feudal society, the Tang Dynasty in China was doubtless a peak in the development of human civilization. The Tang government not only opened up the country to the outside world, allowing foreigners to do business and come to study, but went so far as to allow them in exams for selection of government officials. It was tolerant, and often appreciative, of religions, art and culture from the outside world. Chang'an, the Tang capital, therefore became the center of exchange among different cultures. What is worth special mention is that women of the Tang Dynasty did not have to abide by the traditional dress code, but were allowed to expose their arms and back when they were dressed, or wore dresses absorbing elements from other cultures. They could also wear men's riding garments if they liked, and enjoyed right to choose

---

❶ 语料参考：华梅. 中国服饰［M］. 北京：五洲传播出版社，2004：29.

their own spouse or to divorce him.❶

【赏析】唐代是中国历史上政治、经济、文化发展最为繁荣的朝代之一。本段不仅介绍了唐朝的对外政策与态度，还介绍了唐代文明的开放性与包容性，为读者呈现出国泰民安的盛世图景。译文体现了唐代昌盛开明，运用多种翻译策略使得唐朝盛世跃然纸上。在本段的英译过程中，译者采用了灵活的翻译策略。首先，不拘泥于原文词语的字面含义，进行灵活的选词，如将"外国"一词意译为"outside world"，而没有照搬其中文对应词"foreign countries"。这是出于对当时中国的时代背景的考虑，唐朝以外的疆土在如今并不全是"外国"，用"outside world"一词严谨而又贴切。其次，译者将"外国人"与"外国留学生"译为统一译为"foreigners"，使两个分句合二为一，简洁明了。最后，文章的第二句介绍唐代的开放政策，包含较多信息，由若干分句构成。这种由多个流水分句构成的长句，在进行英译时如果不进行适当的处理就容易使译文冗长、语义不清。译者采用分译法将该句拆分为三个独立的句子，分别描述长安的开放政策、对外来文化的态度以及长安的重要地位，译文表意清楚，简洁易懂。其中间还使用"therefore"一词进行衔接后两个句子，明示前后文的因果关系，逻辑自然严密。

段落2

面妆虽说不是唐女发明的，但奇特华贵、变幻无穷。唐女在脸上广施妆艺，不只是涂上妆粉，以黛描眉，以胭脂涂两颊，以唇膏点唇，还要在额头上涂黄色月牙状（称为"额黄"）饰面，据说是模仿西北少数民族。眉毛被画成不同的形状。眉式也花样翻新，传说唐玄宗曾命画工画十眉图，它们都有不同的名称，如"鸳鸯""小山""垂珠"等。平民也有自己时髦的眉毛风格。此外，双眉中间要有花钿，用鸟羽、螺钿壳、鱼鳃骨，金箔，或直接用颜料画。眉梢则要描上一道"斜红"。嘴唇被涂成当时最流行的形状，在唇角外一厘米处再点上两个红酒窝。盛唐以后，这些酒窝甚至长到鼻子两侧，还变化出钱形、杏桃形、小鸟形、花卉形等。❷

Although facial makeup was not invented by the Tang women, they were quite elaborate and extravagant. These women not only powdered their faces, darkened their eyebrows, rouged their cheeks, put on lipsticks, but also decorated their foreheads with a yellow crescent (called *ehuang*), which was said to be an imitation of the northwestern ethnic minorities. Eyebrows were penciled in different shapes, and the eyebrow styles were constantly renovated. It was said that the Xuanzong Emperor asked his court painter to record the ten eyebrow styles, which all had different names, such as the "mandarin duck" the "small peak" the "drooping pearl". The commoners had their own trendy brow styles as well. Moreover, decorative designs were put between the eyebrows as a finishing touch, made with feathers, seashells, fish bones, pure gold, or just painted on. At the tip of each eyebrow there was a "red slant". Lips were painted into the trendiest shapes of the time, complimented by two artificial red dimples about one centimeter from both sides of the lip. At the

❶ 语料参考：Hua Mei. *Chinese Clothing*［M］．北京：五洲传播出版社，2004：29.
❷ 语料参考：华梅．中国服饰［M］．北京：五洲传播出版社，2004：35.

most prosperous time of the Tang Dynasty, these dimples went so far as to reach the two sides of the nose, in shapes of coins, peaches, birds and flowers. ❶

【赏析】该段落栩栩如生地描绘了唐代女性的面部妆容，精致、高雅，如诗如画，仿佛是艺术品般呈现在读者面前。由于唐女妆容独特，属于中华文化中的瑰宝之一，故涉及较多的专有名词，在进行英译时需要进行适当的处理。译文不仅完整展现了唐女妆容的特征，还运用异化的翻译策略，完整地保留了唐代妆容的意象，给外国读者展现了中国传统独有的妆容内涵。如"鸳鸯""小山""垂珠"分别被译为"mandarin duck""small peak""drooping pearl"，既考虑了译文读者的接受能力，又淋漓尽致地显现了中国古典审美韵味，一举两得。此外，在部分词句中，译者运用替代法，如将第二句的"唐女"译为"these women"，这是因为前文提到了"the Tang women"，所以此处用"these women"来替代它，使上下文衔接更为紧密，为读者营造了生动的文化氛围。此外，译者还对部分句式进行了转换，如将"眉式也花样翻新"译为"the eyebrow styles were constantly renovated"，这是根据中英文语言表达上的差异，将汉语中常用的主动句转换成了英文中常用的被动句，这样更符合英文读者的语言使用习惯，有利于译文读者的接受与理解。

八、翻译练习

（一）请将下列汉语句子译成英文

1．唐代服饰的基本设计理念是历史的必然、时代的选择，对后来服饰的发展有着深远的影响。

2．优秀的设计最能体现本民族传统的特点，在精神和形式上都具有与其他民族不同的强烈个性。

3．那个时期，政局稳定，经济发达，生产和纺织技术有了长足的进步，对外交往频繁，使服装史无前例地繁荣起来。

4．唐朝服饰融合了周朝精心制作的花纹、战国时期的轻松感、汉朝的活泼感以及魏晋的雍容华贵。

5．相当富足的物质基础和相对宽松的社会环境，使得唐代的文化空前发展，诗歌、绘画、音乐、舞蹈等领域群星璀璨。

6．纺织业到了唐代有了长足进步，随着缂丝和染色技术的进步，纺织材料的种类、质量和数量都达到了前所未有的高度，服饰风格的多样化成为时代的潮流。

7．与花枝招展的襦裙装相比，将整套男服穿在身上则别有一番情致。

8．尽管儒家经典中早就规定："男女不通衣裳"，但在唐代绘画和敦煌石窟中，常常可以看到穿着男式服装的女性。

9．在那个年代，无论身份尊卑，无论在家还是外出，许多妇女都是这样穿的。不难想象，唐代对于女性来说是一个相当开放的社会。

---

❶ 语料参考：Hua Mei. *Chinese Clothing* ［M］. 北京：五洲传播出版社，2004：35.

10. 唐代典型男服是头戴幞头，身穿圆领袍衫，腰间系带，脚蹬乌皮靴。

（二）请将下列汉语段落译成英文

隋唐时期，幞头是男子主要的首服。幞头比较软，不太美观，人们便在幞头里垫入衬物，使幞头显得更硬挺，这个衬物称"巾子"。到了五代，幞头发生了较大变化，通常做成方裹型，顶部有两层，前低后高，形似帽子。隋唐时期的帽延用南北朝的式样，纱帽的使用仍很普遍。笠帽在当时也很普遍，以竹篾、棕皮、草葛及毡类等材料编成，形状多为圆形，有宽大的帽檐，用于遮阳、避雨，常用于劳作之人。唐代汉族服饰受西域服饰的影响很大，少数民族所戴的胡帽传入中原，在当时颇为流行。❶

九、译笔自测

唐代女装

盛世唐装中最夺目的要数女装，以及妇女那变幻多样的发髻、佩饰和面妆。唐女讲求配套着装，每套都是一个独具特色的整体形象。人们不是一时心血来潮，而是依据着装所处的社会背景，展示服饰的美。因而，每一种搭配都个性鲜明，又有其令人玩味的文化底蕴。唐女配套装可归结为三种：胡服（或是丝绸之路引入的外来服装）、传统的襦裙装、打破儒家礼仪规范的整套男装。

唐女爱披一件帔子，或是两只胳膊上搭着披帛。两者的区别在于帔子阔而短，一般披在一肩，从出土的唐代女俑上可以看到。传说有一次宫中露天筵席，一阵风起，将杨贵妃的帔子吹到了别人的帽子上。由此来看来，帔子或许是很轻盈的，当然也不排除以厚重毛织帔子御寒的可能性。披帛就是我们通常所说的"飘带"，长长的，一般较窄，从身后向前搭在小臂上，两端自然下垂，中国古典绘画经常有这种美妙的披帛。

与襦裙装相配合的足服，凤头高翘式锦履，也有麻线编织的鞋或蒲草鞋，很精致很轻巧。除了绘画作品为我们提供了形象资料以外，在新疆等地出土文物中可以看到实物。唐女著襦裙装时，头上一般不戴帽子，当然也有花冠等是属于装饰性的，出门时用帷帽遮住脸。这种帷帽从唐初开始流行，至盛唐时，女人们连帷帽也不屑于戴了，干脆露髻骑马出行。当年发式可谓多变，体现着唐女的奢华之风。有30多种高髻、双髻、下髻，这些发髻大多因形取名，也有的是以少数民族的族称取名。在唐代仕女画中可以看得到发髻上插满了金钗玉饰、鲜花或丝。

盛世唐装就是这样散发着耀眼的光芒。尽管人们习惯将对襟袄通称为"唐装"，以其代表中国传统服饰，但那不过是一种以唐代为荣的说法。事实上，现代的唐装远不及唐代的服装璀璨夺目、千姿百态而富有生命气息。当年"万国衣冠拜冕旒"的气势是多么宏大，唐代中国才真正称得上是"衣冠王国"。❷

❶ 该段落语料参考：沈周. 中国红：古代服饰［M］. 合肥：黄山书社，2012：124.
❷ 语料参考：华梅. 中国服饰［M］. 北京：五洲传播出版社，2004：30-38.

十、知识拓展

由于紫色的官服在唐代最为尊贵，所以"紫袍"一词也成为显官要职的代称。绯色的官服即指大红色的袍服，大袖，右衽，衣襟及袖常有镶边。绿袍是唐代六品及七品官的官服，品级相对较低。青袍是唐代官服中等级最低的，后来"青袍"一词多用来代称品级低的官吏。唐代的公服制度对服色的规定虽然很严格，但在具体实行的时候也可以变通。如果一些官吏的品级不够，但遇到奉命出使等特殊情况，经过特许可穿用比原品级高一级的服色，俗称"借紫"或"借绯"（图2-8）。❶

图2-8　唐代男子服装❷

Since purple was seen as a noble color, people who wore purple robe were regarded as those with an important position in Tang Dynasty. Red gown had large sleeves and right lapel with trimmed edges and sleeves. The green robe was for the sixth and the seventh rank of officials who were in low position, whereas officials with cyan gown ranked the lowest among all the moderate positions. Later on, people referred to use the phrase "cyan gown" for low-rank officials. Even though there were strict rules for the official costume in Tang Dynasty, it tended to be flexible when these rules were put into practice. When some lower ranked officials had to be presented on special occasions as being sent on a diplomatic mission, they were authorized to wear a higher ranking costume. It's referred to as "borrowing purple" or "borrowing red".

---

❶ 内容参考：沈周. 中国红：古代服饰［M］. 合肥：黄山书社，2012：124.
❷ 图片来源：华梅. 中国服饰［M］. 北京：五洲传播出版社，2004：33.

第五节　传统裙装

学习目标

1. 了解中国传统裙装的历史及文化内涵
2. 熟悉中国传统裙装的基本语汇及其汉英表达方式
3. 掌握翻译中重构法相关的知识与技能
4. 通过实践训练提高中国传统裙装文化汉英翻译能力

一、经典译言

在演技上，理想的译者应该是"千面人"，不是"性格演员"。

——余光中

二、汉语原文

女裙

　　裙是中国古代女子的主要下装，由裳演变而来。裙是由多幅布帛连缀在一起组成的，这也是裙有别于裳之处。古人穿裙之俗始于汉代，并逐渐取代下裳成为女性单独使用的服饰。那时的女子穿裙，上身要搭配襦或袄。魏晋以后，裙子的样式不断增多，色彩搭配越来越丰富，纹饰也日益增多。宽衣广袖、长裙曳地是贵族服装的主要特点（图2-9）。

图2-9　周昉《簪花仕女图》❶

---

❶ 图片来源：沈从文. 中国服饰 [M]. 上海：世纪出版集团，2011：12.

南北朝以后，裙子才逐渐成为女性的专属服装。两晋十六国时期，一种名为"间色裙"的裙子为流行，它使女性身材显得修长。隋代，裙的式样承袭了南北朝时的风格，间色裙也仍为女性普遍穿用。

唐代的裙子以宽博为尚，裙幅有六幅、七幅、八幅、十二幅不等，裙摆长度也明显增加。女子在穿裙时，多将裙腰束在胸部，甚至到腋下，以使裙子显得修长，这与隋代装饰风格一致。由于曳地宽裙不便于劳作，而且在用料上极其浪费，以致朝廷下禁令加以限制"女子裙不过五幅，曳地不过三寸"。宋代沿袭了唐代的裙装风格，裙身仍然宽博，裙幅以多为尚，折裥很多。

辽金元时期，汉族女性的裙装基本上沿袭了宋代的裙式，少数民族的裙装则保留了民族特点。如辽金时期的契丹、女真族穿裙，颜色多为黑紫色，上面绣全枝花，通常把裙穿在团衫内。

明代的裙式仍然具有唐宋时期裙装的特色。明初，女性以颜色素雅的裙子为尚，纹饰不明显。到了明末，裙上的纹样日益讲究，褶裥越来越密，出现了月华裙、凤尾裙等。传统的裙式在此时被进行了改制，如鱼鳞百褶裙就是在百褶裙的基础上发展并流行起来。

清代有一种朝裙，专供太后皇后及命妇在祭祀、朝贺等典礼上穿用。此外，平民女子所穿的马面裙也是清代女裙中较有特色的样式。❶

译前提示：衣装是人类由野蛮跨向文明的重要标志，中华民族开创的裙装艺术是古代中国优秀文化传统中最具艺术魅力的财富之一。裙装又可分为男裙与女裙，本文正是对我国古代女裙进行介绍。作者从汉代到清代，对每一时期典型女裙的样式以及特征进行了简要描写，对中华女裙历史进行了大致梳理，给读者们呈现了一幅色彩斑斓的"中华女裙历史图鉴"。

三、英语译文

The Women's Skirt

The skirt, evolved from *shang* , was the ancient Chinese women's main garment for lower part of the body. It was made of several connected widths (*fu* in Chinese) of cloth which differed from *shang*. The custom of wearing skirts started in Han Dynasty (206B.C.-220A.D.).It gradually replaced *shang* for women's exclusive use. Women wore skirts to match *ru* or *ao* at that time. After Wei and Jin Dynasties (220-420), the styles and the colors of skirts became diverse. Loose garment with wide sleeves and long skirt down to the ground became the main feature of noble clothing.

Skirts gradually became women's exclusive clothing after Northern and Southern Dynasties. A kind of "skirt of colored strips" prevailed in the Wester Jin, East Jin and Sixteen Kingdoms Period. Women who wore in the skirt looked slender. In the Sui dynasty, the skirt style inherited that of the Northern and Southern Dynasties (420-589), and "skirt of colored strips" was still commonly worn by women.

In Tang Dynasty(618-907), women's skirts varied in widths including six widths, seven widths,

---

❶ 语料参考：沈周. 中国红：古代服饰［M］. 合肥：黄山书社，2012：96.

eight widths and twelve widths. The skirt hem was obviously lengthened. When women wore the skirts, they usually bound the skirt waist around their chests beneath the armpit. In this way, it would make skirts look long and slim. The decoration style remained consistent with that of Sui Dynasty (589-618).Considering the length of the skirt down to the ground wasn't fit for labor work and it was a waste of cloth，this style of dress was restricted by the court. According to the order of the court, the number of widths of women's skirt shall not exceed five and the length of the skirt end on the ground no more than three Cun (Chinese size unit). The skirt style, that is, wide skirt with many widths and lots of folds, was followed in the Song Dynasty.

In Liao, Jin and Yuan Dynasties, women's skirt of Han nationality basically followed the style of the Song Dynasty, while the skirt of ethnic minorities still retained their characteristics. Taking the skirts of Khitan and Jurchen ethnic minorities of Liao and Jin Dynasties as examples, they were mostly in dark purple with a whole embroidered flower and *tuanshan* (a kind of gown) was worn outside.

Skirts in Ming Dynasty retained some characteristics of Tang and Song Dynasties. Women in early Ming Dynasty wore plain skirts with little patterns on it. Until late Ming Dynasty, the number of patterns and folds increased greatly. Lunar corona skirt and phoenix tail skirt appeared. For example, the traditional skirt style was updated. Based on the pleated skirt, fish-scale pleated skirt was developed and flourished.

In Qing Dynasty (1644-1911), there was a court skirt exclusive for the use of the empress dowager, empress and women who were given titles or ranks by the emperor when they offered sacrifice to gods or ancestors. In addition, the horse-faced skirt worn by ordinary women had distinctive features in Qing Dynasty.❶

四、词汇对译

多幅布帛　several connected widths （*fu* in Chinese）of cloth

长裙曳地　long skirt down to the ground

间色裙　skirt with colored strips

裙幅　skirt width

折裥　pleat

裙式　skirt style

少数民族　ethnic minorities

契丹族　Khitan

女真族　Jurchen

黑紫色　dark purple

团衫　*tuanshan*（a kind of gown）

素裙　plain skirt

月华裙　lunar corona skirt

凤尾裙　phoenix tail skirt

鱼鳞百褶裙　fish-scale pleated skirt

朝裙　court skirt

太后　empress dowager

命妇　women who were given titles or ranks by the emperor

马面裙　horse-faced skirt

祭祀　offered sacrifice

❶ 语料参考：沈周著. 中国红：古代服饰［M］. 合肥：黄山书社，2012：96.

五、译文注释

1. 裙是由多幅布帛连缀在一起组成的，这也是裙有别于裳之处。

The skirt was made of several connected widths (*fu* in Chinese) of cloth which differed from *Shang* .

【注释】该句对"裙"的结构进行了介绍，同时指出"裙"与"裳"的区别。裳主要用于冕服、朝服等礼服，它是前后两片的形式。汉译英过程中，译者采用了合译法，将原本由两个分句构成的句子，译为只包含一个分句的译文。译者将原文的第二个分句译为"which"引导的定语从句"which differed from *shang*"，作"skirt"的定语，在句尾对其进行补充说明。合译法一方面使译文句式精炼，另一方面使句子所描写的裙的特征更加突出。

2. 两晋十六国时期，一种名为"间色裙"的裙子广为流行，它使女性身材显得修长。

A kind of "skirt of colored strips" prevailed in the West Jin, East Jin and Sixteen Kingdoms Period. Women who wore in the skirt looked slender.

【注释】该句是对两晋十六国时期"间色裙"及其构成的介绍。在译文中，译者采用了直译策略，将"间色裙"译为"skirt of colored strips"，使译文读者能清楚了解该裙装的色彩特点。此外，为使表意更简洁流畅，译者采用了分译的策略，将原文的一句话分译为两个独立的句子，分别对间色裙的流行时间和特点进行说明，避免了句子冗杂，体现了译法的灵活性。

3. 女子在穿裙时，多将裙腰束在胸部，甚至到腋下，以使裙子显得修长，这与隋代装饰风格一致。

When women wore the skirts, they usually bound the skirt waist around their chests beneath the armpit. In this way, it would make skirts look long and slim. The decoration style remained consistent with that of Sui Dynasty (589-618).

【注释】该句对唐代女子的穿裙方式做简要介绍。原文采用五个分句对唐女穿裙的步骤以及效果进行描写；汉译英过程中，译者运用分译的策略将其译为三个独立的句子：先是对唐女穿衣步骤这一举动具体描写，进而点出其效果，最后将其与隋代装饰风格进行比较。由于原文分句较多，信息涵盖范围广，如果采用直译的方式容易使得译文结构繁杂且无法突出重点。译者采用分译法，将其拆分成三个句子，层层递进、逻辑清晰。

4. 如辽金时期的契丹、女真族穿裙，颜色多为黑紫色，上面绣全枝花，通常把裙穿在团衫内。

Taking the Khitan and Jurchen ethnic minorities of Liao and Jin Dynasties as examples, they were mostly in dark purple with a whole embroidered flower and *tuanshan* (a kind of gown) was worn outside.

【注释】该句对辽金时期契丹族、女真族的裙装进行了介绍。在本句的中文原文与译文的对比之下，英汉语言的差异体现得淋漓尽致。汉语惯用散点句法，作者在对契丹、女真裙装介绍时，不仅描述其颜色，同时对图案加以分析；英文惯用焦点句法，在译文中，译者增加了"they were"二词，使译文主语更加突出。并且在句子末尾，译者巧妙运用同义替换，将"把裙穿在团衫内"译为"团衫穿在外"，准确传达原文语意之时又简化了译文。

"*tuanshan*"是典型的异化翻译，译者在其后对其进行了补充解释，一方面突出中国服饰的特色，另一方面又能够让译文读者轻松地了解其内涵，可谓一举两得。

5. 南北朝以后，男子逐渐不着裙装。

Skirts gradually became women's exclusive clothing after Northern and Southern Dynasties.

【注释】该句对南北朝以后裙装的着装对象进行了介绍。作者在原文中使用否定句式点明男子不再成为裙装的着装对象，换言之，裙装开始成为女子专属服装。下文中，对于裙装的介绍也以女性为主体。译者没有直译原文的否定句式，而是巧妙地使用正译法，调整对着装对象的表述，译为肯定句式，直截了当地点明南北朝以后裙装的穿着对象为女性，更好地衔接了下文，同时又不背离原文语义。

6. 清代有一种朝裙，专供太后皇后及命妇在祭祀等典礼上穿用。

In Qing Dynasty (1644-1911), there's a court skirt exclusive for the use of the queen mother, empress and women who were given titles or ranks by the emperor when they offered sacrifice to gods or ancestors.

【注释】该句对清朝朝裙的使用人群以及使用场合进行了介绍。清朝朝裙专为太后、皇后以及命妇所用。译文中，译者采用增译的翻译策略对"命妇"这一人物进行补充介绍。命妇，是中国古代特有的人物，泛称受有封号的妇女。如果译者不对其内涵进行简要介绍，很容易让英语读者摸不着头脑，从而对朝裙的使用人群产生误解。译者采用定语从句"who were given titles or ranks by the emperor"的方式，解释命妇的内在含义，能够让英语读者体会到汉语特色词"命妇"所包含的传统文化信息。

六、翻译知识——正译和反译

英语和汉语都有肯定式表达和否定式表达，翻译时一般可以直接转换，即原文肯定式直接译为译文肯定式，原文否定式直接译为译文否定式。只要表意准确，译文流畅，直译最好，因为这样与原文表达更为接近。但由于英汉两种语言的差异，有时候直接转换会使表意不准确，表达不流畅，这样就需要转换表达方式，将原文的肯定式译为否定式，即正话反说，也称反译法；或者将原文的否定式译为肯定式，即反话正说，也称正译法。下面将分别举例说明。

（一）反译

翻译时反译分为两种情况。一种情况是原文里的肯定表达在译文中无法使用肯定句式，因而只能将其转换为否定句式，这种情况占正话反说，即反译的绝大多数。另一种情况是原文中肯定表达在译文中既可以使用肯定句式，也可以使用否定句式，但使用否定句式比较地道，更符合目的语的表达习惯，而肯定句式虽然可以达意，但不太符合平时表达习惯，或表达效果不及否定句式，这时还是正话反说，即反译为好。这里的否定或反说除了指使用not, nor, neither, hardly等否定词的句子之外，也包括使用由否定前缀dis-, un-, im- 等构成的否定词的句子。

例1　原文：清代的女子非常重视肚兜，直到新婚之夜才能被丈夫看到。

译文：Women in Qing Dynasty attached great importance to *dudou*. It wouldn't be seen by others except for her husband until the first night of the wedding.

【分析】在中国服饰历史中，不同时期对内衣有不同的称谓，汉代称为"心衣"，唐代出现了一种无带的内衣，被称为"诃子"，宋代出现了"抹胸"，元代出现了"合欢襟"，明代出现了"主腰"，清代的女性内衣则称"肚兜"，是一种极有特色的内衣。肚兜一般做成菱形，也有长方形、正方形、如意形、扇形、三角形等。该句主要介绍清代女子对肚兜的重视程度，原文使用的是肯定句式来说明肚兜的重要性以及相关习俗。汉译英过程中，译者采用了反译法，将肯定句式"直到新婚之夜才能被丈夫看到"译为否定句式"It wouldn't be seen by others except for her husband until the first night of the wedding"，即"肚兜除了在新婚之夜被丈夫看到，其他人都不会看到它"。"not...until"这一英语中惯用否定句式使译文表达地道、流畅，强调了清代女子对肚兜出现场合的重视程度，同时也加强句子语气。

例2　原文：裳，是遮蔽下体的服装统称，无论男女、尊卑都可以穿着。

译文：*Shang*, as a general designation of dress covering lower part of the body, could be worn by anyone disregarded social status.

【分析】该句对中国传统服饰中的裳及其使用人群进行了介绍。裳的形制与后世的裙十分相似，但区别在于裙多被做成一片，而裳则分成前后两片。原文点明任何性别和社会地位的人都可穿裳。汉译英过程中，译者采用了反译法，将穿着裳的人群译为"anyone disregarded social status"，"disregarded"中的"dis"为否定前缀，该词义为"不管、不顾、不理会"，带有否定意义，突出了使用人群不受社会阶级限制的特点。

例3　原文：服饰文化的蓬勃发展，揭开了中国人民生活习俗的神秘面纱。

译文：The vigorous development of costume culture has uncovered the mysterious veil of life and customs of Chinese.

【分析】该句将服饰文化与中国人民的生活习俗进行了关联。原文中，作者强调的是服饰文化给大众展示了我国人民的习俗风貌，属于肯定表达。而在译文中，译者没有采用肯定形式，以"reveal"来译"揭开"，而是采用含否定前缀"un-"的"undercovered"一词来表达揭开之意，"undercover"一词意为"揭开、揭示"，是"cover"一词的否定形式。反过来，虽然"undercover"一词含有否定意义，但在英译汉的过程中也可以被译为肯定形式。

例4　原文：袍衫之内穿裤，裤作为内衣，款式变化很小。

译文：Pants as underwear with little change in style，were worn inside gown。

【分析】该句对唐代男子的袍衫做介绍。唐代男子以袍衫为常服，其着装方式也与其他服装略有不同。原文对款式变化的描述为肯定表达"很小"，但译文将原文的动词"变化"转译为名词"change"，并采用反译方法，用否定词"little"来修饰它，形成否定结构。英文中，表示否定或者少量时，常用"little"，此处使用反译法正是迎合英语读者的语言习惯，有助于其对于文章的理解。

例5　原文：这是一件素色上衣，上面空空如也。

译文：It was a plain coat, with nothing on it.

【分析】该句看似一句简单句，但在英译的过程中，译者同样采用了反译手法。中文里"空空如也"表面上并无明显否定含义，但其实际表明的是什么也没有，其背后暗含否定意义。译者将其译为"nothing"，简单直观，突出该服装的简洁素雅，精准传达原文语义。

（二）正译

正译与反译一样，都是由两种语言的差异造成的，翻译过程中无法直接转换时只好反话正说，即所谓正译。正译也是分为两种情况。一种情况是原文的否定表达方式在译入语里没有，只能转换成肯定式。另一种情况是原文的否定表达方式在目的语中既可用肯定式，也可以用否定式，但用肯定式比较地道，更符合译入语的表达习惯，这时还是转换成肯定式较为合适。另外，翻译中有时还要看上下文的语气和译入语的表达习惯，经权衡利弊后再选择是否正译。

例6 原文：膝裤又称"套裤"，两裤筒不连成一体。

译文： Knee pants are known as "leggings" with two individual pant legs.

【分析】该句是对膝裤形制的介绍。在中国服装史上，传统服装一般为"上衣下裳"，或者上下相连，裤子大多以内衣的形式存在。膝裤没有腰，没有裆，上面系带，可束在胫上，上长至膝，下长至踝，用带子系扎。男女皆可穿，穿时加罩在长裤之外，裤管造型多样。原文采用否定形式表明膝裤裤筒相互分离，以否定词"不"为标志。汉译英过程中，译者采用正译法，反话正说，将否定表达"两裤筒不连成一体"译为不含任何否定词的"with two individual pant legs"，精确说明了两裤筒相互独立的特点，避免了句子冗余。

例7 原文：腰上系着腰带，飘洒自如，给人不入世俗的感觉。

译文： Tied around the waist with sash, danging freely, the wearer looked as drifting beyond the secular society.

【分析】该句是对对襟袒胸衫的介绍。对襟袒胸衫是魏晋时期文人广泛穿着的一种衫，对襟、大袖，衣长至膝下。衫的对襟常有垂褶，随心所欲、半披半曳。原文中"不入世俗"带有否定词"不"字，属否定表达，该词可理解为超脱、游离于规则之外。翻译过程中，译者采用正译法，用肯定形式"drifting beyond the secular society"表示这一意义，其中"beyond"译为"超脱"，也恰如其分地表达了不入世俗之意，而且使得译文更有意境感。

例8 原文：花翎如此高贵，低级官员无法享用。

译文： Peacock feather as a symbol of high rank was exclusive for high rank officials.

【分析】顶戴花翎是清代独特的礼帽，分两种，一种为暖帽，另一种为凉帽。礼帽在顶珠下有翎管，质为白玉或翡翠，用来安插翎枝。翎枝又分为蓝翎和花翎两种。该句主要说明花翎的使用人群，"低级官员无法享用"含否定词"无法"，属否定表达。翻译过程中，译者采用了正译法，将原文转换为肯定表达"exclusive for high rank officials"，意即"只能为高级官员享用"，同义替换了原句句意。在充分传递原文信息的同时，与前句的"高贵"一词相对应，能让译文读者对花翎的尊贵地位产生更为深刻的理解。

例9 原文：明代以后，冕冠不再使用，而使用朝冠。

译文： Emperor crown was replaced by court crown after Ming Dynasty.

【分析】冠是一种首服，在中国古代具有特殊的意义，是"昭名分，辨等威"的工具。古代贵族男子成年后就要戴冠，普通人则裹头巾。戴冠者，还要根据等级和身份的不同，佩戴不同形制的冠。冕冠始于西周，是中国古代最重要的冠式，是帝王、王公、卿和大夫参加祭祀典礼时所戴的等级最高的礼冠。冕冠制度一直为后代沿用，冕冠的基本式样也被历代沿用。原文使用否定表达，说明朝冠取代冕冠这一改变。在翻译过程中，译者并未直接采用原文的否定形式，而是运用正译法，使用"replaced"一词精练地说明二者的更替以及朝冠地位的重要性。例句译文如果采用与原文一致的否定句型，则会显得过于繁杂、拖沓。

例10 原文：衫的袖口不收紧，呈垂直型。

译文： *Shan* had loose sleeves falling straight down.

【分析】衫由深衣转变而来，是魏晋时士人的常用服装，并逐渐成为单衣的通称。衫多由轻薄的纱罗制成，一般采用对襟，衫襟既可用带子系缚相连，又可不系带子，比袍穿用方便，散热性好，适合夏天穿用。该句介绍了衫袖口的形制特点。原文采用否定表达"不收紧"对其进行描述。而翻译过程中，译者采用正译法，使用"loose"一词来替换，意为"松的"，与"不收紧"同义，但却更简洁明了，准确传达原文语义。

七、译文赏析

段落1

"拜倒在石榴裙下"是中国人熟知的俗语，多比喻男子对女子的爱慕倾倒之意。石榴与中国的服饰文化有着密切的联系，古代女子的裙装多为石榴红色，而当时染红裙的颜料也主要是从石榴花中提取而来。石榴裙在唐代非常流行，很多年轻的女子都喜欢穿着，显得格外美丽动人。石榴裙一直流传至明清，久而久之就成了古代年轻女子的代称。相传唐明皇的宠妃杨玉环（又称"杨贵妃"）非常喜爱穿石榴裙。唐明皇过分宠爱杨贵妃导致终日不理朝政，大臣们不敢指责皇上，便迁怒于杨贵妃，对她拒不行礼。唐明皇感到宠妃受了委屈，于是下令：所有文官武将，见了贵妃一律行礼，拒不跪拜者，以欺君之罪严惩。大臣们无奈，只得在每次见到杨贵妃穿着石榴裙走来时，纷纷下跪行礼。于是，"拜倒在石榴裙下"的典故流传至今，成为男子爱慕、崇拜女子的俗语。❶

"Bowing to a lady's pomegranate skirt" is a Chinese well-known saying. It's often a metaphor for a man's love to a woman. The pomegranate had close relations with Chinese costume. Women of ancient times mostly wore skirts in pomegranate-red color, and the dye was extracted from the pomegranates at that time. The pomegranate skirt was very popular in Tang Dynasty (618–907). It was so beautiful and charming that many young women loved wearing it. It remained popular until Ming and Qing Dynasties (1368–1911) so it became the synonym of young ladies in ancient times. According to legend, Emperor Ming of Tang Dynasty loved concubine Yang Yuhuan（also known as "concubine Yang"）who liked pomegranate skirt very much. Emperor Ming was so indulged in his affection to concubine Yang that he had no time to handle the state affairs. Because the ministers did not dare to blame the emperor, they turned their resentfulness to concubine Yang and refused to bow down to her. Emperor felt the wrongs she suffered and released an order: once meeting concubine Yang, all the civil and military officials must bow to her. The one who didn't bow down to her shall be given a severe punishment. Thus, the literary quotation "bowing to her pomegranate skirt" has

---

❶ 语料参考：沈周. 中国红：古代服饰［M］. 合肥：黄山书社，2012：101.

passed down to the present. It becomes the idiom to express a man's affection to a woman. ❶

【赏析】该段落是对中国俗语"拜倒在石榴裙下"的故事起源的精彩描述。一方面，作者对石榴在中国文化里的意蕴进行了介绍；另一方面，作者也向读者生动叙述了唐明皇独宠杨玉环的故事。由于该段叙事语句较多，原文句式也较为零散，英译时译者的灵活性显得尤为重要。英译文既向外国读者展示了石榴在中华文化中的独有内涵，又活灵活现地讲述了唐明皇与杨玉环之间的故事，将俗语"拜倒在石榴裙下"的来龙去脉完整地呈现在读者面前。译者在译文中对句式"so...that..."的运用十分灵活，比如将"石榴裙在唐代非常流行，很多年轻的女子都喜欢穿着，显得格外美丽动人。"译为"it was so beautiful and charming that many young women loved wearing it"。这里，译者没有拘泥于原文的句式结构，而是采用"so...that..."英文句式结构，将"显得格外美丽动人"译作唐女穿着石榴裙的原因，不仅使译文因果关系清晰，同时使表意更加简洁明了。再如，译者将"唐明皇过分宠爱杨贵妃导致终日不理朝政"译为"Emperor Ming was so indulged in his affection to concubine Yang that he had no time to handle the state affairs"。译者同样使用"so...that..."句式直截了当地指出"唐明皇宠爱杨贵妃"和其"终日不理朝政"之间的关系，明示了语言内部的逻辑。此外，译者将"终日不理朝政"译为"had no time to handle the state affairs"，细究译文中的遣词造句，可以发现译者进行了否定转移，即将原文对动词"理"的否定转移到译文中对名词"time"的否定，表达形式发生了变化，但却准确传达了作者的原意。可见，在本段落翻译中，译者在翻译方法和译文表达方面体现出了相当好的灵活性。

段落2

襦裙装，上为短襦、长衫，下为裙，这也许算不上新颖，但唐女将它穿出了新样。如短襦或长衫，在圆领、方领、斜领、直领和鸡心领的交替流行中，竟索性将其开成袒领，这是在前朝未曾出现过的创新之举。最初还主要为宫廷嫔妃、歌舞伎等穿用，但很快便引起了仕宦贵妇的垂青，这说明唐代人思想是非常开放的。儒家经典明确规定要用衣服将身体裹得很严，像唐女这样领子低到能见到双乳上侧露出乳沟的款式，在礼法森严的中国古代社会是空前绝后的。❷

Ruqun, which is made up of the upper jacket and long gown and a lower skirt, may not be novel, but Tang women made a new look by wearing it. The Tang women inherited this traditional style and developed it further. For example, the collar of short jacket or gown, with alternative popularity of round, square, bias, straight and chicken-heart shapes, was opened up as far as exposing the cleavage between the breasts. This was unheard of and unimaginable in the previous dynasties, in which women had to cover their entire body according to the Confucian classics. But the new style was soon embraced by the open-minded aristocratic women of the Tang Dynasty. The style of Tang women, whose collar was so low that it reached the cleavage above both breasts, was unprecedented in ancient Chinese society with strict etiquettes and laws. ❸

❶ 语料参考：沈周著. 中国红：古代服饰［M］. 合肥：黄山书社，2012：101.
❷ 语料参考：华梅. 中国服饰［M］. 北京：五洲传播出版社，2004：30-31.
❸ 语料参考：Hua Mei. *Chinese Clothing*［M］. 北京：五洲传播出版社，2004：31.

【赏析】该段落主要对唐女穿着的襦裙装进行了介绍。作者详细描述了襦裙装领口类型和穿着人群，并将其与唐代文化风气相联系，将服装内涵上升到了社会文明，让读者从襦裙装里领略到唐朝的开放风气。在译文中，译者不仅采用了增译法、分译法的翻译技巧，还在部分语句的英译过程中使用了反译法，使译文流畅自然，更易为外国读者所接受。首先，在对襦裙装领口样式进行介绍之前，译者增加了"The Tang women inherited this traditional style and developed it further"。这句话看似普通，实则起到了承上启下的作用。它既是对前文传统襦裙装样式的总结，又引入后文唐女的创新，恰到好处地发挥了"桥梁"的作用。其次，在对唐女的穿法创新进行说明时，译者将后文"儒家经典明确规定要用衣服将身体裹得很严"，提至此处，对前朝妇女的着衣现象作补充说明"in which women had to cover their entire body according to the Confucian classics"，这样灵活调整原文内容的先后顺序，提前交代文化背景，更有利于译文读者对译文内容的理解。此外，译者对于"空前绝后"一词的翻译采用了反译法。该词在中文语境中含有开创性的意义，译者采用含否定前缀"un-"的"unprecedented"一词来表述，简洁凝练，加强了语气，突出了唐女着装形式之"新"，有助于加深译文读者的印象。

八、翻译练习

（一）请将下列汉语句子译成英文

1. 郁金裙也是以植物色染成的，它既有郁金香的美又散发着香气。

2. 唐中叶时一位公主的百鸟裙，更是中国织绣史上的名作，其裙以百鸟羽毛织成，白天看是一色，灯光下看是一色，正看一色，倒看一色，而且裙上呈现出百鸟的形态，可谓巧夺天工。

3. 石榴与中国的服饰文化有着密切的联系，古代女子的裙装多为石榴红色，而当时染红裙的颜料也主要是从石榴花中提取而来。

4. 皇帝感到宠妃受了委屈，于是下令：所有文官武将，见了贵妃一律行礼，拒不跪拜者，以欺君之罪严惩。

5. 唐代崇尚丰满、浓艳之美，赏花喜欢赏牡丹，人则讲究男无肩女无颈，马也要头小颈粗臀部大。

6. 笼裙是一种桶形的裙子，用轻薄纱罗为料制作，呈桶状，穿着时从头套入。最初常见于西南少数民族地区，隋唐时期传入中原。

7. 百褶裙是指有数道褶裥的女裙，褶裥布满周身，少则数十，多则逾百，常以数幅布帛为料制作。每道褶裥宽窄相等，并于裙腰处固定。隋唐时期，此裙多用于舞伎乐女，宋代时广为流行。

8. 马面裙是清代最为常见且流行的裙式。裙的两侧是褶裥，中间有一段光面，俗称"马面"。

9. 罗裙是以罗为面料制成的裙子，主要流行于唐代。一幅画中的唐代仕女头梳高髻，身穿圆领祖胸罗衫，披红色帔帛，下着双色相间的罗裙，双臂舒展，正翩翩起舞。

10. 凤尾裙是一种由布条组成的女裙，将各色绸缎裁剪成宽窄条状，其中两条宽，余下的均为窄条，每条绣上花纹，两边镶滚金线，或者是缝缀花边。

（二）请将下列汉语段落译成英文

唐代仕女画家张萱、周昉惯画宫中艳丽丰腴的女子。周昉的《簪花仕女图》中，美人着踝肩长裙，上身直披一件大袖纱罗衫，轻掩双乳。由于画家手法写实，既如实地描摹出唐代细腻透明的衣料，又逼真地描绘出女子那柔润的肩和手臂。唐朝的审美看重丰腴和华美，比如花以牡丹为美，男人和女人以脖子和肩膀短为美，马以头小脖子粗背部硕大为美。在唐代绘画中，女性通过手风琴式的褶裥裙来展示她们的柔美，并将腰部一直抬高到腋下，使腰线呈桶形，以显示丰满圆润的身体轮廓。❶

九、译笔自测

男裙

在汉代，裙不限于女性穿用，也是贵族男性的常见装束，如"裙屐少年"成为富家子弟的代称。汉魏六朝时期，男人穿裙子的风气极为流行、时尚。《宋书·羊欣传》里记载说，羊欣从小就喜欢书法，尤其擅长隶书，深得著名书法家王献之的喜爱。羊欣12岁那一年，其父为乌程（今浙江湖州市）县令，王献之为吴兴郡（治所在乌程）太守。这年夏天，羊欣"著新绢裙，昼寝"。这时，王献之来到羊欣家拜访羊欣的父亲。临走时，王献之见羊欣穿的新裙子非常好看，一时兴起，"书'裙'数幅而去"。羊欣醒来后，按照王献之给他书写的几幅"裙"字条幅练习书法，书法造诣"因此弥善"。

公元1368年，明朝建国。新王朝的统治者不遗余力重振汉民族的传统服饰风格，并重新规定了服饰制度。明代的服饰仪态端庄，设计宏美，成为中国近世纪服饰艺术的典范。中国明代曾一度兴起肖像画，艺术家们写实而传神的作品，刻画了当时的服饰形象以至服饰的细节。肖像画中留下最多的是官员和文人的形象，大多戴儒巾或四方平定巾，穿大襟长衫，有的还手拿拂尘。扬州西部明代墓穴出土过一套士人服饰，其中有拖着长长垂角的儒巾，有镶着深色缘边的圆领宽袖斜襟大袖衫，还有高筒毡靴。类似的配套服饰形象被保留在京剧的剧装中，因此我们一看就知道是中国的文人雅士。由于受到肖像画的影响，反映当时人民生活的其他绘画方式也承袭了相同的现实主义风格。

明朝，是中国男人穿裙的最后一个朝代。明朝最后一个皇帝崇祯（1628—1644年在位）在国破家亡之际，命其皇子换上青布棉袄、紫花布枪衣、白布裤、蓝布裙、白布袜、青布鞋、皂布巾，打扮成平民百姓模样，以避战祸。可见当时平民男子是穿裙的。在画家戴进的《太平乐事图》中，骑在水牛背上和步行的农人都穿着短裙，即那种围在腰间，一圈皱褶，长仅及膝的裙子。裙子里有长裤或短裤。如今京剧中丑角店小二的典型服饰形象，还保留着这种皱褶短裙的基本造型。❷❸❹❺

❶ 该段落语料参考：华梅. 中国服饰［M］. 北京：五洲传播出版社，2004：30.
❷ 语料选自：沈周. 中国红：古代服饰［M］. 合肥：黄山书社，2012：96.
❸ 安广禄. 趣话古代的男裙[J]. 文史天地：2007（11）：2.
❹ 臧迎春. 中国传统服饰［M］. 北京：五洲传播出版社，2003：113.
❺ 华梅. 中国服饰［M］. 北京：五洲传播出版社，2004：55.

图一五六　五代一高髻，上襦，公服圆花长裙，拨书盛装，宝琚宴住（摹自南唐顾闳中《韩熙载夜宴图》墨住部分）

图2-10　唐代女子裙装❸

十、知识拓展

对于唐裙的描绘，诗人几乎想尽了绝妙的诗句，除了款式外，还有不少提及裙色。从诗中可以看到，当年的裙色相当丰富，而且官方的束缚少，因而可以尽人所好。仅色彩就有深红、杏黄、深紫、月青、草绿、郁金等，其中以石榴红裙流行时间最长。李白、杜甫、白居易诗中都有关于石榴裙的描述。《燕京五月歌》中记述石榴裙流行盛况，说石榴花开的时候满是浓重艳丽的石榴红，千家万户买石榴花给家中的女子染红裙。可以想象有多么壮观（图2-10）。❷❸

Descriptions of the Tang dresses were found in a vast array of poems, both in terms of style and color. A vast array of colors was found in the poems, because there was no official decree on what color was or was not appropriate. Personal preference was all that mattered, be it deep red, apricot yellow, deep violet, ultramarine, sap green or turmeric. Pomegranate red was popular for the longest time. In poems by Li Bai, Du Fu and Bai Juyi, the most outstanding poets of the time, lady in pomegranate skirt was an enduring image of beauty. *Song of May in Yanjing* had an interesting account of the popularity of pomegranate red skirt. It was so popular that in the season when pomegranate blossoms colored the city in red, every household was buying the flower to dye the dresses of their girls. You can imagine how spectacular it was.

第六节　官服

学习目标

1. 了解中国传统官服的历史及文化内涵
2. 熟悉中国传统官服的基本语汇及其汉英表达方式
3. 掌握汉语流水句翻译的相关知识与技能
4. 通过实践训练提高中国传统官服文化汉英翻译能力

一、经典译言

真有灵感的译文，像投胎重生的灵魂一般，令人觉得是一种"再创造"。

—— 余光中

---

❶ 图片来源：沈从文. 中国古代服饰研究［M］. 上海：世纪出版集团，2005：385.
❷ 本章知识拓展中文内容参考：华梅. 中国服饰［M］. 北京：五洲传播出版社，2004：31-32.
❸ 英文内容参考：Hua Mei. *Chinese Clothing*［M］. 北京：五洲传播出版社，2004：31-32.

二、汉语原文

中国官服

古代中国官员的服饰是很丰富的，各个朝代都有自己的规定，甚至在同一个朝代里也会多次变更。

以袍为朝服，始于东汉（25—220），此前的官服为上衣下裳制。官服必有冠，汉代的文官多戴进贤冠，冠下衬有介帻（一种头巾）；武官戴武弁大冠，与官阶相呼应。秦汉时的男人，不分贵贱都戴帻，只不过官员的帻衬在冠下，平民无冠。魏晋南北朝时，官员戴漆纱笼冠，它的制作方法是在冠上用黑色丝纱编织丝笼，笼上涂漆水，使之高高立起，而里面的冠顶还隐约可见。汉代的冠式，都前高后低。以后逐渐改制，到了魏晋时期，改"高山"冠使之卑下，此后冠式就逐渐改为平式或前俯后仰式，到了明代已基本看不到汉代冠式的痕迹（图2-11）。

图2-11 官服❶

唐代官员和士庶都戴幞头。幞头初期是以一幅罗帕裹在头上，较为低矮。后在幞头之下另加头巾，以桐木、丝葛、藤草、皮革等制成，犹如一个假发髻，以保证裹出固定的幞头外形。中唐以后，逐渐形成定型的帽子，名字仍叫幞头。幞头之脚，或圆或阔，犹如硬翅而且微微上翘，中间好似有丝弦，因其有弹性，这一类叫作"硬脚"。

直脚幞头是宋代官员首服的独有式样，幞头两侧的直脚各向左右长长地展开，为什么要那么长，有一种说法是为防止官员上朝站班时交头接耳。

---

❶ 图片来源：网易网。

　　明代官员的首服由唐宋幞头演变为乌纱帽，其间的样式并无多大差别，只是原为临时缠裹，后定型为帽子。"乌纱帽"在汉语里也成了官位的代名词。唐、宋、明三代官袍的样式变化不大，官员品级的高低一般以服色区分，有明确的规定，其间曾稍做调整，并沿用到清王朝退出历史舞台。

　　唐代时女皇武则天曾赐百官绣袍，以文官绣禽，武官绣兽。明朝对此加以仿效，开始在官服前襟饰以有图案的补子来区分文武官员的品级。明代官员朝服为盘领，袖宽三尺，袍色分三种，一品至四品穿绯袍（绛红色袍），五品至七品穿青袍，八品、九品穿绿袍；未入流杂职官与八品以下相同。而官员的常服为团领衫束带。这对京剧服饰中的官吏服饰有着直接的影响。

　　对于格外重视宫廷朝仪的清朝初期的统治者来说，能够彰显身份阶层的朝服官服相当重要。他们制定了中国历史上最为繁杂的衣冠制度，无论色彩、纹饰、款式均出章入典、规定严谨，并以图示说明，要求后世子孙也能"永守勿愆"。朝廷甚至设立督造官服的织造局，慎选织工绣手专事官服制作，清宫服饰也格外讲究织工的细致、刺绣的华美、饰件的齐备。

　　马蹄袖、马褂是清代官服的主要元素，但朝服和常服胸前绣"补子"的做法却直接取自前代明朝。文官绣禽类，武官绣兽类，并依品级的高低绣制不同的飞禽走兽，以此显示不同级别的职位和权力。与明朝不同的是，禽兽的花样与明朝略有差异，常常配以精致的花边，突出了装饰效果。样式上，清朝的补子绣在袍服外面的对襟大褂上，前襟补子也随之分为两块。明朝的乌纱帽到了清代换成了花翎，用孔雀毛上的"目晕"花样的多少区分级别。官员的朝服和常服，里三层外三层，行袍、行裳、马褂、坎肩、补服，重重叠叠，还要佩戴各种朝珠、朝带、玉佩、彩绦、花金圆版、荷包香囊等等，朝珠又有翡翠、玛瑙、珊瑚、玉石、檀木的等级限定，连丝绦都有明黄、宝蓝、石青之分，服饰的等级之别到了高度细密的程度。❶

　　译前提示：官服，又称"公服"，是中国古代官吏或侍从在处理公务时所穿的服装。衣冠服饰常常用于彰显个人的社会身份和地位。特别是在等级森严的封建社会，统治者制定各种着装规范，以此来标识官员等级和庶民地位，在维护统治秩序的同时，在客观上增加了中国服饰的多样性。本文梳理了中国古代官服发展的时间脉络，简要介绍了各历史时期中国官服的主要特点。

三、英语译文

Chinese Official Uniform

　　Chinese official uniform is very complex . Each dynasty had rules of its own, which could be changed many times even within a dynasty.

　　It was not until the Eastern Han Dynasty (25-220) before the *pao* replaced the two-piece costume as the official uniform. There is no official uniform without the cap. In Han Dynasty, civil officials wore *jinxian* caps with a *ze* or kerchief lining the cap, while the military officials wore a hat

---

❶ 语料参考：华梅. 中国服饰［M］. 北京：五洲传播出版社，2004：59-66.

designed for their ranks. Men of all social strata in Qin and Han Dynasties all wore a kerchief, the only difference being that the commoners wore no cap or hat above it. In Wei, Jin, and Southern and Northern Dynasties, officials wore gauze hats. To make a gauze hat, black silk gauze was used in the weaving process of a silk cage where lacquer was applied to make it stand high. Though coated with black silk gauze, the top of the crown was still visible inside. The crown of the Han Dynasty which is high in the front and low in the back changed as time flied. During the Period of Wei and Jin, the *gaoshan* crown symbolizing modesty was promoted. From then on, crowns became either flat or low in the front and high in the back. The Han style vanished without a trace in the Ming Dynasty.

Both officials and the common people in the Tang Dynasty wore *futou* or turbans. In comparison with the later style, the *futou* or turban of this time is relatively lower at height. The early *futou* was just a close-fitting piece of cloth wrapped around the head. Later on a kerchief was put under the turban padded with wood, silk, grass or leather to form the shape of hair buns. Only in this way could craftsmen make sure that fixed shape of *futou* was made. After the mid Tang Dynasty, the cap finally took shape, still given the name *futou*. The two corners of the *futou* are round or wide, tipping slightly upwards, which easily reminds people of hard wings. They seem to have strings inside. Due to their elasticity, they were named as "hard tips".

The Song Dynasty *futou* was unique in that it had two long and straight tips protruding at both sides. Why were these straight tips so long? Some say, they were made for the purpose of preventing officials getting their heads too close to whisper when in court.

Later on the official headdress evolved from the *futou* of Tang and Song Dynasties to the black gauze cap of the Ming Dynasty. There were no significant changes to the style. The difference was that the turban, which had been temporarily wrapped, became a cap of a fixed shape. "Black gauze cap" also became a synonym for the government official status, used until the present day. No significant changes were made in the Tang, Song and Ming official gowns. There were clear rules specifying the appropriate color for each of the ranks, with slight modifications made in each dynasty. This system was passed down until the Qing Dynasty ended in history.

In Tang Dynasty the woman emperor Wu Zetian had all officials wear embroidered gowns, specifying that civil official gowns were embroidered with birds and military official gowns with beasts. The Ming dynasty followed this tradition, distinguishing types and ranks of officials with *buzi*, embroidered pieces attached to the chest and back of gowns indicating the wearers' ranks. The official dress of the Ming Dynasty is characterized by the round collar and the three-feet-wide sleeves. In terms of colors, official uniforms can be divided into three categories. Officials from top four ranks were required to dress in scarlet robes. Those from the following three ranks were asked to wear green robes. Officials of the lowest ranks were left with no choice but to wear indigo robes. The round collar and belt attached to the informal official uniforms inspired the design of costumes in Peking Opera.

To the ruling class of early Qing Dynasty who was intent on defining rigid court ceremonies, the official attire was used as an important instrument to distinguish social status. The Qing rulers invented the most complicated system of official attire in Chinese history strictly defining the color, decorative patterns and style of official uniform in books with clear illustrations, intended to be passed down to all generations to come. The court even set up supervisory office ensuring that all rules are followed in the making of official uniform, and all court attire is complete with the most refined weaving and embroidery, and complemented with complete set of ornaments.

The most distinguishing elements of the Qing official uniform are the horse-hoof shaped sleeve and the Mandarin jacket style. However, the use of *buzi*, or ornament patches, was borrowed directly from the previous Ming Dynasty. The court insignia badges clearly distinguished the civil and military officials with embroideries of birds or beasts. Emblems with different animals were used to further distinguish the ranks and authority of these officials. The emblems embroidered on the decorative patches were however different from the Ming dynasty in that they were much more decorative, often accentuated with an elaborately embroidered border. In terms of style, the Qing *buzi* was embroidered on the outer jacket worn over the gown with front closure, and the front embroidery was done in two pieces at each side. The black gauze cap of the Ming Dynasty was replaced by the *hualing* or feathered cap. The number of "eyes" on the peacock feather was used to differentiate each different rank. The official court uniform and daily uniform were both worn in different layers of robe, jacket, gown, vest, and decorative patches, complemented by court beads, court belt, jade ornaments, colored silk ribbons and perfumed sachet. Officials wore court beads made of jade, agate, coral, or sandalwood, and silk ribbons of bright yellow, turquoise or azure, all according to their ranks. Strict rules were enacted and followed to make sure that officials from different ranks can be differentiated by their clothing.❶

四、词汇对译

官服	official uniform	马褂	the Mandarin jacket
上衣下裳	two-piece costume	标志	insignia
进贤冠	*jinxian* cap	徽章，标志	badge
头巾	kerchief	徽章	emblem
薄纱	gauze	精巧地	elaborately
高山冠	*gaoshan* crown	对襟大袍	the gown with front closure
幞头	*futou*	花翎	the *hualing* or feathered cap
包头巾	turban	朝珠	court beads
发髻	hair bun	朝带	court belt
乌纱帽	black gauze cap	玉佩	jade ornaments
官袍/服	official gown/ attire/ uniform	彩绦	colored silk ribbons

❶ 语料参考：Hua Mei. *Chinese Clothing*［M］. 北京：五洲传播出版社，2004：59–66.

修改，修正，改变　modification

绣袍　embroidered gown

补子　*buzi*

说明，插图　illustration

装饰品　ornaments

马蹄袖　the horse-hoof shaped sleeves

香囊　perfumed sachet

玛瑙　agate

白檀，檀香木　sandalwood

青绿色　turquoise

蔚蓝色　azure

五、译文注释

1. 其实，古代中国官员的服饰是很丰富的，各个朝代都有自己的规定，甚至在同一个朝代里也会多次变更。

In reality, Chinese official uniform is very complex. Each dynasty had rules of its own, which could be changed many times even within a dynasty.

【注释】原文为一句，包含很多流水短句，并且中间还发生了主语的变化。译者在翻译过程中采用了最为简单的处理方式，那就是断句，即译为两句，让带有不同主语的成分独立成句。这种方式比较简单快捷，在翻译实践中也最为实用，是避免犯错的很好的方式。此外，原句中"甚至在同一个朝代里也会多次变更"部分省略了主语"规定"。译者借助这个核心词汇，将前后两个分句的关系搭建了起来，以"rules"为先行词，将"甚至在同一个朝代里也会多次变更"处理成"which"引导的后置定语从句，使译文表意清晰，结构严谨。

2. 汉代的文官多戴进贤冠，冠下衬有介帻（一种头巾）；武官戴武弁大冠，与官阶相呼应。

In Han Dynasty, civil officials wore *jinxian* caps with a *ze* or kerchief lining the cap, while the military officials wore a hat designed for their ranks.

【注释】原句结构清晰易懂，整体由四个分句构成，其间一个分号，明确地将文官与武官划分开来。在本句的翻译中，译者选用了"while"一词来代替中文中的分号，"while"前后形成轻微转折，构成对比关系，使译文同样结构清晰，语义明确。在翻译"冠下衬有介帻（一种头巾）"这个分句时，译者借助了介词短语结构"with+名词"，构成"进贤冠"的后置定语。

3. 官员品级的高低一般以服色区分，有明确的规定，其间曾稍做调整，并沿用到清王朝退出历史舞台。

There were clear rules specifying the appropriate color for each of the ranks, with slight modifications made in each dynasty. This system was passed down until the Qing Dynasty ended in history.

【注释】原句表达符合汉语相对自由的表达方式，译者翻译时将结构进行了重新调整。在翻译前两个分句时，译者找到了可以串联两个分句的核心词汇，那就是"规定"，在译文中以它为核心，利用"there be"句型结构进行翻译。而"其间曾稍做调整"部分，作者采用了介词短语的结构，使其与前面句子相衔接。在最后一个分句的处理上，译者直接另起一句，把"the system"作为主语进行展开。

4. 明朝对此加以仿效，开始在官服前襟饰以有图案的补子来区分文武官员的品级。

The Ming dynasty followed this tradition, distinguishing types and ranks of officials with *buzi*, embroidered pieces attached to the chest and back of gowns indicating the wearers' ranks.

【注释】在翻译本句时，译者将"明朝对此加以仿效"这一分句作为主句，然后用分词状语形式翻译后面的分句。在翻译的过程中，译者充分考虑到读者反应，认为读者可能在"补子"的理解上存在困难。尽管如此，译者没有放弃将本真的中国服饰文化传递出去，所以在翻译时采用了音译的方式来翻译"补子"。而考虑到读者的阅读困难，译者紧接其后用"embroidered pieces"作为同位语来对"*buzi*"进行解释说明。

5. 他们制定了中国历史上最为繁杂的衣冠制度，无论色彩、纹饰、款式均出章入典、规定严谨，并以图示说明，要求后世子孙也能"永守勿愆"。

The Qing rulers invented the most complicated system of official attire in Chinese history strictly defining the color, decorative patterns and style of official uniform in books with clear illustrations, intended to be passed down to all generations to come.

【注释】词句翻译中，译者没有拘泥于原文的形式，而是遵循英语语言的形合特征进行了调整。译者在翻译中将第一个分句作为主句。在翻译第二个分句时，选用了主句中的"the Qing rulers"作为逻辑主语，将"无论色彩、纹饰、款式均出章入典"，译为分词短语"defining the color, decorative patterns and style of official uniform…"作为定语修饰，限定前面的"the most complicated system"，使译文结构清晰，焦点突出。同时，在翻译"并以图示说明"这一分句时，译者通过"with"的介词结构将其处理成了"books"的后置定语。

6. 样式上，清朝的补子绣在袍服外面的对襟大褂上，前襟补子也随之分为两块。

In terms of style, the Qing *buzi* was embroidered on the outer jacket worn over the gown with front closure, and the front embroidery was done in two pieces at each side.

【注释】这句话的结构非常简单，但是在具体翻译上由于涉及服饰领域部分较多，并不十分容易翻译。首先，译文开头"In terms of"是在另起话题时常用的表达方式，可以使文章显得逻辑严密、结构清晰。其中一个翻译难点就是"对襟大袍"的翻译，此处译者将其译为了"the gown with front closure"，表述简洁明了。译者将"前襟补子"译为了"the front embroidery"，是意译，使用"embroidery"以避免与上一个分句中的"*buzi*"重复。

六、翻译知识——汉语流水句的翻译

汉语是意合的语言，往往由流水短句组成。究其原因，汉语在行文表达过程中经常会采用并列结构，一个句子中可以存在几个并列分句，一个主语可以带出几个并列谓语，从而形成由一连串流水短句组成的复合句。根据汉语的表达习惯，不使用连词的时候远比使用连词的时候多，各流水短句之间的逻辑关系秘而不宣；而英文表达则完全不同，需要使用连词以及其他词汇或语法手段将隐含逻辑关系明示出来。

对初学翻译者而言，汉语流水句的英译往往是一个难题。在翻译过程中，我们应该厘清原文的意思，重构逻辑关系，在把握中英文语言差异以及原文文本大意的基础上，使译文符合英文形合语言的特点。对于比较复杂的汉语流水句，除了灵活利用英语中的从句结构外，还可以通过巧用诸如分词短语、介词短语等短语结构来译出流畅达意的英译文。

例1 原文：苗族服饰中的诸多奇幻刺绣、织锦纹样，蕴含着人类童年时代形成的神秘观念，具有若干鲜明的原始文化特征。

译文： The fantastic designs of embroidery and brocade for Miao dresses and adornments reflect mysterious ideas formed in the childhood of mankind and have distinct characteristics of a primitive culture.

【分析】 "蕴含着人类童年时代形成的神秘观念，具有若干鲜明的原始文化特征"两个分句共享一个主语即苗族服饰中的刺绣和织锦纹样，构成十分明显的并列结构。汉语在语言表达中常常会采用并列结构，但是对于连词则是能省就省。英语则不然。作为形合语言的英语极其重视结构的完整性。即便是看上去显得可有可无的简单连词"and"，在英语中也扮演着重要的角色，是必不可少的。而"and"这样的词，却经常被不熟悉英汉两种语言差异的人所忽视，翻译汉语流水分句的时候应加以重视。

例2 原文：从龙的故乡，旗袍谱写出了一曲美丽的乐章，颂扬着东方服饰文化深厚底蕴的精髓，牵系着每一颗热爱中国的心。

译文： In the homeland of the Loong, *qipao*, the traditional Chinese dress, is just like a beautiful legend singing praises of the quintessential, profound connotation in Eastern culture of clothing and accessories, and links together all the Chinese people who love China.

【分析】 原句由四个部分组成，以逗号相隔。第一部分由"从"字引导，在句中充当状语。第二部分为全句的核心，结构也相对完整。其后顺接的两个分句，从结构上看像是与其形成并列之势，在句意理解上也可将"旗袍"作为其共同的主语来对待。但同时，也可以将旗袍所谱写出的乐章作为后两分句共同的主语。由于中文作为意合语言的特性，在句子结构的解读上具有一定的灵活性，但这并不妨碍读者对于句意的理解、把握。但英文重视句子结构的完整性。译者翻译时，针对原句流水短句间无连词或省主语的情况，进行了相应调整。译者将"旗袍谱写出了一曲美丽的乐章"作为主句，将其后的"颂扬着东方服饰文化深厚底蕴的精髓"处理为表主动的现在分词短语形式，将其置于"legend"后作定语。同时译者使"牵系着每一颗热爱中国的心"与前面的"旗袍谱写出了一曲美丽的乐章"并列存在，其间用连词"and"相连。

例3 原文：不直接用绣线刺之于布面作绣，而是将绣线编成粗于单支线的辫子状带子，盘锁成图案，称为编带绣或辫绣、辫子股绣。

译文： Unlike common embroidery with threads pulled through the fabric, braided embroidery involves braiding embroidery threads into plait-likes bands thicker than single threads and arranging them into designs.

【分析】 该句原文由一系列汉语的流水短句组成，这些短句有一个共同的主语，而这个主语在原句中却被省略掉了。根据句意来推断，此处的主语应该是"这种刺绣"。汉语与英语最大的区别之一就是，即便是省略主语，不用连词，只要意思完整，也是合情合理的。但是英文表达中，一定要尊重语法规范，必要的句子成分都不能缺少。译者在翻译本句时用"编带绣"（"辫绣"或"辫子股绣"）作为主语，将"称为编带绣或辫绣、辫子股绣"这一分句拆解省译，使其意义融入全句当中。

例4 原文：明光铠是一种在胸背装有金属圆护的铠甲，腰束革带，下穿大口缚裤。

译文： *Mingguang* suit is one with round metal plates protecting the chest and the back, worn with a leather belt and wide trousers.

【分析】本句主要是对明光铠的定义和介绍，由三个流水分句组成，其间没有连词相连，逻辑上处于并列关系。但是本句在翻译时如若按照并列关系处理，效果并不好。如果译为三个关系并列、结构独立完整的句子会使整个句子显得冗长啰嗦。译者在翻译本句时打破了原文这种并列关系，进行了逻辑关系的重构。译者找到了本句的核心即明光铠，或者说是金属圆护铠甲，第一个分句自然而然译为主句，后面两个分句则合并处理为分词作状语。

例5 原文：服饰是人类文明的重要组成部分，是划分族群的外在标识。

译文： Clothing is part of human civilization and represents the identity of one nation and its people.

【分析】本句较为简单，由两个分句构成，两个分句共同享有一个主语，那就是"服饰"。译者在翻译此句的过程中，抓住了两句间的并列关系，在英文译文中用连词"and"来体现。在英文表达中，两个谓语拥有共同的主语"clothing"，由"and"连接。在翻译"划分族群的"这一定语时，译者用"of"构成介词短语，展现了族群和身份标识之间的关系。

例6 原文：这套中山装是蓝黑色，立领精致剪裁，样式更加简洁，是对传统"中山装"的重新设计，非常吸人眼球。这是中国男性参加正式宴会的典型服饰。

译文： The eye-catching dark blue suit, slim-cut with a standing collar, is a simplified and redesigned "*zhongshan* suit", or "*mao* suit" — a typical formal garment for Chinese men.

【分析】原句主要介绍现代改良中山装，该句包含多个流水分句，构成相当长的句子，句中各分句之间是并列关系，共享一个主语"这套中山装"。译者在翻译时打破原句结构，重构了译文句子，将这些原文中的并列分句处理成了译文中的不同成分。第一个分句中本来作表语的"蓝黑色"与第五个分句中"非常吸人眼球"的表述，共同构成了译文中"suit"的定语，第一个分句由此整体转化为主语，第二个分句变成了"dark blue suit"的后置定语，在句中被放在插入语的位置，第三个分句与第四个分句合并，以系表结构形成了译文句子的谓语部分。经过对汉语原文流水分句的调整与重构，英译文句子变得主干清晰，表意明确。

例7 原文：深衣的材料多为白色麻布，祭祀时则用黑色的绸，也有加彩色边缘的，还有的在边缘上绣花或绘上花纹。

译文： Material used for making *shenyi* is mostly linen, except black silk is employed in garments for sacrificial ceremonies. Sometimes a colorful decorative band is added to the edges, or even embellished with embroidered or painted patterns.

【分析】例句中文原句是一个典型的由流水分句构成的长句，主要介绍了深衣的面料。每个分句各介绍一种情况，四个分句共用一个主语即"深衣的材料"。译者在翻译此句时，先进行了断句，将介绍颜色和材质的前两个分句划分为一句，将介绍装饰的后两个分句划分为另一句。随后在译文中补充了原文所缺少的连接词，如"except""or"等，使句子更连贯，逻辑更清晰。

例8　原文：中国历史上有个"黄袍加身"的故事，说的是公元959年，一位皇帝病死，由他年幼的儿子即位，第二年，掌握兵权的将军赵匡胤被手下将士披上黄袍拥为皇帝，立国号为宋。

译文：In Chinese history there is a story of "dressed with yellow robe" that occurred in 959 A.D., one year after a young emperor took over the throne at the death of his father, the old emperor. In the next year, Zhao Kuangyin, a general in command was dressed with the royal yellow robe by his supporters and made emperor. That was the beginning of the Song Dynasty.

【分析】例句内容丰富，是一连串流水分句组成的长句，主要讲述了"黄袍加身"的故事。从中文原句结构来看，各成分处于一种平等的地位，且之间并没有连词相连。所有的逻辑关系都含在语义中，而非彰显于结构中。英文与中文不同，十分重视通过结构来展现语言逻辑的严谨性。所以译者在翻译时根据逻辑结构对语言结构进行了重新构建。首先，译者将句子进行了适当拆分。译文第一句引出"黄袍加身"的故事，说明其发生时间。第二句主要讲述"黄袍加身"这个故事的主要情节内容。最后一句则是说明了故事对应的朝代。

例9　原文：冕服包括冕冠，上有一块板，做成前圆后方形，戴在头上时后面略高一寸，使冕冠呈向前倾斜之势，以示帝王向臣民俯就，就是真心惦记臣民、尊重臣民的意思。

译文：*Mianfu* is a set of garments including the *mianguan*, a crown with a board, round in the front and square in the back, that leans forward, as if the emperor is bowing to his subjects in full respect and concern.

【分析】汉语原文中前四个分句主要介绍冕冠的样子，第五个分句中的"使"和第六个分句的"以示"有表目的的含义，而第七个分句的"就是"有解释说明的含义。英译文的主句部分是"*Mianfu* is a set of garments including the *mianguan*"，"冕冠"和"冕服"的翻译都采用了直译。为了让读者明白"冕冠"的含义，译者在"*mianguan*"后安排了一个同位语"a crown"来进行解释说明，并且将冕冠上有前倾板的情况通过介词短语和定语从句的方式加以说明。后半句译者用"as if"作衔接，解释了"使冕冠呈向前倾斜之势"的用意。整个译文句子结构层次分明，表意清楚。

七、译文赏析

段落1

顶戴花翎是清代独特的礼帽，分两种，一为暖帽，一为凉帽。礼帽在顶珠下有翎管，质为白玉或翡翠，用来安插翎枝。翎枝又分为蓝翎和花翎两种，蓝翎为鹖羽（鹖：今名褐马鸡）所做，花翎为孔雀羽所做。蓝翎只赐予六品以下的官员或在皇宫和王府当差的侍卫佩戴，也可以赏赐建有军功的低级军官。花翎在清代是一种高品级的标志，非一般官员所能戴用，一般被罚拔去花翎的官员一定是犯了重罪。❶

Official Hat with tail feather was a unique style in Qing Dynasty(1644-1911) including winter hat and summer hat. Feather duct (*lingguan*) was made of jade and emerald jade (used as a holder for

---

❶ 语料参考：沈周. 中国红：古代服饰［M］. 合肥：黄山书社，2012：55.

the feather) beneath the top bead. There were two types of feather, blue feather (feather of brown eared-pheasant dyed blue) and peacock feather. Blue feather was offered to the officials below the sixth rank. In addition, it's granted to guards serving in the imperial palace and the low rank military officials who performed meritorious deeds in battles. Peacock feather as a symbol of high rank was exclusive for high rank officials. Punishment of removing the peacock feather was regarded a felony.

【赏析】段落原文详细介绍了顶戴花翎这种清代独特的礼帽，可分为暖帽与凉帽两种，其上翎枝与官阶等级相关。译文语言流畅、表意准确，完整地呈现了原文信息。在文化特色词汇的英译上，译者根据具体情况进行了灵活的处理。在翻译"顶戴花翎"时，译者采用了意译的方式，将其译为"official Hat with tail feather"即"带尾羽的官帽"。而在"翎管"的翻译上，译者将其直译为"feather duct"，同时又将它的音译版本"lingguan"附在后面，使译者在理解其内涵的同时充分接触到了原语文化。原文中还有许多中文特有的流水短句，译者通过句式调整，使其更加符合西方读者的阅读期待。例如："翎枝又分为蓝翎和花翎两种，蓝翎为鹖羽（鹖：今名褐马鸡）所做，花翎为孔雀羽所做。"一句为典型的流水句，全句由三个部分组成，第一句总体叙述翎枝的分类，第二、三分句分别介绍蓝翎、花翎为何物，三个分句句子完整，结构上独立于彼此，中间亦无连词使之相连。译者在处理时将第一分句作为重点，利用"there be"句型说明翎枝有两类，并将"blue feather and peacock feather"作为同位语放在"two types of feather"之后。译者将第二分句的内容整体作为注释，备注在"blue feather"后的括号中，同时直接将"花翎"意译为"peacock feather"，也就省去了翻译第三分句的必要。从篇章的角度来看，全文多用被动语态，符合说明文对于客观性、科学性的要求。

段落2

清代的补子基本沿承明代，但也有所变化：明代的补子施于袍上，而清代补子用于褂上；明服为团领衫，前胸补子为完整的一块，而清服是对襟褂，前胸的补子被一分为二；明代的补子大约40厘米见方，清代的补子稍小，约30厘米见方；明代的补子多以红色等为底，金线绣花，而清代的补子则是以青、黑、深红等深色为底，五彩织绣；明代的补子只饰于前胸后背，清代宗室的圆补有的不仅饰胸，还饰于两肩之上。❶

Buzi of Qing Dynasty(1644-1911) basically evolves from Ming Dynasty with some changes. Dress of Ming Dynasty has a round collar. *Buzi* of Ming Dynasty is a whole piece of 40cm square cloth attached to fore breast of gown. It has flower embroidery with gold thread against a red background. Whereas the costume of Qing Dynasty is opposite front pieces jacket with two halves of *buzi* which is 30cm square against dark red, black and cyan background. Colorful embroidered *buzi* is not limited to decorate fore breast and back of the dress, it's also on two shoulders.

【赏析】段落原文从样式、色彩、位置等方面比较明、清两个时代的补子，列举了两者间的差异。译文表述清晰，内容丰富，完整地传递了原文信息，向读者展现了明清官服之

---

❶ 语料参考：沈周. 中国红：古代服饰［M］. 合肥：黄山书社，2012：51.

美。在文化特色词汇"补子"的翻译上，译者采用了音译的方式，将最原汁原味的表达呈现给读者。在具体句子的翻译上，译者也处理得十分灵活，充分考虑到了中西方语言的差异。例如"明代的补子多以红色等为底，金线绣花"一句，译者考虑到作为动态语言的中文与作为静态语言的英文之间的差异，在翻译时没有生硬地按照字对词的翻译，而是巧妙地利用"with""against"这样的介词来表意。此外，译者对文段的篇章结构进行了很大的调整。原文从样式、颜色、位置等方面入手，分别对明、清补子进行比较。而在翻译时，译者则选择先整体描述明代补子，再对清代补子进行介绍。这样的叙述方式，对于没有相关知识储备的译入语读者而言更为友好，也更有助于理解。

八、翻译练习

（一）请将下列汉语句子译成英文

1. 在古代中国，着装规范不仅是民间习俗，更是国家礼制的一部分。

2. 历朝历代都有各种条文、律令，对服装的材质、色彩、花纹和款式作详尽的规定，将皇族、文武官员和普通百姓的服饰严格区分开来。

3. 官员的女眷服饰也精雕细琢到了巨细无遗的程度，镶边被大量使用，在镶滚之外还在下摆、大襟、裙边和袖口上缀满各色珠翠和绣花。

4. 折裥之间再用丝线交叉串联，连看不到的袜底、鞋底也绣上密密的花纹。

5. 纵观中国古代官服，尽管有许多讲究，但是最能体现服饰与权力关系的还是补子。

6. 补子的图案很有意思，有的文官补子的图案来源于现实世界的动物，如仙鹤、锦鸡、孔雀、云雁、白鹇、鹭鸶、黄鹂、鹌鹑等，有的却非实有之物，如练雀，它的形状有点像鹭鸶，又有点像孔雀。

7. 武官的补子，也各式各样，有狮、虎、豹等实在之物，也有想象出的动物。不同的动物代表不同的官阶。

8. 宋代崇尚文治，冠服制度渐趋繁缛，也曾经多次修改。

9. 清规定禁穿明代官服但明代的补子为清代继续沿用，图案内容大体一致，略有改动。

10. 宋代没有鱼符的制度，但官员仍佩有鱼袋。

（二）请将下列汉语段落译成英文

凉帽为官员夏季所戴，呈圆锥形，清初时崇尚扁而大，后流行高而小的形状。通常用藤竹、篾席、草麦秸编结帽体，外裱绫罗，内衬红色纱罗，沿口镶滚片金缘，顶部装饰红缨、顶珠、翎管和翎羽。根据清代礼冠制度，每年春季三月将暖帽换成凉帽，八月将凉帽换成暖帽。❶

---

❶ 语料参考：沈周．中国红：古代服饰［M］．合肥：黄山书社，2012：56．

九、译笔自测

官服与等级

官服，又称"公服"，是中国古代官吏或侍从在处理公务时所穿的服装。早期的官服是一种单衣，两袖窄小，便于从事公务，这也是有别于祭服、朝服之处。据史书记载，它作为官吏所穿官服的主要式样，一直沿用至隋代。

到了唐代，官服制度较为完善，官服的形制采用袍制，两袖仍比较窄小，以服色、纹样、佩饰区分官吏的等级身份，对后世官服产生了深远的影响。宋代官服款式为圆领、大袖袍服，腰束革带。元代官服沿袭宋代，但又有所创新，在官服上绣以花卉图案，以图案品种、大小区分品级。明代的官服为袍式，盘领，右衽（衽即衣襟），袖宽三尺，多用苎丝、纱、罗等材料制成。根据公服的服色、绣花的花种和大小以及腰带的材质区分品级。一至四品为绯色；五至七品为青色；八至九品为绿色。清代的官服废除了服色制度，无论职位高低，颜色都是蓝色，只在庆典时方可用绛色。清代官服由袍、褂组成，袍均为圆领，右衽。

官服的颜色是区分官吏等级的标准之一。以唐代为例，唐贞观四年（630年），官服的颜色被定为四等：一品至三品服紫，四品至五品服绯，六品至七品服绿，八品至九品服青。唐末又规定：三品以上仍旧用紫色，四品用深绯，五品用浅绯，六品用深绿，七品用浅绿，八品用深青，九品用浅青。

由于紫色的官服在唐代最为尊贵，所以"紫袍"一词也成为显官要职的代称。绯色的官服即指大红色的袍服，大袖，右衽，衣襟及袖口常有镶边。绿袍是唐代六品及七品官的官服，品级相对较低。青袍是唐代官服中等级最低的，后来"青袍"一词多用来代称品级低的官吏。

唐代的公服制度对服色的规定虽然很严格，但在具体实行的时候，也可以变通。如果一些官吏的品级不够，但遇到奉命出使等特殊情况，经过特许可穿用比原品级高一级的服色，俗称"借紫"或"借绯"。

官服的配饰也是其品级的体现。以唐代为例，官吏按照品级高低，穿用不同材质的腰带，如金、玉、犀、银、石、铜、铁，并以腰带上的饰物区分等级。

中国古代官服的配饰有很多，其中最具代表性的就是佩绶。佩，指佩戴于身的玉饰，如大佩、组佩等；绶，指用来悬挂印、玉佩的丝带。以佩绶区分尊卑是我国古代服饰制度的显著特征。佩绶在秦代以前就已出现，并作为一种官服制度流传下来，但绶带的色彩规格时有变化。到了清代，这种佩绶制度不再使用，取而代之的是顶戴制度。

玉佩也是中国古代贵族和官员们礼服上必不可少的一种装饰，受森严的等级约束，玉佩的形制、佩带的方法及部位也都据佩玉者的身份有明确的规定。❶

十、知识拓展

忠靖服是明代职官退朝闲居时所穿着的服装，配忠靖冠，含"进取尽忠，退思补过"之意，交领，右衽，大袖，上下相连，衣长过膝，常以深青色纱罗为料。不同的品级官员以不

---

❶ 语料参考：沈周. 中国红：古代服饰［M］. 合肥：黄山书社，2012：40-52.

同的图案或素色来区别地位等级。忠靖冠是明代的官员退朝闲居时所戴的帽子，制作时用铁丝围成框架，用乌纱、乌绒包裹表面。冠的形状略呈方形，中间微突，前面部分装饰有冠梁，并且压有金线。后部的形状像两个小山峰。冠前的梁数根据官职的品级而定（图2-12）。❶

Loyal Peace (*zhongjing*) Dress is the home garment for officials of Ming Dynasty (1368–1644), together with Loyal Peace Hat. Loyal Peace means keeping forging ahead and mending mistakes. The so-called Loyal Peace Dress is a knee-length garment with crossed collar, large sleeves and right front piece and is usually made of dark cyan

图2-12 忠靖冠❷

cloth. Ranks vary in different patterns and plain colors. Officials of Ming Dynasty wore Loyal Peace Hats (*zhongjingguan*) at home.The framework of this kind of hat is made of wires and covered by black gauze and black velvet. It is in square shape with a projecting part in the middle. There is a beam pressed by gold thread on the front part of the hat. The shape of the rear looks like two hills. The number of beams on front part is judged by the rank of the official.

第七节 戎装

学习目标

1. 了解中国传统戎装的历史及文化内涵
2. 熟悉中国传统戎装的基本语汇及其汉英表达方式
3. 掌握汉语无主句翻译的相关知识与技能
4. 通过实践训练提高中国传统戎装文化汉英翻译能力

一、经典译言

把作品从一国文字转变成另一国文字，既不能不因语文习惯的差异而露出生硬牵强的痕迹，又能完全保存原有的风味。那就算得入于"化"境。

—— 钱钟书

二、汉语原文

早期甲胄

在中国的神话传说中，认为甲是由被后世称为"战神"的蚩尤发明的（距今约5000年前）。那个年代正是中国从部落联盟到国家创建的时期，社会动荡、战争频繁。甲胄的出现

---

❶ 内容参考：沈周. 中国红：古代服饰［M］. 合肥：黄山书社，2012：42.
❷ 图片来源：沈周. 中国红：古代服饰［M］. 合肥：黄山书社，2012：42.

当然是战争的产物。在氏族社会时期，为了抵御石箭木斧的攻击，利用藤木皮革制作保护身体的防护工具，是完全有可能的。

早期盔甲只遮住头、胸等人体的要害部位，后来的铠甲则主要由甲身、甲袖、甲裙组成。根据出土实物来看，殷商时已有铜盔；周代已有青铜盔和胸甲，胸甲是遮护前胸的，用犀牛皮或水牛皮做成。从文字记载中可以看到，周代已有专门负责甲胄的官，周代时的铜铠甲多以正圆形的甲片为主，且七片为一组，甲上加漆，以使之呈现出白、红、黑等各种颜色。穿铠甲出征时，一般要罩上精美的绣袍以示军威军仪，在战场上厮杀时才解下罩袍。

图2-13　秦代甲胄❶

从秦始皇陵兵马俑坑和石甲胄陪葬坑的文物资料看，秦代的铁质甲胄已占相当比例，但同时也使用着大量皮甲，说明秦代正处于战国至汉代甲胄质料发展转变的过渡阶段，这也是中国古代甲胄发展史上承上启下的关键时期（图2-13）。甲胄质地由皮革到铁质的改变，主要缘于战国至汉代进攻性武器由青铜转变为更锋利的铁兵器，迫使作为防护兵器的甲胄随之逐步由皮质转变为铁质。

大批秦始皇陵兵马俑的出土，为人们提供了较为完整的中式铠甲的形象资料。出土的秦代兵俑分为步兵俑、军吏俑、骑士俑、射手俑等，他们的铠甲服饰装束表现出严格的等级制度，军官和骑士戴冠，普通士兵无冠。虽然不是实物，但是由于陶俑塑造得精致细腻，铠甲的结构可以看得很清楚。秦兵俑中最为常见的铠甲样式，即普通战士的装束，有这样一些特点——胸部的甲片都是上片压下片，腹部的甲片都是下片压上片，以便于活动。从胸腹正中的中线来看，所有甲片都由中间向两侧叠压，肩部甲片的组合与腹部相同。肩部、腹部和颈下周围的甲片都用连甲带连接，所有甲片上都有甲钉，钉数或二或三或四不等，最多者不超过六枚。甲衣的长度，前后片相等，其下摆多呈圆形，不另设缘饰。目前所发现的秦代甲胄资料显示，同一类型的甲胄之间，其形制、尺寸、结构以及甲片的数量等基本相同，甚至其相同部位的甲片亦几无差异，说明秦代甲胄的尺寸、形制等在秦始皇统一度量衡的大背景下已趋于统一，同时也说明甲胄是由官府统一组织制作的，而非私造。

秦代甲胄的日趋成熟和完善，绝非偶然，而是有着多方面原因的。一方面，当时各国之间的战争使甲胄在制作工艺和质量上有所提高；另一方面，从甲胄自身的发展阶段来看，经过原始社会末期至秦代两千多年的漫长发展，皮甲胄的制作工艺已经相当完善，与汉代皮甲胄逐步减少的状况相比，秦代可以称为皮甲胄发展的最高阶段；同时，铁甲作为新型戎装也有所发展。❷

❶ 图片来源：百度网。

❷ 语料参考：华梅. 中国服饰［M］. 北京：五洲传播出版社，2004：67-71.

译前提示：古代戎装在中国传统服饰中扮演着举足轻重的角色。本文简要介绍了古代戎装的起源（追溯到远古蚩尤的时代），战国以及秦代的戎装发展。以秦朝出土文物为依据，详细介绍了秦代普通士兵甲胄的基本样式，归纳了秦朝甲胄趋于成熟和完善的原因，认为秦朝皮甲胄已发展到了巅峰。

三、英语译文

Armor of Early Times

In ancient Chinese mythology, Chi You, the "god of war" (from 5000 years ago), invented the armor. That period of transition from tribal allegiance to the state was a period of volatility and frequent wars. The emergence of the armor was inseparable from the appearance of wars. To guard against the attack of stone arrows and wooden axes, people of the tribal period were very likely to use protective instruments made from canes, wood or leather.

The early armor suits only covered the head and the chest, whereas later they developed into separate pieces of the body shield, the shoulder shields and the leg shields. Judging from artifacts excavated in early times, the bronze helmet appeared in as early as the Shang Dynasty. In Zhou Dynasty, bronze helmet and chest shield made of rhinoceros or buffalo hide were used in wars. It was also recorded in early history that at that time there were officials in charge of armored suits. The copper armor in Zhou Dynasty were made with round pieces in groups of seven, painted in white, red and black. An elaborately embroidered robe was worn over the armored suit to display the dignity of the army, removed only when the actual fighting began.

Judging from the terracotta warrior burials and the accompanying stone burial armors, it is apparent that armor suits made of iron were already prevalent in the Qin Dynasty, although leather was also used very often. It seemed that the Qin Dynasty was a transitional period (from the Warring States Period to the Han Dynasty) for armor suit materials, and it was also a crucial linking period in Chinese ancient armor development. The move away from leathered armor suits towards iron was primarily due to the replacement of bronze weapons by much sharper iron weapons in the period between the Warring States Period and the Han Dynasty. Accordingly, the material of the armor suits— the protective devices of that period also changed from leather to iron.

The excavation of a large number of terracotta warriors in the Emperor Qin Shi Huang Mausoleum has provided us a complete set of visual images of the Chinese armor suits of that time. The unearthed soldier figures included foot soldiers, army clerks, riders and archers, all with armors that strictly reflect their ranks and statuses. Generals and riders wore hats, while ordinary soldiers did not. Although they were not real artifacts of armor suits, the fine artifacts put into these clay figures were so meticulous that the structure of the armor was clearly seen. The most common armor style, the style for common soldiers, had one distinct feature—all metal chips were covered like fish scale by the piece on top of them at the chest, and in the reverse direction at waist level, a design intended for easy movement. Looking from the central line, all chips cover the next outwards. Construction of shoulder chips was similar to the waist. Chips at the shoulders, waist and below the neck were connected with belts and nails, from two nails to four and no more than six. The length of the armor is equal at front and back, rounded at the lower edges with no additional decoration. Materials we have today on Qin armor indicate that armor of the same type is similar in style, measurement, construction and number of chips, and chips in the same part of armors are identical. This unification trend in the size and style of armor suits of the Qin Dynasty can be seen as the result of measurement unification promoted by Emperor Qin Shi Huang, and shows that the production of armor was centralized instead of privately done.

The gradual maturity and perfection of the Qin armor was no coincidence. On one hand, the fine quality was a direct result of frequent wars among states. On the other hand, the development of armor itself had gone through over two thousand years of history from the late primitive period to the Qin Dynasty and the craft of leather armor was already quite advanced. Contrasted to the lessening numbers of leather armors in the Han Dynasty, the leather armors in Qin Dynasty reached its height, and the new style iron armors gradually gained development.[1]

四、词汇对译

防护用具	protective instruments	绣袍	embroidered robe
藤条	cane	战国时期	The Warring States Period
盔甲	armor suits	兵马俑坑	the terracotta warrior burials
甲身	body shield	步兵	foot soldier
甲袖	shoulder shields	军吏	army clerk
甲裙	leg shields	射手	archer
头盔	helmet	金属甲片	metal chips
胸甲	chest shield	鱼鳞	fish scale
犀牛皮	rhinoceros hide	统一度量衡	measurement unification
水牛皮	buffalo hide	皮甲胄	leather armor

五、译文注释

1. 在中国的神话传说中，认为甲是由被后世称为"战神"的蚩尤发明的（距今约5000年前）。

In ancient Chinese mythology, Chi You, the "god of war" (from 5000 years ago), invented the armor.

【注释】战神蚩尤是中国特色文化中的人物。在翻译"战神"和"蚩尤"时，译者分别采用了直译和音译的方式，以最大限度地展现中国神话传说的本真魅力。在英文译文中，"god of war"被用来作为"Chi You"的同位语，以插入语的形式放在"Chi You"后面对其

---

❶ 语料参考：Hua Mei. *Chinese Clothing*［M］. 北京：五洲传播出版社，2004：67-71.

进行解释说明。

2．从文字记载中可以看到，周代已有专门负责甲胄的官，周代的铜铠甲多以正圆形的甲片为主，且七片为一组，甲上加漆，以使之呈现出白、红、黑等各种颜色。

It was also recorded in early history that at that time there were officials in charge of armored suits, which were made with round pieces in groups of seven, painted in white, red and black.

【注释】译者在翻译此句时，没有生硬地照搬或者套用中文原句结构，而是借用主语从句来搭建句子的主框架，为了避免句子结构头重脚轻的问题，用"it"作为形式主语置于句首。

3．穿铠甲出征时，一般要罩上精美的绣袍以示军威军仪，在战场上厮杀时才解下罩袍。

An elaborately embroidered robe was worn over the armored suit to display the dignity of the army, removed only when the actual fighting began.

【注释】例句原文的逻辑主语是出征的将士，在结构上省略了主语。原文采用无主句是因为这个主语容易推测出来，同时也并非句子的表达重点。本句的重点在于对服饰的说明，所以译者在翻译此句时采用了被动式，将"精美的绣袍"作为主语。这样译的另一个原因是，倘若希望用一个条理清晰的长句去整合原句的所有元素，就需要找到能统领各分句的关键词汇，而在本句中这个可以用作主语来统领全句的就是"精美的绣袍（an elaborately embroidered robe）"。

4．从秦始皇陵兵马俑坑和石甲胄陪葬坑的文物资料看，秦代的铁质甲胄已占相当比例，但同时也使用着大量皮甲胄。

Judging from the terracotta warrior burials and the accompanying stone burial armors, it is apparent that armor suits made of iron were already prevalent in the Qin Dynasty, although leather was also used very often.

【注释】在翻译"秦代的铁质甲胄已占相当比例"这一分句时，译者没有生硬地采用词对词的翻译方式，而是运用意译的方法来进行翻译。"占相当比例"在句中即大量使用、普遍流行的意思。"prevalent"一词不但可以充分表达这个含义，而且可以使整个句子显得简洁。

5．甲胄质地由皮革到铁质的改变，主要缘于战国至汉代进攻性武器由青铜转变为更锋利的铁兵器。

The move away from leathered armor suits towards iron was primarily due to the replacement of bronze weapons by much sharper iron weapons in the period between the Warring States Period and the Han Dynasty.

【注释】与中文不同，英文属静态语言，这类中英文表达上的差异在翻译过程中要特别注意。原文中"由青铜变为更锋利的铁兵器"突出的是动态的变化，在转换到英文中时，译者对这个"变为"进行了名词化，用"replacement"一词来表达，体现了英文的静态表述趋向。

6．肩部、腹部和颈下周围的甲片都用连甲带连接，所有甲片上都有甲钉，钉数或二或三或四不等，最多者不超过六枚。

Chips at the shoulders, waist and below the neck were connected with belts and nails, from two nails to four and no more than six.

【注释】例句原文用主动的句子结构表示被动的含义。相较于中文，英文使用被动句的频率要高得多。译者在翻译此句时，考虑到中英文语言表达在被动语态方面的差异，采用被动结构进行翻译，使译文更加符合英语语言规范，更适合英语国家读者的阅读习惯。

六、翻译知识——汉语无主句的翻译

汉语无主句是指只有谓语部分而没有主语部分的句子。在汉语的使用中，句子中有时不需要出现主语，或是在没有主语的情况下，整个句子依然可以达到表意完整。无主句是中文的特色，特别是在口语中大量存在。与汉语不同，英语的句子结构严格遵循语法要求，主要成分必须完整，尤其是主语，大多数情况下是句子中必不可少的成分。所以在汉语无主句的翻译过程中，我们需要结合上下文语境，进行适当的增补或调整。

在汉英翻译实践中，针对汉语无主句，可以采用灵活的方法来翻译。比如：可以补充主语。汉语中有些表达类似警句格言，泛指性很强，采用无主句表达很适当。但是，英语语言具有形合特点，要求结构完整。所以，在汉译英过程中，应当根据实际情况补充出主语。可供选择的主语有表示泛指的代词、名词，或者使用非人称主语来进行补充。在翻译实践中，有时还可以将汉语无主句译成英语被动句、"there be"结构、倒装句、祈使句等。

例1 原文：没有文字，就用刺绣来表达其对宇宙起源的认识。

译文： Ethnic Miaos without written language, use their embroidery to express their view of the origin of the universe.

【分析】例句原文是典型的无主句，比较偏向于口语表达。结合原文句子的上下文语境，读者可以毫不费力地推测出本句逻辑上的主语应该是苗族人民。译者在翻译本句时采用了最常规的方式即补充出主语，来保证英译文句子结构的完整性。

例2 原文：按照一定的纹理，由外向内将辫成的辫带平盘绕织盖在剪纸上，用同色彩线将辫带固钉。

译文： The braided silk bands are wounded on the paper cut on cloth from outside to inside before stitched to the cloth with the silk lines of the same color.

【分析】例句原文同样是一个无主句。原句的逻辑主语是人，是苗族制作辫绣的手工艺者。这是显而易见的，即使不借助上下文，单独看这句话，读者也可以轻易推测出本句的主语，了解句子所要传达的含义。译者在翻译时补充了主语，但是并没有使用原句的逻辑主语，而是采用被动结构，以谓语动词动作的接受者"辫带"作主语，使译文更符合英语语言的表达特点。

例3 原文：由于底布和纹饰主调均采用相近的暖色，显出一派热烈的喜气。

译文： The similar warm colors of the ground fabric and the design evoke an exuberantly festive mood.

【分析】例句原文的前半句是一个原因状语从句，后半句由动词"显出"开头，缺少主

语。从字里行间推断，与谓语"显出"相搭配的主语应该是句子所描述的服饰作品。译者在翻译时，并没有按常规的直接根据原文的谓语增补主语的方法，而是采用了一个更为灵活的方式。译者选择"暖色（warm colors）"作主语，谓语动词相应地变成"evoke（激发）"，使译文的主谓结构自成一体，但却充分传递了原文的含义。

例4　原文：被罚拔去花翎，一定是犯了重罪。

译文：Punishment of removing the peacock feather was regarded a felony.

【分析】例句原文两个分句都没有主语。结合上下文，可以推断出原文的逻辑主语为官员。原句中虽然省略了主语，但并不妨碍读者的阅读、理解。英文则不然，大多数情况下，缺少必要成分的句子只能被视为不符合语言规范的病句。为使译文句子结构完整，译者在翻译时补充了主语，但是译者并没有选用"官员"作主语，而是巧妙地以"punishment（惩罚）"作主语，采用被动结构来传达原文的意义。可见，在符合原文语义和目标语语言规范的基础上，译者可以灵活选用主语，重构译文句式。

例5　原文：根据清代礼冠制度，每年春季三月将暖帽换成凉帽，八月将凉帽换成暖帽。

译文：According to the institution of Qing Dynasty, wearing summer hats started in March every year and the summer hats were replaced by winter hats in August.

【分析】例5也是一个无主句，根据上下文可以推测出原文的主语为"官员们"或其他泛称。译者在处理时，为两个由"and"连接的并列分句选用了不同的主语，第一个分句以戴凉帽这个动作为主语，采用主动语态；第二个分句以"凉帽（the summer hats）"作为主语，采用被动语态，两个主语的补充和相应谓语的选择使得译文表意充分，结构清楚。

例6　原文：多以红色作为明代补子的底，用金线绣花。

译文：It has flower embroidery with gold thread against a red background.

【分析】例6是典型的无主句，译者在翻译时补充出了主语，因为英文具有形合的特点，需要保证其结构的完整性。译者将"明代补子"作为主语，又因为"明代补子"在前文有所提及，所以在此处用代词"It"来代替。

例7　原文：通常用藤竹、篾席、草麦秸编结帽体，外裱绫罗，内衬红色纱罗。

译文：Its frame is woven by vines, thin bamboo strips and grass with mounted silk and red gauze lining.

【分析】例7原句是一个无主句。译者在英译过程中，根据中文原句上下文的逻辑关系，在译文中补充出主语"Its frame"，并采用被动结构来翻译此句。另外，译者将后两个分句译为介词短语"with+名词"，用作状语，使译文主次分明，表意明确。

例8　原文：面料的选择上除传统的提花锦缎外，还增加了棉布、麻、丝绸等更为轻薄的品种。

译文：In terms of materials, besides the traditional jacquard woven brocade, more light and thin fabrics with printing such as cotton cloth, linen and silk were used.

【分析】例8原句中的逻辑主语显然应该是制作者，但制作者并非这个句子的重点，无须突出强调，因而在原文中被省略掉了。中文无主句并不会为读者阅读构成障碍，反而使句子表述清晰、简洁。但在英文表达中，结构是重中之重的，语法是必须遵守的。所以译者在英译时做了适当处理，将"轻薄面料"作为主语，采用被动式来进行翻译，符合英语行文规范。

七、译文赏析

段落1

由于盔甲易腐烂的特质，中国战争中盔甲的历史和演变研究很难有确凿的定论，但留传下来的文字描述，诸如壁画、陶俑一类的艺术品，以及幸存的盔甲金属部件，可以帮助我们重现这重大发展历程。什么人需要穿盔甲，又是在什么时候穿盔甲，是另一个讨论的焦点。战国时期（公元前475-公元前221年）的兵书中提到，任何级别的军官都穿盔甲。同样的资料中还提到指挥官将盔甲存放在储存袋中，分发给部队战士。但至少一些应征入伍的普通步兵可能不得不自己准备盔甲。❶

The history and evolution of armour in Chinese warfare is difficult to ascertain with certainty, given its often perishable nature, but text descriptions and appearances in art, such as in wall paintings and on pottery figurines, along with surviving metal parts can help reconstruct major developments. Just who wore armour and when is another point of discussion. Military treatises of the Warring States Period (475B.C.-221 B.C.) suggest that all officers of any level wore armour. The same sources contain references to commanders keeping armour in storage bags and distributing it to troops, but at least some of the ordinary conscripted infantry probably had to provide their own.

【赏析】段落原文探讨中国盔甲的历史渊源，虽难下定论，却好在有据可考，同时作者说明了盔甲穿着者身份，穿着场合等问题，并提到了一些兵书上所记载的战国时期的情况。译文信息丰富、语言流畅、结构清晰、句式多样。在文化特色词汇的翻译上，译者采用了异化的方式，使译入语读者能够有机会更贴近原文。译者将"战国时期"直译为"the Warring States Period"。这样的翻译方式在形式和内容上都贴近原文，且译入语读者也可以借助这样的翻译迅速了解到战国这个各诸侯国混战的时代。译者为适应英语表达习惯，对句子也进行了适当调整。例如"什么人需要穿盔甲，又是在什么时候穿盔甲，是另一个讨论的焦点。"一句，由三个部分组成，第一个分句结构相对完整，第二个分句则是一个省略主语的句子，其主语应为"这些人"，因紧接前文，构成语义上的顺联，因而中文表达中此处主语可以省略。而第三个分句同样也缺少主语，它的主语是前两分句所呈现的问题。译者在翻译时，将第一、二个分句分别处理为由"who"和"when"引导的主语从句，并对"when"所引导的内容进行省略，与"who"引导的句子进行合并，使其整体作为译文的主语。最后，篇章中代词的灵活使用，在避免了词汇的重复，简化了语言的同时，使译文句子之间的逻辑更加清晰，联系更加紧密。

---

❶ 语料参考：*Armour in Ancient Chinese Warfare*，略有改动。来源：世界历史百科（Wold History Encyclopedia）网。

段落2

特别是从汉朝（公元前202年—公元220年）开始，随着铁质兵器的广泛运用，铁这种材质越来越多地应用于盔甲的制作。制作者通过缝合或铆接的方式将小板连接在一起，形成具有一定灵活性的束腰外衣，这种外衣也保护了上臂的外侧。与此同时，铁也被用来加固盾牌和制作头盔。这一时期的头盔呈兜帽状，下垂一个保护颈部的部分。但即便在汉代的军事文献中能够找到关于铁面罩的记载，这样的头盔仍然不能保护脸部。❶

Especially from the Han Dynasty (202 B.C.-220 A.D.) onwards, with the wider use of the iron weapons, iron was increasingly used in making armour. Small plates were stitched or riveted together to form a somewhat flexible tunic which also protected the outer upper arms. Iron was at the same time used to strengthen shields and to make helmets. Helmets of this period took on a hood-like shape with a hanging part to protect the neck. However, they still offered no protection for the face, even if there are references to iron face-masks in Han military treatises.

【赏析】段落原文详细介绍了汉代的戎装，提及铁这种原材料以及缝合、铆接工艺在该时期戎装制作上的广泛应用。译文语言流畅，表述清晰且详细，内容丰富。首先，译者用词准确，体现出其对于原文及其所承载的文化的深刻理解。例如，在翻译"铁面罩"时，译者采用的是直译的方式，选用"face-mask"一词来对应面罩，而非"mask"。"mask"可用于指隐藏长相的面具、阻挡有害气体的面具。而"face-mask"指的是阻隔污染空气以及细菌的口罩，在危险环境中保护面部免受伤害的面罩。此处的"铁面罩"是战场上的军用防护设备，用"face-mask"更为准确。其次，译者也根据中西语言差异对个别句式进行了具体调整。例如"制作者通过缝合或铆接的方式将小板连接在一起，形成具有一定灵活性的束腰外衣，这种外衣也保护了上臂的外侧"这句话，译者在翻译时将"small plates"作为主语，化主动为被动，同时提炼出原文后两分句的共同成分即"束腰外衣"，通过限制性定语从句的形式，将两分句进行合并处理。从整个篇章来看，"at the same time""however""even if"等词的灵活使用，使文章显得条理清晰。

八、翻译练习

（一）请将下列汉语句子译成英文

1. 当时的史书记载中，有用数领精制铁甲换取一领纸甲的记载，想必性能不会太差。

2. 绸护领有护颈及护耳作用，绣有纹样，并缀以铜或铁钉。

3. 由于战乱不断，魏晋南北朝时期的戎装在原来基础上有了很大发展。

4. 铠分前后两大片，遮住前胸后背，类同于背心式样，长至腹下。

5. 明光铠是一种在胸背装有金属圆护的铠甲，腰束革带，下穿大口缚裤。

6. 裲裆铠的结构比前代有所进步，甲身由鱼鳞等形状的小甲片编制，长度延伸至腹部，取代了原来的皮革甲裙。

7. 有十三种铠甲作为正式军服，包括铜、木、皮、布等各种材质。

---

❶ 语料参考：*Armour in Ancient Chinese Warfare*，来源：世界历史百科网（略有改动）。

8．用于实战的，主要是铁甲和皮甲。绢布甲用绢布一类纺织品制成，外形美观，只作为武将平时的服饰和仪仗用装束。

9．由于采用冷锻法加工，当甲片冷锻到原来厚度的三分之一以后，其末端留下像筷子头大小的一块，隐隐约约，像皮肤上的瘊子，故名。

10．到了明朝，军队开始大量装备一种棉甲，这与当时火器大量运用于战场的情况相适应。

（二）请将下列汉语段落译成英文

针对特定类型的士兵需要设计专门的盔甲。一般会有两三个士兵布设在战车之中，他们的运动量相对较小，因而他们的盔甲也会更加沉重，但这样的盔甲的确能为士兵提供更好的保护。在成功覆盖身体的各个部位的同时，使手臂有足够的空间可以自由地使用长矛和戟（斧头和矛的混合体）等武器。与此同时，步兵只有较短的束腰外衣和更为基础的护腿保护，这使他们能够在战场上快速移动。骑兵从公元前4世纪开始取代战车。传统骑兵都是手持戟和弓，轻装上阵，为的是方便他们在行进中自由移动，骑在马上展开攻击。正因如此，他们的衣服必须轻便，不束缚手脚。

九、译笔自测

中国秦代盔甲

秦朝是在中国战国时代之后出现的朝代。经过常年征战，秦国征服了敌对国家，秦朝第一位皇帝、专制君主秦始皇统一中国，建立了秦朝。

在秦朝，中国的战士总是身穿由200多甲片构成的制作精良的盔甲。历史学家对这种盔甲的了解大多来自那7000多个真人大小的秦始皇陵兵马俑，这些兵马俑似乎都是比照各不相同的个体战士制作的。1974年在西安市附近发现的兵马俑包括装甲步兵、骑兵、弓箭手和古战车驾驶员。从对这些数据的分析入手，人们可以了解很多有关中国古代军队的情况。

从公元前221年到公元前207年，秦朝统治了现在的甘肃和陕西两省。这种状态是通过战国时期几次成功的征服战争实现的，秦始皇借此巩固了他的王国。这样的战果，使得秦朝骁勇善战的战士扬名四海。普通士兵以上的士兵需要穿着由薄皮革或金属板制成的特殊盔甲。步兵穿着覆盖肩膀和胸部的套装，骑兵穿着覆盖胸部的套装，而将军则穿着附带丝带和头饰的盔甲套装。

盔甲似乎先是用铆钉固定在一起，然后通过捆绑或缝合连接在一起的。这些薄片是由皮革或金属制成的小板（大约2英寸×2英寸，或2英寸×2.5英寸），每块板上都有一些金属钉。一般来说，较大的板被用来覆盖胸部和肩部，较小的板被用来保护手臂。置于其他部位的保护，一些战士除了会在外套下穿裤子外，还会在大腿上穿额外的衣服。还有一些战士会佩戴护膝，弓箭手因为射箭的需要，有时可能需要采用跪姿，出于保护膝盖的需要，会佩戴护膝。

起初匠人们通过涂漆的方式为兵马俑的衣服涂上了包括蓝色和红色在内的明亮颜色。不幸的是，暴露在外界环境中，受到空气和火的侵蚀，兵马俑衣服的最初的鲜亮颜色最终脱落或褪色。褪色的斑点保留了下来。历史学家无法确定秦始皇的士兵是否真的穿得如此鲜艳。

还是说兵马俑身上的鲜亮颜色只是出于装饰的目的而有意画上去的。

秦军的盔甲款式设计相对简单。无论是覆盖胸部、肩膀、手臂的盔甲，还是只是覆盖胸部的盔甲，它都是由小而重叠的甲片构成的。为了将自己与低级别的士兵区分开来，秦朝的军事将领在自己的脖子上系上带状装饰。一些军官戴着平顶便帽，而将军们则戴着像野鸡尾巴的头饰。

没有一个兵马俑是手持盾牌的。然而，历史学家认为，秦朝士兵是使用盾牌的。士兵们使用各种各样的武器，包括弓、长矛、剑、匕首、战斧等等。甚至连剑也可以区分出很多种类，有的直如大刀，有的弯如弯刀。这些武器大多是用青铜制成的；其余的则是由含有铜和其他元素的合金制成的。

秦军士兵将头发梳得十分整齐，胡须也打理得很精致，他们将头发中间分缝儿，编成精致的发辫，或是在自己头顶右侧盘成顶髻，有时秦兵会佩戴皮帽，这种帽子在秦朝骑兵身上最为常见，士兵们没有佩戴头盔的习惯。骑兵们用马鞍，但不用马镫，他们在紧身的裤子外面穿一种据历史学家称要短于秦步兵外套的衣服。

将军们将带子系成花结，并将其别在外套的不同位置上。花结的数量以及其所处的位置表明了一个军官的军衔、等级；在服装上可能就是一个小小的差别，但它所代表的官阶上的差距可能就相当于四星上将和五星上将之间的差别。

十、知识拓展

清代是中国古代戎装发展中变化最大的一个时期。一是满族作为统治者，对汉族军戎服装加以改造，二是火枪、火炮的运用导致了戎装的变革。清代的铠甲分为甲衣和围裳。甲衣肩上装有护肩，护肩下有护腋；另在胸前和背后各佩一块金属的护心镜，镜下前襟的接缝处另佩一块梯形护腹，名叫"前挡"。腰间左侧佩"左挡"，右侧不佩挡，留作佩弓箭囊等用。围裳分为左、右两副，穿时用带系于腰间。在两副围裳之间正中处，覆有质料相同的虎头蔽膝。一般的盔帽，无论是用铁或皮革制品造，都在表面髹漆。盔帽前后左右各有一梁，额前正中突出一块遮眉，盔顶正中竖有一根插缨枪、雕翎或獭尾用的铁或铜管。盔后下沿垂石青等色的丝绸护领，有护颈及护耳作用，上绣有纹样，并缀以铜或铁钉。到了清朝末年，水兵、陆军、巡警等服装，已明显带有西欧军服的特征（图2-14）。❶❷

Qing Dynasty, however, was a period with the most

图2-14 乾隆戎装❸

---

❶ 本章知识拓展中文内容参考：华梅. 中国服饰［M］. 北京：五洲传播出版社，2004：75-76.

❷ 英文内容参考：Hua Mei. *Chinese Clothing*［M］. 北京：五洲传播出版社，2004：75-76.

❸ 图片来源：搜狐网。

significant changes in the development of ancient martial attire. The Manchu, as the ruling people, made their own reforms of the Han military wear. The use of guns and cannons led further to this transformation. The Qing Dynasty armor suits were divided into coat of mail and *weishang*. On each shoulder of the coat of mail, there was a protective shoulder pad, under which there was an armpit guard. In addition, metal chest plates were attached on front and back, and a trapezoid shaped belly protector was added as well. The left side of the body was protected, while the right remained open for carrying bow and arrows. The double width *weishang* was used to protect the sides, fastened around the waist when needed. Helmets, whether made of iron or cattle hide, were painted on the surface. On all four sides of the helmet, there were vertical ridges, a brow protector and metal tubing for attaching the decorative feather, tassel or animal fur. A protective silk collar was attached for shielding the neck and the ears, decorated with fine embroidery and metal tacks. By the end of the Qing Dynasty, army uniform for the navy, the infantry and the police patrol already had distinct Western features.

第八节　旗袍

学习目标

1. 了解旗袍的历史及文化内涵
2. 熟悉旗袍的基本语汇及其汉英表达方式
3. 掌握与被动句相关的翻译知识与技能
4. 通过实践训练提高旗袍文化汉英翻译能力

一、经典译言

直译的意义若就浅处说，只是"不妄改动原文的字句"；就深处说，还求"能保留原文的情调与风格"。

——茅盾

二、汉语原文

20世纪的旗袍

随着第一次世界大战的爆发，西方女权主义运动开始萌芽，妇女不再甘心做男人的附属品和家庭的牺牲品，不少妇女尝试一直是男人在做的工作，开始穿长裤、剪短发。这股风潮与席卷中国的"新文化运动"合流，女性在追求科学、民主、自由风气的影响下，纷纷走出家庭接受高等教育，谋求经济独立，追求恋爱婚姻自由。

留洋女学生和中国本土的教会学校女学生率先穿起了"文明新装"——上衣多为腰身窄小的大襟衫袄，衣长不过臀，袖短及肘或是喇叭形的露腕七分袖，衣摆多为圆弧形，略有纹饰；与之相配的裙，初为黑色长裙，裙长及踝，后渐缩至小腿上部。这种简洁、朴素的装扮成为20世纪一二十年代最时髦的女性形象。而对西方审美眼光的推崇，也影响到了中国女性整体形象的重塑。欧美的化妆品、饰品进入中国市场，美白皮肤、养护头发、向上翻翘眼睫毛、涂抹深色眼影、剪掉长发、烫发，以及戴一朵夏奈尔式的茶花或一条长长的绕颈珍珠项链、拎一只皮毛质地的手提包、脚穿丝袜和高跟鞋……构成了时髦女性的日常形象。

图2-15　旗袍❶

　　而今天的人们津津乐道的旗袍（图2-15）也是在这个时期不断改良，成为一种具有现代意义的时装。所谓"旗袍"，即旗人之袍，而"旗人"，是中原汉族人对满族人的称谓。旗袍原本腰身平直，而且很长。1921年，上海一批女中学生率先穿起了长袍。初兴的式样是一种蓝布旗袍，袍身宽松，廓形平直，袍长及踝，领、襟、摆等处不施镶滚，袖口微喇，看上去严冷方正。这种式样的服装一经走上街头，就引起了城市女性的极大兴趣并竞相仿效。此后的旗袍不断受到时代潮流的影响，在长度、腰身、衣领、袍袖上多有变化。

　　20世纪20年代中期，旗袍的袍身和袖子有所减短，腋下也略显腰身，但袍上面仍有刺绣纹饰。20年代末期，袍衣长度大幅度缩短，由原来的衣长掩足发展到衣长及踝进而缩至小腿中部。腰身更加收紧，大腿两侧的开衩也明显升高。30年代以后，改良旗袍的变化称得上日新月异。先是时兴高领，待高到双颊时，转而以低领为时髦，低到不能再低时，又突兀地将领子加高以显示时尚。袖子也是这样，长时可以遮住手腕，短时至小臂中部，继而露出肘部，至上臂中部，后索性去掉袖子。下摆也是忽而长可曳地，忽而短至膝上。除了两侧以外，有的开衩还被设计在前襟，并使下摆呈现弧形。面料的选择上除传统的提花锦缎外，还增加了棉布、麻、丝绸等更为轻薄的品种，采用印花图案，色调以素雅为美，领、袖、襟等部位也用镶滚，却并不烦琐。中国传统的服饰形象并不突出腰身，但随着20世纪女性服饰追求身体曲线美的倾向越来越鲜明，旗袍成了展现女性性感身材最理想不过的装束。❷

　　译前提示：旗袍被视为华人女性的传统服饰，形成于20世纪20年代，素有"中国国粹"和"女性国服"之称。本文简要介绍了改良旗袍的起源，与之相伴的社会、文化背景，以及20世纪旗袍的主要特点。

三、英语译文

Qipao in the Twentieth Century

With the burst of the First World War, western feminist movement began to sprout. Women were

❶ 图片来源：中国环球电视网。
❷ 语料参考：华梅. 中国服饰 [M]. 北京：五洲传播出版社，2004：89–92.

no longer reconciled to be the accessories and victims of the family, so quite a number of women tried to pursue a career that had been previously occupied by men. They began to wear long trousers and cut short hair. This social tide converged with Chinese "New Civilization Movement" that had spread across China. Under these social influences of seeking for science, democracy and freedom, numerous women started to walk out of the family to receive higher education. They sought for both financial independence and freedom in love and marriage.

Female students who studied overseas and students from local mission schools took the lead in wearing "civilized new dress"— upper outer jacket was mostly jackets with tight waistline, big front garment pieces not lower than hips, elbow-long sleeves or 70% sleeves in the shape of horn. The clothing hem was mostly in arch shape and decorated with patterns. The matching skirts originally were ankle-long black skirts, and gradually the length of skirt rose to the upper shank. This style of simple and plain dress became the most fashionable female image in 1920s and 1930s. The esteem towards the western aesthetics also influenced the remoulding of the general image of Chinese women. Cosmetics and adornments from Europe and America entered into Chinese market. To whiten the skin, nourish hair, curl the eyelash, sweep dark eye shadow, cut short hair, curl hair, wear a Chanel style camellia or a very long pearl necklace around neck, carry a fur handbag, wear stockings and high heel shoes... formed the daily images of Chinese fashionable women.

Qipao that is well received nowadays was also amended and improved in this period and became a modern fashionable dress. *Qipao* in Chinese is called "banner robe", which means the banner people's robe. And the banner people are how the middle land Han people referred to Manchu people. In 1921, a group of female high school students in Shanghai took the lead to wear long robes. At the beginning, the prevailing style was a kind of blue cotton cloth *qipao* with loose clothes body, straight and flat outline and bell-mouthed sleeves. The robe was ankle-long with no edgings or lacework in collar, front garment piece and hem parts. The robe looked very serious and formal. This style of dress aroused great interests of city women and was the rage once it appeared on the streets. Later under constant influences of modern tide, *qipao* showed changes in length, waistline, collars and sleeves.

In middle years of 1920s, the clothes body and sleeve of *qipao* shortened, and the oxter part tightened. The robes were still decorated with embroidery and patterns. At the end of 1920s, the length of the robe shortened greatly, rising from the foot to ankle and then to the middle part of shank. The waistline tightened and the vents on both sides of the thigh part heightened also. After 1930s, *qipao* improved and changed constantly. First high stand-up collar style was the rage, and then when it was high to reach the cheeks, low collar style began to prevail. When the collar couldn't be lower anymore, suddenly it was again heightened to look fashionable. The design of sleeves was the same case. It rose from wrist, to the middle of lower part of arm, to elbow, to the middle of the upper part of arm, and then finally there were no sleeves. The lower hem of robes sometimes was long enough to reach floor and sometimes was knee-high. Besides the vents on the sides, there might be a vent designed in the front garment piece and the lower hem was in the shape of arch. In terms of materials, besides the traditional jacquard woven brocade, more light and thin fabrics with printing such as cotton cloth, linen and silk were used. The colors chosen were usually simple but elegant. The collars, sleeves and garment pieces were decorated with edgings, but didn't look complicated with trivial details. Traditional Chinese costumes didn't highlight the waistline, but with the more and more distinct tendency of seeking the curve beauty of body shape in lady costumes, *qipaos* have become the most desirable dress to show the sexual body shape of women. ❶

四、词汇对译

旗袍	*qipao*	袖口微喇	bell-mouthed sleeves
文明新装	civilized new dress	镶边	edging
紧腰	tight waistline	花边	lacework
配饰	accessories	刺绣	embroidery
衣襟	front garment piece	立领	stand-up collar
袖长及肘	elbow-long sleeves	低领	low collar
下摆	hem	开衩	vent
弧形	arch shape	提花织锦	jacquard woven brocade
裙长及踝的黑色长裙	ankle-long black skirt	棉布	cotton cloth
夹眼睫毛	curl the eyelash	亚麻	linen
眼影	eye shadow	体形	body shape
轮廓	outline	曲线美	the curve beauty

五、译文注释

1. 而今天的人们津津乐道的旗袍也是在这个时期不断改良，成为一种现代意义的时装。

Qipao that is well received nowadays was also amended and improved in this period and became a modern fashionable dress.

【注释】在原文中"旗袍"是"人们津津乐道"的对象，这对关系中，"人们"是施事者，而"旗袍"则是受事者，但这个动作的接收者才是整句论述的重点，而作为施动者的"人们"，即便是被隐去，读者也依旧可以凭借语义、常识以及英语表达习惯精准无误地推测出施事者是谁。译文采用被动式，可以在隐去不必要的施事者的同时，突出句子的重点内容。"旗袍也是在这个时期不断改良"，"改良"一词在中文原文中在形式上没有被动标记，却能够结合语境表达被动的含义。英文是重结构形式的语言，它的时态、语态对于其意思的表达十分重要，所以在这部分的英译过程中，译者采用了结构被动式"be动词+动词过去分词"，使句子表意更清楚，结构更符合英文规范。

---

❶ 语料参考：Hua Mei. *Chinese Clothing*［M］. 北京：五洲传播出版社，2004：92–94.

2．面料的选择上除传统的提花锦缎外，还增加了棉布、麻、丝绸等更为轻薄的品种。

In terms of materials, besides the traditional jacquard woven brocade, more light and thin fabrics with printing such as cotton cloth, linen and silk were used.

【注释】中文原句采用了省略主语的形式，即汉语中常见的无主句，若将句子成分补充完整，则为"面料的选择上，除传统的提花锦缎外，（旗袍制作者）还增加了棉布、麻、丝绸等更为轻薄的品种"。英语中经常使用被动语态，汉译英时可选择将主动句转换为被动句，尤其针对汉语无主句，译为英语被动句是一个常用的方法。例句译文正是按照英语的表达习惯，将原文的宾语"棉布、麻、丝绸等更为轻薄的品种"译为译文中的主语，使之变为被动句，表意清晰，符合英文的表达习惯。

3．领、袖、襟等部位也用镶滚，却并不烦琐。

The collars, sleeves and garment pieces were decorated with edgings, but didn't look complicated with trivial details.

【注释】首先需要理解的是，原句想要表达的意思是"领、袖、襟等部位也用镶滚来装饰"。中文读者可以根据语境补充原文，推断出语句的意思。但是英文与中文存在着较大差异，英文重结构，主干结构必须完整。中文更倾向于主动式的表达，往往会用无主句或主语省略句来保持语句的主动形式。而英语则倾向于被动式，这使得相应句子在英译时必须做出适当调整。例句译文保留了原文主语，在充分理解原文含义的基础上用"were decorated with"被动结构来译，地道、清楚地传递了原文的信息。

4．妇女不再甘心做男人的附属品和家庭的牺牲品，不少妇女尝试一直是男人在做的工作。

Women were no longer reconciled to be the accessories and victims of the family, so quite a number of women tried to pursue a career that had been previously occupied by men.

【注释】中文是意合语言，在一个长句中，可以有许多具备完整意思、完整结构的流水短句存在，它们之间甚至不需要用连词来梳理逻辑，其文字背后暗含的含义、逻辑将它们紧密地联系在一起。比如此例句原文为两个分句，中间没有任何连词，但其隐含的因果关系却可以被中国读者轻易从字里行间中解读出来。而英语则不然，它的句子需要严格遵守其语法规则，框架结构非常重要，它甚至是表意的重要一部分，句中各成分之间的逻辑关系需要有明确的标记。因此，例句译文补充了连词"so"，来使译文前后分句间的逻辑关系明晰化，方便英文读者更好地理解原文所传递的信息。

5．初兴的式样是一种蓝布旗袍，袍身宽松，廓型平直，袖口微喇。

At the beginning, the prevailing style was a kind of blue cotton cloth *qipao* with loose clothes body, straight and flat outline and slightly bell-mouthed sleeves.

【注释】"初兴的式样"是指最初流行的式样，原文中"初"与"兴"是作为一个整体，放在"式样"前面做定语，译文在处理上比较灵活，把"初"化为时间状语，用"At the beginning"来翻译，而"兴"则继续做定语，译为"prevailing"放在"style"之前。这样的调整使译文更加流畅，也更加符合英语的表达方式。

6．这股风潮与席卷中国的"新文化运动"合流。

This social tide converged with Chinese "New Civilization Movement" that had spread across

China.

【注释】汉语里定语一般前置，如果定语过长，可用一串紧随其后的流水短句来表达，抑或另起一句。英文中稍短的定语会放在被修饰限定的单词之前，而稍长的定语则会借助介词短语、不定式结构、从句结构，甩到被修饰的单词之后。译者在翻译原句中的"席卷中国的'新文化运动'"时就采用了这样的方式，利用定语从句来进行结构上的调整，将原本前置的定语"席卷中国的"，译为以that引导的后置定语从句"that had spread across China"。

六、翻译知识——被动句与翻译

汉语和英语中都有被动语态，但两种语言在被动语态的使用方面存在差异，在翻译中需要谨慎对待。被动语态在英语中是一种常见的语法现象，是英语学习者必须掌握的语法点。英语中使用被动语态或是出于对于施事的隐藏，认为施动者无需或无法指明；又或是出于句法或修辞的考虑，认为使用被动语态可以提高上下文衔接的流畅程度以及全文句式的丰富程度。另外，从文体的角度来看，一些特定的文体更适于通过被动语态来表述，以体现客观、正式等特点。这些文体包括科技文体、新闻文体、公文文体、论述文体等。

汉语被动句的使用无论在语义上还是在形式上都受到一定程度的限制。就语义而言，多数情况下，汉语中的被动句通常用来表达不如意的事。而且按照汉语的表达习惯，如果原文不需要提及施动者，且不会带来歧义或造成理解障碍，一般不会采用结构被动式，而是用意义被动式来表达。因此，汉译英时，往往存在将原文的主动结构译为译文被动结构的情况。在翻译中掌握被动语态的用法是翻译的基本功。译者只有在透彻理解中英文被动结构使用差异的基础上，才能翻译出达意、流畅的好译文。

例1 原文：纹样用黑布剪出来，贴在红布底上，再用粗约2厘米的梗线环扣轮廓边。

译文：The design was cut from black cloth and pasted to red ground fabric before 2-cm-thick threads are stitched around the borders.

【分析】汉语常常用意义被动式而非结构被动式来表达被动的概念，这一点与英语存在较大差异。一些汉语中表示行为的动词没有形态变化，但却既能表达主动又能表达被动意义。例句原文中"纹样用黑布剪出来，贴在红布底上"。"用""剪""贴"，三个动词虽然在形式上无法体现被动含义，但是放在"纹样"之后，读者自然可以体会到其被动意义。英文则不然，作为一种形合语言，句法结构是其极为重要的组成部分。英语中如果要表达被动意义，就必须使用被动语态的结构（be动词+动词过去分词），即如译文"was cut...and pasted..."

例2 原文：人们称花溪型现代挑花为新花。

译文：Modern *huaxi* type couching is called "new couching".

【分析】英文表达中有时会隐去施动者，可能因为这个施事者是显而易见的，无需特殊强调，可能是为了彰显一种客观、理性的姿态，也有可能是因为不便将施事说得太过明了。例句原文中的"人们"是泛称，属于显而易见的施事者，隐去并不会干扰读者理解。不仅如此，译文中隐去施事者，使用被动语态，将原文的受事者"花溪型现代桃花（Modern *huaxi* type couching）"译作主语，有助于将重心转移到受事者身上，方便读者抓住重点，符合英语的表达习惯。

例3 原文：20世纪90年代以前，干部普遍穿着中山装，将其作为无产阶级团结的一种标志。90年代起，中山装多为西装所取代。

译文： *Zhongshan* suit was regularly worn by cadres as a symbol of proletarian unity until the 1990s when it was largely replaced by the Western business suit.

【分析】 例句选自一篇介绍中山装、普及中国文化常识的文章。在英文表达中，这样的文体适合用被动语态来彰显其客观、理性、公正等特点。原文结构为主动语态，译文则顺应文体需要采用被动语态，将中山装作为主语，使表达清楚客观，同时也体现出英文的物称倾向。

例4 原文：从文字记载中可以看到，周代已有专门负责甲胄的官，周代时的铜铠甲多以正圆形的甲片为主，且七片为一组，甲上加漆，以使之呈现出白、红、黑等各种颜色。

译文： It was also recorded in early history that in Zhou Dynasty officials were in charge of armored suits, which were made with round pieces in groups of seven, painted in white, red and black.

【分析】 汉语中，当不需要或不可能说出施事者时，常常采用无主句或者主语省略句来保持句子的主动形式。例句中隐含的施事者应该是"人们"或是"我们"一类，即"（我们/人们）从文字记载中可以看到"，这样的施事者易于理解，因而可以省略。英语注重句子结构完整，句子不能没有主语，在同样情况下只能采用被动式或者其他结构完整的句式。另外，这是一篇说明类文章，采用被动语态显得更加客观、严谨。同时，译文在隐藏施事者后，用"it"作为形式主语，将实质主语置后，可以使句子结构更加平衡，避免头重脚轻。

例5 原文：通常用藤竹、篾席、草麦秸编结帽体，外裱绫罗，内衬红色纱罗。

译文： Its frame is woven by vines, thin bamboo strips and grass with mounted silk and red gauze lining.

【分析】 例句译文打破了原文的无主句的句式结构，运用被动语态来进行翻译，相较于按照原文语序直译并补充出原文隐去的施事者，例句译文的翻译方式更贴近英语国家的表达习惯，也更能为读者所接受。

七、译文赏析
段落1

由于中国执行改革开放的经济政策，国门大开，这为服装的多样化流行和观念的变化，提供了有利的条件。国际潮流信息及时通过服装的经贸交流传递到国内，新科技、新思潮、新时装对国内市场震动很大，女性着装更加具有现代意识。时装已经成为市场经济的重要组成部分，极大地影响着女装更加向国际化的方向发展。一种新的价值观念，深入中国人的生活之中，深刻地影响着人们的审美意识、购买行为和消费观念。这个时期的服装，越来越成为个人对生活的态度、兴趣和消费方式的象征。与此同时，被冷落的传统旗袍，几乎成为特定环境下的特殊装束。表现在：少数人用作礼仪着装；戏剧、电影中用作回顾服装；饭店中用作礼宾招待服装。因此，这时的旗袍对市场已经起不到影响作用了。偶尔能见到一些锦缎、丝绒、盘花扣，甚至精致的镶、绲传统工艺制作的旗袍，很受中、老年人喜爱。因此，它是礼宾场合的"常客"。而大多数受新潮思想冲击的青年人，对旗袍只是抱着改良主义意愿、旁观的态度，很少问津。❶

China's economic policy of reform and opening up to the outside world has provided conditions feasible for the prevalence of multiple styles of apparel and also changes in fashion concepts. Through economic and commercial exchanges in clothing and accessories, knowledge of international trends have passed into China. The new technology, new ideological trends, new fashions have all made a great impact on China's market, especially in modernizing womenswear. Fashionable dressing has become an important part of the market economy, greatly influencing the development of China's clothing industry in directions of internationalization. The new value introduced into the lives of the Chinese people has greatly influenced the people's aesthetic purchasing behavior and changed consumption patterns. One's way of dressing has more and more begun to express one's attitudes, interests in life as well as modes of consumption style. By then, *qipao*, as traditional dress has nearly become specialized attire for certain specific occasions: as ceremonial fashion for a few during rituals; as costumes in plays, operas and movies that look back on old days; as uniforms for waitresses of the high-class hotels and restaurants. Therefore, at that time, *qipao* has less influence on the market. Occasionally, middle-aged and elderly people like to buy and use traditionally tailored *qipao* made of silk, brocade and satin with floral frog closure or piping. So it shows up only at special ceremonial occasions. As for the young people, most of them only have the attitude to reform *qipao* instead of wearing it.

【赏析】段落原文介绍了旗袍在现代生活中由于全球化的发展导致工作、生活方式以及审美意识方面的诸多改变，而遭受冷遇的情况。译文正确传达了原文作者所要表达的信息，同时实现了语言的流畅表达。首先，在翻译特色词汇"旗袍"时，译者采用了音译的方式，将原语的外在形式及其文化内涵完整呈现在读者面前。至于"镶""绲"，由于该信息在本段中并非重点，且没有较多篇章可以用于这两项复杂工艺的解释与说明，而照实译出还可能会增加读者阅读理解的困难，所以译者在处理时，用"traditionally tailored"对应"传统工艺制作"，而将传统工艺的具体例子（"镶""绲"）进行了省译。其次，为了适应英语读者阅读、表达的习惯，译者也在翻译过程中进行了适度调整。原文中"偶尔能见到一些锦缎、丝绒、盘花扣，甚至精致的镶、绲传统工艺制作的旗袍，很受中、老年人喜爱"是无主句，这是汉语中一种较为常见的句式。译者在充分理解了原句句式特征的基础上，根据英语句法特征的需要，结合上下文以"中、老年人（middle-aged and elderly people）"作为主语，使译文结构完整，并且呼应最后一句中的"young people"，突出与下文结构上的对应、内容上的对比。最后，译文也在很多细节处体现了英语语言结构性强的特点，比如使用"as for"这样的连接词；又如在列举旗袍的特殊用途时，每一个用途前都使用介词"as"，可见英译文结构之严谨。

---

❶ 语料参考：袁杰英. 中国旗袍［M］. 北京：中国纺织出版社，2000：67-68.

段落2

从20世纪50年代起，为了适应现代生活需要，服装设计者对中国的旗袍开始不断地进行改良，使这种民族服装既具有东方特色，又符合世界时装的流行趋势，具有优雅、贤淑气质的旗袍，已经得到了国际公认。近20年来，所见到的改良旗袍，受国际时装流行思潮影响颇大，一时间低领、无袖、紧腰、高开衩、超短、袒胸、裸背等各种形式变化无穷，珠片、刺绣、毛皮饰边、织物印花等工艺装饰大放异彩，颜色绚丽、跳跃、浓重、柔和，大胆突破了旗袍的旧有模式。改良旗袍既保留了原有的特点，又融入了创新意识。从此，传统的满装又被注入了时代的血液，赋予了青春的活力。❶

From the 1950s in order to conform to the demands of the modern life, *qipao* was continually improved not only in its Eastern characteristics, but also in conformity with world fashion trends. *Qipao* with its elegant and virtuous disposition was generally acknowledged internationally. In the recent 20 years, international fashion trends have greatly influenced and improved *qipao*, which has appeared with countless changes, such as in styles with low collar, sleeveless, tight waistlines, high slit openings, ultra-mini, robe décolletée, and bareback. It has been augmented with pearl ornamentation, embroidery, fur hems and woven prints. The stereotyped *qipao* has thus boldly broken through with deep, vibrant, gorgeous and mild colors. *Qipao* after modifications has not only integrated new conception, but also preserved its original characteristics. Since then, the old and classic Manchu *qipao* has rejuvenated the blood of the times and been given a new youthful vitality.

【赏析】段落原文介绍了设计师为使旗袍适应现代生活而针对它进行了系列改良的情况。译文完整传递了原文的文字信息与文化内涵，语言流畅、条理清晰，基本再现了原文的行文风格。首先，在翻译特色词汇"满装"时，译者采用了异化的翻译方式，"Manchu"与"qipao"都是音译。同时，"装"的翻译在含义上又体现出意译的成分。译者没有简单地将其直译为"costume""clothing""attire"一类的词，而是结合上下文理解出其所指是旗袍，并用"qipao"将其翻译出来，使该句与上文衔接得更加紧密，从而降低了读者阅读难度。其次，原文中"具有优雅、贤淑气质的旗袍，已经得到了国际公认"采用的是主动句式，而在该句的翻译过程中，译者采用"be acknowledged"的结构，将其译为了被动式。相较于中文，英文更偏向于在特定文体中使用被动式以体现客观性与科学性。在篇章结构上，译文总体上顺应了原文顺序，忠实于原文的行文风格。同时，这也体现出，在全球化程度不断加深的今天，中西方现代语言体系在相互交流、学习的过程中逐渐趋近于彼此。

八、翻译练习

（一）请将下列汉语句子译成英文

1. 旗袍的基本样式是：立领，右大襟，全身较宽松，长袖，上下直线剪裁，下摆宽大，不开衩。

---

❶ 语料参考：袁杰英. 中国旗袍［M］. 北京：中国纺织出版社，2000：52–53.

2．旗袍是一种高领紧身连衣裙，裙子在半边开衩。

3．20世纪20~30年代，旗袍被视为旗人之袍，后来上海的社会名流和名媛佳丽针对旗袍进行了现代化改造。

4．旗袍，在广东话里被称为"长衫"，是上下一体的裙装，起源于17世纪满族统治下的中国。

5．从龙的故乡，旗袍谱写出了一曲美丽的乐章，颂扬着东方服饰文化深厚底蕴的精髓，牵系着每一颗热爱中国的心。

6．清代男子袍衫用色十分丰富，以月白、雪青、湖蓝、银灰和枣红色为多，浅竹布色（淡蓝）也极为流行。

7．在20世纪20年代的上海，旗袍进行了现代化改造，之后在名人和上流社会中流行起来。

8．满族统治时期的原始旗袍是宽松、肥大的。

9．传统旗袍由丝绸制成，上有复杂精美的刺绣。

10．现代旗袍是一种上下一体式连衣裙，贴身合体，单边或双边开衩。

（二）请将下列汉语段落译成英文

服装在20世纪20年代的流行，已模糊了满、汉之间的民族界限。男性此时的常服有下列四种：欧式西装；中山装、学生服；长袍、马褂或长袍、坎肩；衫袄、长裤。女性在"民国"初年，多流行上衣、下裙，旗袍也有服用者，但不普遍。而到20世纪20年代中期，旗袍又一度盛行起来。这种一衰一兴的现象，正是服装循环起落流行规律的反映。男式改良的旗袍，大大精简了烦琐的装点修饰。腰身宽松，袖口宽大，长度适中，领前低（5.5厘米），领后高（8厘米），袍面加里，加宽下摆，提高开衩，显露缎面长裤，以示荣华风姿。20世纪20年代前期袍身较长；中期袍身逐渐减短，有的短至膝下6~12厘米，腰身变窄，下摆收小，同时开衩变低，并去掉长裤。领型也相继降低，领前低（1.5厘米），领后高（2.6厘米）；后期产生了新的变化，女性以突出领型为时尚，流行以10.5厘米的"马鞍型"掩面护颊高领，显示女性柔美容颜。❶

九、译笔自测

旗袍

旗袍，顾名思义，是指清朝满人入关前后八旗妇女的衣袍，即以满、蒙为主体的关外妇女的常服。其基本样式是：立领，右大襟，全身较宽松，长袖，上下直线剪裁，下摆宽大，不开衩。通常在领口、大襟、袖边、下摆处，镶饰刺绣花边或其他颜色的边。这种旗袍主要在北方流行，南方妇女仍多数沿袭明朝风俗习惯穿着较长的上褂，下露长裙。三百多年来，除满族已汉化外，至今蒙古族的妇女仍着长袍，只不过是为骑马方便而加缚腰带罢了。

辛亥革命后，虽然新潮女青年多数穿着白色或绿色上衣、黑色短裙，但一般家居妇女无论南北方，普遍穿着较简化的旗袍，尤其是棉旗袍已成为老少必备、人人宜穿的冬日服

---

❶ 语料参考：袁杰英. 中国旗袍［M］. 北京：中国纺织出版社，2000：52–53.

装了。

记得我在北平读小学和初中时，那是20世纪20年代，学校的制服是白衣、黑裙，可是在家时，特别是假日有客人来或跟随父母出门做客时，都要换穿旗袍。这种长袍下摆较大，不开衩，成宝塔形，袖口也较宽，呈喇叭形。若是素色，则在领口、大襟、下摆处镶小花边。当时已能买到较廉价的机织花边了，无须刺绣。若是有花纹的衣料，则周身滚深色细边。母亲和长辈中的中年妇女依时尚也有穿长裙和齐腰、圆下襟短上衣的，但多数仍穿旗袍。

此后，旗袍的剪裁样式有了很大变化。20世纪30年代我家移居上海，经常可以从画报电影传媒中看到服装时尚。还有月份牌、香烟画片、广告等，都以时装美人为题材。旗袍的样式逐渐有了很大改变：衣袖从原来的长袖过肘变成短袖到肘以上，再后来更短，甚至到肩，成为无袖了。衣长都在膝以下，最长直到脚跟，只有穿高跟鞋才能不拖地。最大的变化是剪裁，依腰身曲线完全贴身，下摆窄，左右开衩到膝部。这首先是流行于上海十里洋场，无疑是受西方晚礼服裸露曲线的影响。

1934年我在南京"中大"艺术系就读时，开始喜欢穿自己设计、自己手制的服装。记得我在夏天就穿过自制创新的旗袍，取消了高领和大襟，露颈的圆领绲边，肩部开口，肩下短袖，膝下开衩，既凉爽，又缝制省事、省料。直到20世纪90年代的今天，无论在国外或国内，这仍然是常见的最简单的夏日便服样式。20世纪40年代后的旗袍还曾流行衣长短到膝以上，这也是受了迷你（MINI）裙的影响。

第二次世界大战以后，随着国家间的频繁交往、对话和文化交流，服装潮流已经成为世界性的。近年以意大利、法国为首的服装展和欧美流行的时装趋势，很明显可以看出中国旗袍的影响，晚礼服的流行基本形态就是紧身长袍式。连便服长裙也流行窄下摆在后面或侧边开衩了。

在中国香港和中国台湾，半个多世纪以来旗袍始终是中国妇女的标准服装而且成为婚丧、社交等正式场合的礼服。在样式上比起20世纪30~40年代的旗袍除了在襟领上的设计花样有变化以外，普遍的大变化就是旗袍的开衩提高到几乎接近臀部。国内从20世纪80年代改革开放以来，凡晚会典礼仪式的主持人和礼仪小姐以及茶楼酒馆的接待人员，也多流行这种旗袍。

中国旗袍从清代样式发展到如今已经相去甚远。相信我国新兴的服装业和新一代优秀的设计师，会以这基本样式为基础，吸收中国各民族装饰艺术的精华，创制出更美、更新、更合理的中国旗袍。

如前所述，一种民族的服装样式也是一种文化的包装，就像语言是思想感情的表现一样，它担负着与人交流沟通的任务。如同母语会使人产生特别的亲切感，中国旗袍的基本样式也能将人引向中国文化的意趣，使人产生如母语般的亲切感。❶

---

❶ 语料参考：袁杰英. 中国旗袍［M］. 北京：中国纺织出版社，2000.

十、知识拓展

旗袍作为一种光滑、合身的服饰，展现了穿着者内在的优雅，体现了中国女性的身份认同与文化传承。它起源于17世纪的清朝（1644—1911）。满族男性的传统服装是"长袍"，而女性穿的则是"旗袍"，当时的"旗袍"与我们今天所知道的"旗袍"具有很大差异。

今天为我们所熟知的具有标志性的中式服装起源于20世纪20年代的上海。受到西方的影响，满族旗袍在这一时期开始了它的现代化历程。它在上海的名流圈以及上层社会中很受欢迎，甚至成为一种身份的象征。当时，每一件旗袍都是为顾客量身定制的。1929年，官方认定旗袍为当时的一种国服，旗袍也就成为时尚，为各行各业的女性所追捧。除了传统的丝绸外，棉布、羊毛、亚麻和缎子织锦等织物也被用于生产新式的现代旗袍，这使旗袍成为人人都买得起的时尚服装（图2-16）。

现代旗袍更加合身，它可以比以前的旗袍更短，长度保持在膝盖以上，侧面开衩也可以一直到大腿。再加上那令人惊叹的刺绣，旗袍简直是成为美丽、优雅和激情的艺术体现。❶

图2-16　旗袍设计图❷

The *qipao* is a dress that brings out the inner elegance of the wearer and embodies the cultural identity and heritage of Chinese women in one sleek and fitting dress. It has its origins in the 17th century, under the Manchu emperors of the Qing Dynasty (1644–1911). The traditional clothing for a Manchu male was a long robe (*changpao*), while a woman wore a *qipao*, a gown that was very different from the one we know today.

The iconic Chinese dress we're familiar with today originated in Shanghai in the 1920s. The Manchu *qipao* underwent its modernization during this period, owing to a Western influence. It became popular among Shanghai's celebrities and upper classes. It was even something of a status symbol, as each piece had to be custom-made to flatter the female wearer's body. In 1929, on becoming one of the official national dresses at that time, *qipao* became fashionable for women from all walks of life. In addition to the traditional silks, other fabrics—cotton, wool, linen and satin brocade—were used in the production of the new, modern *qipao*, making it an affordable and fashionable clothing item for everyone.

The modern version of *qipao* is more form-fitting; it can also be shorter than its predecessor, reaching to above the knee, with side slits that reach up to the thigh. The addition of stunning embroidery has made *qipao* an artistic embodiment of beauty, elegance and passion.

---

❶ 内容参考：*Travelogue: The qipao, China's most iconic dress*，来源：中国环球电视网。
❷ 图片来源：高光网。

第九节 中山装

学习目标

1. 了解中山装的历史及文化内涵
2. 熟悉中山装的基本语汇及其汉英表达方式
3. 掌握与形合与意合相关的翻译知识与技能
4. 通过实践训练提高中山装文化汉英翻译能力

一、经典译言

我们在翻译的时候，通常胆子太小，迁就原文字面、原文句法的时候太多。要避免这些，第一要精读熟读原文，把原文的意义，神韵全部抓住了，才能放大胆子。

—— 傅雷

二、汉语原文

中山装

服饰是人类文明的重要组成部分，是划分族群的外在标识。我们的实际着装类型和数量受诸多因素影响，比如性别、身体属性、地理以及社会环境。一种服饰风格的形成、完善以及最终固定需要历经数个阶段以及漫长岁月的考验，因为服饰风格根植于一个民族的文化传统。在中国，一些经典的服饰依旧占有重要的地位，比如中山装（图2-17）具有独特的形制和奇妙的内涵，被越来越多的人所熟知并使用。尽管现如今年轻一代多追逐更为现代和新潮的服饰，但是中山装因为蕴藏着非凡的象征含义，仍然受到不少人的青睐。

中山装是中国一种现代男装，其名字来源于孙中山先生。"中华民国"建立之初，孙中山便在国内推广了这种具有明显政治内涵的服饰。当时，服饰类型多基于满族服制，这是由于清朝曾强制推行满族服制，并将其作为社会统治的一种形式。早在"民国"建立之前，就有一些进步人士不满于中国落后的服饰，并将中国长衫与西洋帽子结合在一起，形成了一种全新的着装风格，推动了中国服饰的发展。中山装的设计极力适应当代需求，同时又尽量避免对西方风格的全盘吸收。

孙中山先生积极地参与了中山装的设计，将自己旅日的生活经验融入了中山装的设计之中：日式制服成为中山装的原型。当然，也有很多修改。西服有三个隐藏的口袋，但中山装的四个口袋却是在外面的，并且是以中国平衡对称概念指导分布的。另外中山装也是有一个口袋是隐藏的。随着时间的推移，中山装的样式也发生了些许细微的变化，比如，中山装的扣子由原来的七粒减少到了五粒。中山装的四个兜源自《管子》中的礼、义、廉、耻，中国人认为这是为人处世的基本原则。正中间的五颗纽扣代表"中华民国"根据《宪法》设立的五个院（立法、监察、考试、行政以及司法）。袖口的三颗纽扣代表孙中山的"三民主义"：民族、民权、民生。孙中山先生逝世之后，传言中山装被赋予了爱国与改革的非凡意义。

后背不破缝，表示国家和平统一之大义。

The back has no break seam, having the meaning of peaceful reunification of the nation.

倒山字形"笔架盖"象征崇文兴教。

The Reverse-mountain-shape "Brush-stand Covers" Represent the Thought of Respecting Culture and Promoting Education.

五粒扣代表"行政、立法、司法、考试、监察"五权宪法。

The five buttons represent the five rights in constitution: administration, legislation, judieiary. examination and supervision.

袖口上的三粒扣表示"民族、民权、民生"的三民主义。

The three buttons at sleeve opening mean the Three People's Principles: Nationalism. Democracy, the People's Livelibood.

四个口袋寓意"礼、义、廉、耻"四大美德。

The Four Pockets Have the Morale of the Four Virtues about "Sense of Propriety, Justice, Itonesty and Honor"of Chinese nationality.

口袋上的四粒扣表示人民拥有的"选举、罢免、创制、复决"的四权。

The four buttons on the pocket mean the four rights enjoyed by the people: election, dismissal, making and review.

图2-17 中山装❶

1949年中华人民共和国成立，中山装成为领导人的必备服装，象征着无产阶级团结。此后，中山装在中国掀起一波又一波的热潮。当前，中山装曾经的政治性已经逐渐被现代时尚性所取代。与常见的西装相比，中山装更能体现东方男性的内向气质，从而成为非常适合中国男性的服装。

译前提示：中山装是孙中山先生在广泛吸收欧美服饰的基础上，综合了日式学生服装（诘襟服）与中式服装的特点，设计出的一种立翻领有袋盖的四贴袋服装。我国国家领导人、著名影星和艺术家在出席重大活动时，时常穿着中山装。中山装也逐渐成为一种极具中国特色的标志。本文详细介绍了中山装的起源、其背后的文化内涵以及后续发展，指出中山装经久不衰的符合东方审美的时尚性。

三、英语译文

Zhongshan Suit

Clothing is part of human civilization and represents the identity of one nation and its people. In fact, the type and the amount of clothing we wear are dependent on various bases. They may be dependent on gender, physical attributes, and social and geographical considerations. It involves

---

different process and takes years to perfect and to formally stabilize the types since they are anchored in our traditions and culture in general. In China, there are some classic clothes that remain to be important. For example, the *zhongshan* suit has its exceptional style and wondrous cultural meaning, so it has been known and worn by more and more Chinese people. Though the young generation prefers new and modern fashion styles, *zhongshan* suit is still loved by many people for its remarkable symbolic associations.

The *zhongshan* suit, named after Mr. Sun Zhongshan, is a style of modern Chinese male attire. Sun Zhongshan introduced the style shortly after the founding of the Republic of China as a form of national dress with a distinctly political and later governmental implication. At that time, the style of dress worn in China was based on Manchu dress, which had been imposed by the Qing Dynasty as a form of social control. Even before the founding of the Republic of China, older forms of Chinese dress were becoming unpopular among the elite and led to the development of Chinese dress which combined the *changshan* and the Western hat to form a new dress style. The *zhongshan* suit was an attempt to cater to contemporary sensibilities without adopting Western styles wholesale.

Mr. Sun Zhongshan was actively involved in the design of the *zhongshan* suit, providing inputs based on his life experience in Japan: the Japanese uniform became a basis of *zhongshan* suit. There were some modifications as well: instead of the three hidden pockets in Western suits, the *zhongshan* suit had four outside pockets to adhere to Chinese concepts of balance and symmetry; an inside pocket was also available. Over time, the suit had some minor stylistic changes, for example, the number of the buttons was reduced from seven to five. The four pockets were said to represent the Four Cardinal Principles cited in the classic *Guanzi* and understood by the Chinese as fundamental principles of conduct: propriety; justice; honesty; a sense of shame. The five center-front buttons were said to represent the five branches of government (legislative, supervisory, examinational, administrative and judicial) established by the Republic of China in accordance with the constitution and three cuff-buttons to symbolize the Three Principles of the People: nationalism; democracy; people's livelihood. Long after Sun Zhongshan's death, the *zhongshan* suit was assigned a revolutionary and patriotic significance.

After the establishment of the People's Republic of China in 1949, the *zhongshan* suit became widely worn by government leaders as a symbol of proletarian unity and an Eastern counterpart to the Western business suit. Since then, *zhongshan* suit has been popular again and again in China. At present, the political implicature of the *zhongshan* suit has been gradually replaced by its modern fashion implicature. Compared with common western suits, the *zhongshan* suit can better reflect the introverted temperament of oriental men, therefore they are very suitable for Chinese men to wear.❶

---

❶ 语料参考：*Zhongshan Suit History*，来源：东方良品（Good Orient）网；*Zhongshan Suit*，来源：华鼎旅游（Top China Travel）网；*Zhongshan Suits Make Comeback*，来源：沪江网（略有改动）。

四、词汇对译

中山装　*zhongshan* suit

男装　male attire

"中华民国"　the Republic of China

民族服装　national dress

满族服装　Manchu dress

日式制服　the Japanese uniform

修改　modification

暗袋　hidden pocket

对称（性）　symmetry

里袋　inside pocket

礼、义、廉、耻　propriety, justice, honesty and a sense of shame

袖扣　cuff button

三民主义　the Three Principles of the People

民族　nationalism

民权　democracy

民生　people's livelihood

爱国的　patriotic

中华人民共和国　the People's Republic of China

无产阶级大团结　proletarian unity

商务套装　business suit

五、译文注释

1. 服饰是人类文明的重要组成部分，是划分族群的外在标识。

Clothing is part of human civilization and represents the identity of one nation and its people.

【注释】中文原句由两个分句组成，共用一个主语。原文主要说明服装的重要性，两个分句间没有连词，凭借内在逻辑连接在一起，构成并列分句。而英译文中则使用连词"and"将两个谓语连接起来，构成一个句子。其原因在于，英文是重视结构体系的形合语言，两个谓语动词间必须要有衔接标志，即便是"and"也不能省略。

2. 1949年中华人民共和国成立，中山装成为领导人的必备服装，象征着无产阶级团结，成为中式西装。

After the establishment of the People's Republic of China in 1949, the *zhongshan* suit became widely worn by government leaders as a symbol of proletarian unity and an Eastern counterpart to the Western business suit.

【注释】中文原句的核心在于"中山装成为领导人的必备服装"这一分句。句首"1949年中华人民共和国成立"铺垫出背景，句尾"象征着"，"成为了"两个分句则是对核心信息的补充说明。"中华人民共和国成立"和"中山装成为领导人必备服装"两件事之间存在先后顺序，所以增加介词"after"来明示这个时间顺序。句尾补充成分的翻译中，译者使用连词"as"引出补充内容，这样的处理方式使译文结构明晰，语言简练，符合英语的表达习惯。

3. "中华民国"建立之初，服饰类型多基于满族服制，这是由于清朝曾强制推行满族服制，并将其作为社会统治的一种形式。

When the Republic of China was founded in 1912, the style of dress worn in China was based on Manchu dress, which had been imposed by the Qing Dynasty as a form of social control.

【注释】在本句的翻译中，译者发挥了一定的自主性，对原文句子成分的功能进行了适当调整。原文中"这是由于"引出原因，表明前后分句之间构成因果关系。汉译英过程中，译者使用了转换法，用"which"引导的后置定语从句来修饰先行词"Manchu dress"，补充说明清王朝推行满族服饰来统治社会的信息，从而将原文中表原因的状语转换为了译文中的后置定语。

4. 孙中山先生积极地参与了中山装的设计，将自己旅日的生活经验融入中山装的设计之中：日式制服成为中山装的原型。

Mr. Sun Zhongshan was actively involved in the design of the *zhongshan* suit, providing inputs based on his life experience in Japan: the Japanese uniform became a basis of *zhongshan* suit.

【注释】中文原句中前两个分句之间没有明显的连接词衔接，两分句在逻辑上也可以简单地视为自然顺联。英文是形合语言，注重结构的严谨性，分句间的逻辑关系需要通过连接词或词性变化等方式进行明示。译者在翻译前两个分句时未使用连词，而是将原文第二分句以现在分词"providing"译出，使其成为状语，修饰前文的"参与"，说明孙中山先生参与中山装设计的方式。

5. 正中间的五颗纽扣代表"中华民国"根据《宪法》设立的五个院（立法、监察、考试、行政以及司法）。袖口的三颗纽扣代表孙中山的"三民主义"：民族、民权、民生。

The five center-front buttons were said to represent the five branches of government (legislative, supervisory, examinational, administrative and judicial) established by the Republic of China in accordance with the constitution and three cuff-buttons to symbolize the Three Principles of the People: nationalism; democracy; people's livelihood.

【注释】译者在翻译"代表"时用的是"were said to represent/ symbolize（据说代表/象征着）"这样的被动结构。被动结构具有一个优势，就是在文字说明、新闻报道等特定文体中彰显文章撰写者公正客观、小心严谨的态度，进而使文章内容显得更加真实可信。此外，译者在翻译例句时采用了合译法，将原文的两个句子合二为一，用"and"连接，并在"and"后的谓语结构"to symbolize"之前省略了与前面重复的"were said"，使译文更加简练，前后连贯。

六、翻译知识——形合意合

形合和意合是英汉语言之间的重要区别性特征。形合是指用语言形式手段（如关联词）连接句中的词语或分句，来表达语法意义和逻辑关系。意合是指通过句中词语或分句的含义而非语言形式手段来连接句中的词语或分句，来表达句中的语法意义和逻辑关系。英语以形制意，而汉语以意驭形。英语造句主要采用形合法，而汉语造句则相反，主要采用意合法。

由于语言与思维模式之间相互制约，西方形式逻辑的思维模式要求语言依赖于各种连接手段承上启下。英语句子常用各种形式手段连接词语、分句或从句，注重句子形式和结构完整，以形统意。其主要连接手段包括：关系词和连接词、介词、形态变化形式等。汉民族重内省和体悟，不重逻辑，因而语言简约、意义模糊。汉语句子少用形式连接手段，注重隐形连贯和逻辑事理顺序，以意统形。汉语句子常运用语序、反复、排比、对偶、对照、句子紧缩以及四字格等方式，将语法意义与逻辑联系隐藏于字里行间。汉英翻译过程中，要注重意合语言特征向形合语言特征的转变。下面将结合汉英翻译实践说明翻译中如何协调形合语言与意合语言的关系。

例1 原文：隋朝（581—618年）时，中国重新统一，汉人的服装规范再次推行开来。

译文： When China was reunited in the Sui Dynasty(581-618), the Han dress code was pursued again.

【分析】本句原文为典型的意合形式，分句间的逻辑关系隐含在字里行间。译文则体现了形合语言的特点，由逗号分隔开来的前后两部分形成了明显的主从句关系。译文句子的后半部分为主句，前半部分则是由"When"引导的一个时间状语从句。译文借助时间状语从句这种语言形式手段，厘清了本句所描述的事件之间的逻辑关系。

例2 原文：旗袍具有独特魅力，许多女性通过穿着它来彰显自身独有的优雅气质。

译文： Because of *qipao*'s unique charm, many women wear it to show their special grace.

【分析】译文通过"Because of"的使用，在两分句间构建起了因果逻辑关系，前因后果通过外在形式体现得一清二楚。后半句中"to do"不定式结构清楚说明了女性穿着旗袍的目的。

例3 原文：旗袍不仅强调女性身姿的自然美感，而且使得女性的腿部显得更加修长。

译文： Not only does it lay stress on the natural beauty of a female figure, but also makes women's legs appear more slender.

【分析】译文中"Not only...but also"是英文表达中常用的一组关联词组。在它的连接下，前后句间体现出一种并列且轻微递进的关系，句子重心略微倾向于后半句。英语作为形合的语言，充分借助结构与形式，达到了以形表意的目的。

例4 原文：在此期间，上海成为闻名四海的国际通商口岸，化为中国敞开的门户，受到西方的影响。

译文： During this time, Shanghai became a well-known international treaty port, which opened the door for Western influences.

【分析】例句英译文以形统意，层次清楚。"During this time"作为时间状语置于句首，主句紧随其后，"which"引导后置定语从句修饰"treaty port"。整个句子结构清晰，焦点突出，句子含义与逻辑一目了然。

例5 原文：苗绣苗锦特殊的生存环境和传承方式，使一些支系的纹饰保存着许多远古文化和原始艺术的特征。

译文： Thanks to the uniqueness of the milieu of Miao embroidery and brocade as well as the way in which they have been handed down, the designs of some branches retain many features of ancient cultures and primitive arts.

【分析】例句汉语原文含有潜在的因果联系。英译文使用"thanks to"一词将这种因果联系明确表现出来，使译文逻辑关系更为清楚。而中文则不同，其表达习惯前因后果，原因一般置于句子前面部分，结果紧随其后，这样符合汉语的表达习惯，即使没有因果关联词语，读者也能自然领会其间的逻辑关系。

例6 原文：妇女们常把小型的绣件卷放在身上，走到哪里做到哪里，一有空就拿出来做一些，连放牛时也做着刺绣。

译文：Women are in the habit of carrying rolled-up small embroideries about and working on them whenever they have spare time, including the time when they are grazing cattle.

【分析】汉语采用意合形式来进行表达，一句话中往往出现一连串的流水分句，分句与分句之间没有明显的关联词相连，仅凭意思相互勾连。比如，例句中的汉语原文由四个流水分句组成，形散神聚，一气呵成。汉译英时，译者借助连词"and"和分词"including"将整个译文句子衔接起来，结构紧凑、层次分明，体现了形合语言的特点。

例7 原文：唐代时女皇武则天曾赐百官绣袍，以文官绣禽，武官绣兽。

译文：In Tang Dynasty the woman emperor Wu Zetian had all officials wear embroidered gowns, specifying that civil official gowns were embroidered with birds and military official gowns with beasts.

【分析】例句原文为三个流水分句组成，以意统形，逻辑关系隐含在字里行间。比如"文官绣禽，武官绣兽"在汉语表达中采用了对照的方式，词句整齐、匀称。汉语中，如果采用了反复、排比、对偶、对照等修辞方法，这样的句子往往不需要关联词，关联词的添加反而会破坏其对称结构的美感。汉译英过程中，译者在充分传递原文信息的基础上，采用分词结构"specifying that"，明示原文第一个分句和后面两个分句之间的逻辑关系。同时利用"and"使原文后两个分句形成并列结构，并省略后一分句中与前面重复的谓语"were embroidered"，使前后衔接更加紧密、连贯。

例8 原文："乌纱帽"在汉语里也成了官位的代名词，沿用至今。

译文："Black gauze cap" also became a synonym for the government official status, used until the present day.

【分析】王力曾经指出："西洋语的结构好像连环，虽则环与环都联络起来，毕竟有联络的痕迹；中文的结构好像天衣无缝，只是一块一块的硬凑，凑起来还不让它有痕迹"，西洋语法呆板、强硬，而中国文法富于变化，具有弹性。例句原文由两个各自独立的分句构成，但前后语义连贯，读起来自然流畅。英语语法要求句子结构焦点突出，句中的逻辑关系必须借助词汇或词性变化明确表达出来。因此，在例句的翻译过程中，译者借助过去分词"used"，将原文的第二个分句翻译为第一个分句中"乌纱帽"的后置定语，使译文结构严谨，层次清楚，符合英语形合语言的表达要求。

例9 原文：20世纪70年代，中国执行改革开放的经济政策，国门大开，为服装的多样化流行和观念的变化提供了有利的条件。

译文：In the 1970s, China's economic policy of reform and opening up to the outside world has provided conditions feasible for the prevalence of multiple styles of apparel and also changes in fashion concepts.

【分析】中文是意合的语言，句子的衔接，逻辑关系的传递往往依靠句子本身的含义。相较英语，中文对逻辑连接词，以及句法结构的依赖程度较低。时间观念以及事物状态也不依靠动词的时态变化来表达。比如本例句原文，没有任何连接词，但表意依然流畅。汉译英过程中，译者根据英语语言的表达需要进行合理的调整，重构英文句子的主干结构，并合理安排各修饰成分的位置，形成了结构凝练的英译文，重点突出，层次清楚。

例10　原文：20世纪90年代，受西方的影响，大众不再像过去一样时常穿着中山装，它却依旧是中国领导人出席国内重要场合时的重要正装。

译文：Although the *zhongshan* suit fell into disuse among the general public in the 1990s due to increasing Western influences, they are commonly worn by Chinese leaders during important state ceremonies and functions.

【分析】例句原文没有使用太多逻辑连词，但隐含在字里行间的逻辑使得句子表意清楚，自然流畅。汉译英时，译者在句首增译了连词"although"，突出了句子前后部分的让步关系；增译了表示因果关系的词组"due to"，明示了西方影响和大众不再热衷于中山装之间的因果关系。增译这些凸显文中逻辑关系的表达，使得译文层次清楚，逻辑连贯，符合形合语言的特点。

七、译文赏析

段落1

"中华民国"于1912年成立，但当时的服饰风格依旧基于受清王朝影响很大的满族服饰的形制（女性穿旗袍，男性则穿长衫）。清王朝的统治未能使中国免于帝国主义的侵略，其统治下的中国在科学、技术等层面也远远落后于西方。这些原因促使很多中国革命者起来推翻清政府统治。早在"中华民国"成立之前，老式的中国服装就已经不再受中国先进分子的欢迎。长衫搭配洋帽形成了一种全新的着装方式，促进了中国服饰的发展。中山装事实上也是中西服装风格结合的产物。❶

Although the Republic of China was founded in 1912, Chinese wearing style was still based on Manchu dress (*qipao* for women, *changshan* for men) at that time, which had been deeply influenced by Qing dynasty. Qing dynasty failed to defend China against the aggression of imperialists, besides, the Qing stood low in technology and science compared to the West, which made the majority Han Chinese revolutionaries overthrow the Qing Dynasty. The old Chinese dress was unwelcomed by the elite even before the founding of the Republic of China. *changshan* matching with western hat, as the development of Chinese dress, became a new dress. The *zhongshan* suit was exactly a combination of western and eastern styles.

【赏析】段落原文介绍了中山装诞生的时代背景，清朝服饰风格逐渐过时，西方服饰风格受到追捧，而中山装则是中西结合的产物。译文语言简洁、流畅，逻辑清晰，内容丰富，完整地向读者传递了原文信息。首先，在翻译文化特色词汇时，译者多采用异化的策略，例如在翻译"满族服饰"和"中山装"时采用了音译与直译相结合的方式，在翻译"旗袍""长衫"时直接采用了音译的方式，旨在将原汁原味的中国传统文化展现给译入语读者。其次，中文是意合的语言，而英文是形合的语言，相较于中文，英文更注重结构的完整。例如原文句子"清王朝的统治未能使中国免于帝国主义的侵略，其统治下的中国在科学、技术等层面也远远落后于西方"，由两个分句组成，每个分句各有一个主语即"清王

---

❶ 语料参考：*Zhongshan Suit*，来源：中国华鼎旅游（Top China Travel）网。

朝的统治"和"中国",中间没有连接词,全凭语义相连。在英译过程中,译者在两个分句中间增译连词"besides",发挥衔接作用。同时,译者将两个分句主语分别定为"Qing Dynasty"和"the Qing",使两个分句从形式到意义上都衔接得更为紧密。此外,译者基于文章内涵,对句子进行了拆分与合并,使译文逻辑更清晰,表达更顺畅。例如原文"这些原因促使很多中国革命者起来推翻清政府统治"本为独立的句子,但译者根据上下文的语义联系,将其译为前一句的非限制性定语从句,使译文在结构和语义上更为紧凑。

段落2

中山装深受孙中山先生的喜爱,他也是穿中山装的第一人。曾有规定,一定级别的官员在就职仪式上要穿着中山装以显示自己效忠于孙中山先生。随着中国改革开放政策的推行,西服以及其他服装逐渐取代中山装,变得流行起来。尽管普通民众基本上不再穿着中山装,但是中国的国家领导人依旧喜欢穿着中山装出席各种重要的国内活动。❶

Zhongshan suit was favored by Mr. Sun Zhongshan, who was the first one wearing this kind of suit. There was a rule that a certain class civil official must wear the *zhongshan* suit in inaugurations to show obedience to Mr. Sun. With the promotion of China's reform and opening-up policy, western suits and other garments gradually replaced *zhongshan* suit and became popular. Although *zhongshan* suit faded in folk, Chinese leaders were still fond of wearing the *zhongshan* suit to attend important state events.

【赏析】段落原文介绍了中山装过去与现在的使用范围。译文语言简洁、流畅,逻辑严谨、信息丰富,从形式和内容上再现了原文。首先,段落原文所涉及的文化特色词汇相对较少,译者在处理时基本采用了异化的翻译策略。例如在中山装的英译上,译者采用了音译加直译的方式。中山装借用了西装以及日本学生装元素,融合中国元素发展而成,与西装在样式和功能上具有很多相似之处。译者选用"suit"一词可以使读者迅速联想到西装,从而了解到中山装的大致样式与功用。而"zhongshan"则显示出其中国特色,表明这是中国特有的服饰类型,与西装区别开来。译文"zhongshan suit"既展现了中国特色,又起到了辅助读者阅读理解的作用,在中西方文化之间达到了一种相对的平衡。其次,译者也充分考虑到了中英语言表达上的差异,针对部分句子结构进行了相应调整。例如原文"中山装深受孙中山先生的喜爱,他也是穿中山装的第一人"这一句,由两个分句组成,分句成分完整且相对独立,中间并无起衔接作用的连词。而英语不同于中文,是形合的语言,需要依靠严谨的结构来表意。译者在翻译这句话时,捕捉到了前一个分句中的"孙中山先生"与后一分句中的"他"之间的联系,从而将后一分句译成以"who"引导的非限制性定语从句,用于修饰前一分句中的"Mr. Sun Zhongshan",使译文句子结构紧凑,焦点突出。

八. 翻译练习

(一)请将下列汉语句子译成英文

1. 55年来,红都作为一个历史悠久的国内时尚品牌,一直为中国历代高级官员以及外

---

❶ 语料参考:*Zhongshan Suit*,来源:中国华鼎旅游(Top China Travel)网。

国政要提供中山装定制服务。

2. 假如你走在几十年前中国的大街上，你很可能会看到男女都穿着中山装。这是一种中国版本的商务套装，也被称为"毛装"。

3. 1949年，中华人民共和国成立，中山装成为无产阶级团结的象征，作为东方的商务西装，在男性和政府领导人中广泛使用。

4. 中山装，也被称为毛式制服，由1911年辛亥革命的革命领袖孙中山先生首次引入，中华人民共和国成立后，在毛泽东的推动下，得到推广。

5. 中国在国际舞台上的重要性和影响力与日俱增，越来越多的人回归中国文化和中山装。

6. 中山装象征着领导人所需要的品质——庄重、慷慨和稳重，在中国被视为最适合领导人穿着的服装。

7. 这套中山装是蓝黑色，立领精致剪裁，样式更加简洁，是对传统"中山装"的重新设计，非常吸人眼球。

8. 相反，那是一套改良版的中式立领服装。整个设计符合中国风格，但某些细节处又具有现代剪裁特色。

9. 这些裁缝来自浙江宁波，20世纪初因其缝制中山装和西服的技艺而闻名。

（二）请将下列汉语段落译成英文

中山装在1911年辛亥革命过程中因其简洁、实用而与西装齐名。国民政府后来颁布相关规定，将中山装定为正装。中山装在样式上进行了些许细微的调整。中山装也被赋予了革命与爱国的含义。中山装的四个口袋，正中间的五粒扣子，包括袖口的三颗纽扣都具有特殊的含义，这些意义都取自中国传统经典《管子》、"中华民国"《宪法》与孙中山的先生提出的"三民主义"。

九、译笔自测

中山装

中山装是孙中山先生在广泛借鉴西式套装、日本学生装以及中国本土服饰的基础上设计出来的，在"中华民国"时期非常流行。1925年孙中山逝世后，中山装被赋予了一层革命和爱国的理念。

中山装基本成型于20世纪20年代早期。自20年代起，尽管中山装的样式也有几次调整，但整体变化不大。最主要的变化是中山装前门襟的扣子数量从起初的七粒变成了五粒，去掉了背面的带子和斜缝。中山装有四个口袋，前门襟有五粒扣子，每个袖子上又各有三粒小扣子。孙中山先生赋予中山装重要内涵。后背不破缝，象征着中国的和平统一。封闭式翻领象征着严谨治国的理念。四个口袋代表《管子》中的"四维"：礼、义、廉、耻；正中间的五颗纽扣代表"中华民国"根据《宪法》设立的五个院：立法、监察、考试、行政以及司法；袖口的三颗纽扣代表孙中山的"三民主义"：民族、民权、民生。中华人民共和国成立之后，材料和人力稀缺，针对传统中山装进行了简化和改进。升级版的中山装也随后推广到了全国。

中山装有很多优势，主要包括它平衡对称的板型，典雅端庄的外观，极高的舒适度，凸显男性沉稳的特点，适用范围极其广泛（正式或非正式场合皆可）。改良后的中山装色彩丰富，除却黑、白、灰等常用色彩外，还有驼色、蓝色等。基于不同场合的需要，人们会选择不同的色彩。用于正式场合的中山装色彩应当庄重素雅一点。用于非正式场合的中山装色彩可以鲜亮活泼一些。

近年来，中山装变得越来越流行。很多成功人士都喜欢穿着白色或是米黄色的中山装出席大规模商务场合，这样的打扮显得穿着者沉稳而时髦。中山装的复兴得益于许多著名的电影明星以及艺术家。在很多重大场合，国际巨星成龙都会穿着中山装借助媒体出现在观众面前。这将中山装及其文化推广到了国际舞台。著名导演张艺谋也喜欢穿着中山装。他经常穿着一身黑色的中山装出席国际电影节活动。著名歌星刘德华经常穿着黑、白、灰三种颜色的中山装在台上表演。

中国有很多传统服饰。时至今日，人们向文化生活投入越来越多的关注。这些服装被赋予了更多的重要内涵以及新鲜活力，传统文化也通过多种样式得到保存和展现。❶

十、知识拓展

"民国"初年，许多青年学生到日本学习，带回了日本的学生装。这种服装式样给人朝气蓬勃、庄重文雅之感，深受广大进步青年的喜欢，还衍生出了典型的现代中式男装中山装。

中山装的特殊之处是对衣领和衣袋的设计。高矮适中的立领外加一条反领，效果如同西装衬衣的硬领；上衣前襟缝制了上下四个明袋，下面的两个明袋由压褶处理成"琴袋"式

图2-18　中山装❷

样，以便放入更多物品，衣袋上再加上软盖，袋内的物品就不易丢失。与之相配的裤子前襟开缝，用暗纽，左右各插入一大暗袋，而在腰前设一小暗袋（表袋）；右后臀部挖一暗袋，用软盖。这种由"中华民国"创始人孙中山（1866—1925）率先穿用的男装，较之西装更为实用，也更符合中国人的审美习惯和生活习惯，虽然采用了西式的剪裁、西式的面料和色彩，却体现了中装对称、庄重、内敛的气质。自1923年诞生以来，中山装已成为中国男子通行的经典正式装（图2-18）。❸❹

In early years of Republic of China, many young students went to study in Japan, and brought Japanese student uniforms back. This clothing style showed a touch of youthful spirit, sobriety and refinement, thus becoming popular among

---

❶ 语料参考"闲话'中山装'"，来源：听力课堂（Tingclass）网；*History of Traditional Chinese Suit –Zhongshan Suit*；来源：新汉服（New Hanfu）网。

❷ 图片来源：搜狐网。

❸ 本章知识拓展中文内容参考：华梅. 中国服饰［M］. 北京：五洲传播出版社，2004：87-89.

❹ 英文内容参考：Hua Mei. *Chinese Clothing*［M］. 北京：五洲传播出版社，2004：91-92.

many young men. Later, it was transformed into the typical modern Chinese men's uniform—the *zhongshan* suit.

The special features of the *zhongshan* suit are in the design of collars and pockets. A reversed stand collar of fitted height has the effect of the wing collar of the western style shirt. There are four out pockets in the front garment piece. The lower two ones are pressed and tucked into the style of "*qin* pocket" so that they can hold more stuff. Soft covers are designed above the pocket to prevent articles from losing. The corresponding pants has a front opening with hidden buttons and a big hidden pocket on each of the left and right side. There is a small hidden pocket (watch pocket) on the front waist and a hidden pocket on the right rear hip with a soft cover. This set of men's suit designed by the founder of Republic of China, Mr. Sun Zhongshan, is more practical than western suits, and fits more to the aesthetics and life customs of the Chinese people. Even though it adopted the western cut, materials and color, it showed the qualities of symmetry, solemnity and restraint of Chinese dress. Since it came into being in 1923, the *zhongshan* suit has become a prevailing classical formal dress for Chinese men.

第三章 传统服饰品汉译英

第一节 头饰

学习目标

1. 了解中国传统头饰的历史及文化内涵
2. 熟悉中国传统头饰的基本语汇及其汉英表达方式
3. 掌握语篇翻译中增、删、重组方法的相关知识与技能
4. 通过实践训练提高中国传统头饰文化汉英翻译能力

一、经典译言

想译一部喜欢的作品要读到四遍五遍，才能把情节、故事，记得烂熟，分析彻底，人物历历如在眼前，隐藏在字里行间的微言大义也能慢慢琢磨出来。但做了这些功夫是不是翻译条件就具备了呢？不。因为翻译作品不仅在于了解和体会，还需要进一步把我所了解的体会的，又忠实又动人的表达出来。

——傅雷

二、汉语原文

明清金银首饰

首饰原本通指男女头上的饰物，俗称"头面"，之后又成为全身装饰品的总称。古代首饰已有发饰、颈饰、冠饰、佩饰等。我国早在商代就已经制作出套、组金首饰。1977年，北京平谷刘家河商墓出土的金笄、金耳坠、金臂钏就是采用铸造、锤揲技法制成的，它们造型朴拙浑厚，含金量已达到85%。首饰是随时代的不同、人们生活习惯和服饰的变化而发展的。唐代是我国金银饰品制作和使用的鼎盛时期，人们把金、银、珍珠、宝石相互搭配，发挥不同材料的特点，充分展示出饰品绚丽多姿、豪华精美的风采。这种独具匠心的设计制作对以后的金银细作有着深厚的影响。

北京是元、明、清三朝的都城，制作金银器有它独特的地域优势。明代是中国黄金制造史上的又一个鼎盛时期，有着更加完善的管理体系。对金银矿的开采、熔炼实行严格的控制，专门设立宫廷内府银作局，监管供帝王、后妃们需要的金银器的制作。这些从全国各地抽调上来的工匠，都是制金工艺的高手，他们把各自的智慧和技术特长相互融合，制作出许多绝世精品。所以明代的金银饰品极为丰富，其数量之多、工艺之精美，都是空前的。万贵夫妇合葬墓中出土的金银饰品极为丰富精美，其中一对錾花金凤簪，簪柄扁尖，簪头一只飞

凤立于祥云之上，伸展双翅，凤尾细长向后飘拂，上面錾刻细密的麦穗纹。另一件为嵌宝石葵花型金簪，分三层纹饰，采用锤、拉丝、盘烧、錾刻、镶嵌、焊接等多道工艺技法制成；花瓣是用纤细如发的金丝盘绕而成，均匀厚重，中心镶嵌的黄色碧玺晶莹剔透，周围环绕着16颗红蓝宝石，三色宝石交相辉映，更展示出金簪的华丽富贵。这些金饰都是宫廷银作局所作，原本为后妃用品却成了万贵妃之父死后的葬品。明代金银饰品追求纤细、豪华、实用的风格与当时宫廷内后妃们的服装、发饰有着密不可分的联系。

清代的制金工艺在继承明代传统技法上又有了新的发展，出现了"点蓝""点翠"新工艺。"点蓝"是把一种矿物质釉料点烧在饰物上，成为一种玻璃状的蓝色釉，这种蓝色釉和黄金、白银相辉映，显得清丽华贵。"点翠"是把翠鸟的羽毛依设计要求剪裁备用，把剪裁好的羽毛用胶粘于金、银饰品上。图3-1中点翠凤形银簪，采用拉丝、累丝、錾刻、镶嵌、焊接等技法外，凤鸟的羽毛采用的就是点翠工艺，蓝白相间，素雅相宜。由于羽毛是粘贴的不易保存，所以留存的较少。❷

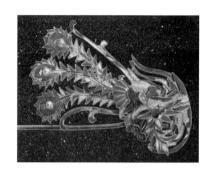

图3-1　点翠凤形银簪❶

译前提示： 种类繁多、制作精美的古代头饰在我国服饰文化中留下了浓墨重彩的一笔，头饰的演变不仅蕴含着文化因素，而且也蕴藏着审美和技艺因素。本文从赏析明清两代比较有代表性的金银首饰文物入手，对历史上制金工艺进行赞美。首饰文化随历史发展在不断演变，制金工艺追求精益求精，都揭示出充满智慧的古代先民关于美的探索之路。

三、英语译文

Gold and Silver Head Ornaments in Ming and Qing Dynasties

Originally, ornament refers to the adornment on the head, also called "*toumian*" in proverb. Afterwards, it refers to adornment from head to foot, containing hair ornaments, neckwear, crown ornaments and other accessories. As early as in the Shang dynasty, our ancestors have created gold ornaments in set or group. In 1977, gold hairpins, gold earrings and gold armlets, made with molding and hammering techniques, were unearthed from a Shang tomb at Liujiahe of Pinggu, Beijing. The shapes of those unearthed ornaments were austere and simple, of which gold amounts to 85%. Ornaments developed along with the times, human habits and dress styles. The Tang Dynasty was the vigorous period of making and using gold and silver ornaments. People matched the usage of gold, silver, pearls and precious stones, giving full play to the characteristics of different materials and displaying the gorgeous, luxurious and exquisite style of accessories. This kind of unique design and production has a profound impact on the later gold and silver works.

❶ 图片来源：《北京文物鉴赏》编委会. 明清金银首饰［M］. 北京：北京美术摄影出版社，2005：100.
❷ 语料参考：《北京文物鉴赏》编委会. 明清金银首饰［M］. 北京：北京美术摄影出版社，2005：10-11.

As the capital of Yuan, Ming and Qing, Beijing had special geographical advantages of making gold and silver works. Ming, with more complete management systems, was a vigorous period for gold manufacturing in Chinese history. The exploitation and melting of gold and silver mines were strictly controlled. The Palace Silver Ware Bureau was established to supervise the production of gold and silver ware needed by the emperor and empresses. Craftsmen, all of whom were masters of gold-making techniques, gathered to the bureau from all over the country. They combined their wisdom and technical expertise to produce many priceless elaborate works. Therefore, in the Ming Dynasty, there were massive amounts of gold and silver ornaments, which were unprecedented not only in quantity, but also in technological exquisiteness. The joint burial tomb of Wangui and his wife contained rich funeral objects, including gold and silver articles with distinct features. One of them was a pair of gold hairpin with engraved design and a flat and sharp-pointed handle. At the head of the hairpin, there was a phoenix flying in the propitious cloud, the wings outspread and the slim tail, with thick ear-of- wheat veins, fluttering backward. The other one was a sun-flower-shaped hairpin inlaid with gems. It had three layers of ornamentation, made by hammaring, wire-twisting, pan-firing, engraving, inlaying, welding and other techniques. The petals were made by spinning gold strings as slim as hair, well-distributed and thick, with a crystal-like yellow tourmaline in the center surrounded by 16 red and blue gemstones. The gems of three colors complemented each other and showed the gorgeous luxury of the gold hairpin. The pursuit of fineness, luxury and practicable characteristics of the Ming gold and silver ornaments had a close relation to the dress and hair style of those empresses in the palace at that time.

The gold and silver making technology of Qing Dynasty had a new development on the basis of Ming achievements, including the new enameling craft and kingfisher craft that made their first appearances. Enameling means to burn a kind of mineral glaze on the ornaments to form a kind of glass like blue enamel. The radiance of gold, silver and enamel makes the ornament beautiful and luxurious. The kingfisher craft means to cut kingfisher's feather into requested shape, then stick them to the gold or silver ornament with glue. The phoenix-shaped silver hairpin of kingfisher craft in the picture of this text was made by the wire-twisting, filigreeing, engraving, inlaying and welding techniques. The feather of the phoenix was made by kingfisher craft with a harmonious match of blue and white, simple and elegant. Because it's hard to preserve the kingfisher's feather, ornaments made by this technique are hardly found now.[1]

四、词汇对译

金笄	gold hairpin	拉丝	wire-twisting
金耳坠	gold earrings	盘绕	spin
金臂钏	gold armlets	蓝色釉	blue enamel

[1] 语料参考：《北京文物鉴赏》编委会. 明清金银首饰［M］. 北京：北京美术摄影出版社，2005：12–13.

锤揲　hammaring

金银细作　gold and silver ware

鼎盛时期　the vigorous period

陪葬品　funeral objects

祥云　the propitious cloud

麦穗纹　ear-of-wheat veins

嵌宝石葵花型金簪　sun-flower-shaped hairpin inlayed with jewels

金银丝编织物 filigree

焊接　welding

点翠工艺　the kingfisher craft

点蓝工艺　the enameling craft

凤形银簪　phoenix-shaped silver hairpin

镶嵌　inlay

錾刻　engraving

五、译文注释

1. 首饰原本通指男女头上的饰物，俗称"头面"。

Originally, ornament refers to the adornment on the head, can also be called "*toumian*" in proverb.

【注释】"头面"是一个汉语词汇，释义为头和脸，旧时指首饰，妇女头上戴的装饰品。文中采取音译的方法，将"头面"译为"*toumian*"，最大限度地保留了原文文化意象。而且译文的前半句已经对"头面"进行了解释，为方便读者理解提供了背景知识，音译不会造成读者理解困难。

2. 1977年，北京平谷刘家河商墓出土的金笄、金耳坠、金臂钏就是采用铸造、锤揲技法制成的。

In 1977, gold hairpins, gold earrings and gold armlets, made with molding and hammering techniques, were unearthed from a Shang tomb at Liujiahe of Pinggu, Beijing.

【注释】"笄"即是簪子，秦汉以后才被称为簪。它是中国最古老的一种发簪，用来插在挽起的长发上，或插在帽子上。商周时期的男女都用笄，女子用来固定发髻，男子用来固定冠帽。金笄的前端留下一长约0.4厘米的榫状结构，原来可能镶有其他的装饰。金臂钏即佩戴于上臂的镯环，是将金丝其两端捶扁成马蹄状，这种形制主要出现在燕山地区。例2原文中将金笄译为"gold hairpin"，采用归化的翻译方法，将这里的"笄"意译为目标语读者容易理解的词汇"hairpin"，最大限度地淡化了原文的陌生感。译文将"金耳坠""金臂钏"按照字面意义直译为"gold earrings""gold armlets"，求得内容和形式的基本对等，可以帮助读者更好地理解这些配饰的意思。

3. 唐代是我国金银饰品制作和使用的鼎盛时期。

Tang Period was the vigorous period of making and using gold and silver ornaments.

【注释】"鼎盛"，语出南朝刘勰《文心雕龙·时序》："经典礼章，跨周轹汉，唐虞之文，其鼎盛乎"，是具有中华文化意象的词汇，"鼎"因有"正当""正在"的意思，鼎盛时期才用来表达时代的兴盛、昌盛。在西方读者没有相应文化背景知识的情况下，对"鼎盛时期"这样的词语，不用深究其词语背后的文化意象及内涵，只需直译出成语整体的含义即可，以"vigorous"来传达其内涵，可为西方读者阅读扫清文化内涵理解方面的障碍。

4. 对金银矿的开采、熔炼实行严格的控制，专门设立宫廷内府银作局监管，供帝王、后妃们需要的金银器的制作。

The exploitation and melting of gold and silver mines were strictly controlled. The Palace Silver Ware Bureau was established to supervise the production of gold and silver ware needed by the emperor and empresses.

【注释】例4的翻译采用了语态转换法，原文句子是省略了主语的主动语态，而且动词较多，有"控制""设立"和"提供"。而英文一般要求完整的句子结构，被动语态使用的频率也高，译文以"the exploitation and melting of the gold and silver mines"为主语，采用了被动语态来译，把"控制"作为谓语动词，这样更符合英文句式的特点，加强上下文的连贯和衔接。"设立"一词的翻译也同样采用了语态转换法。

5. 所以明代的金银饰品极为丰富，其数量之多、工艺之精美，都是空前的。

Therefore, in the Ming Dynasty, there were massive amounts of gold and silver ornaments, which were unprecedented not only in quantity, but also in technological exquisiteness.

【注释】把一个句子由汉语翻译成英语时，首先应从整个句子所反映的思想内容出发做逻辑上的考虑。因为句型是一个句子的骨架，选用的句型恰当与否直接影响到译文的可读性和通顺程度。原句是典型的汉语表述方式，多流水句，一个小句接着一个小句，结构较为松散。而英文的句式结构注重"形合"，在对这句话进行翻译时，译者抓住原句的重点信息，根据原句的语义联系对句子结构进行了调整，将后两个分句译为由"which"引导的后置定语从句，并利用"not only...but also..."结构将句子连接起来，使译文读起来结构严谨、逻辑清晰、语言通顺。

6. 万贵墓中出土的一对錾花金凤簪，簪柄扁尖，簪头一只飞凤立于祥云之上，伸展双翅，凤尾细长向后飘拂，上面錾刻细密的麦穗纹。

One of them is a pair of gold hairpin with engraved design and a flat and sharp-pointed handle. At the head of the hairpin, there is phoenix flying in the propitious cloud, the wings outspread and the slim tail, with thick ear-of- wheat veins, fluttering backward.

【注释】例句原文是对"錾花金凤簪"这一中国古代发式用的一种簪子的描写，在翻译过程中既要完整地传递文化信息，又要方便读者理解和接受。例句中将"錾花金凤簪"译为"gold hairpin with engraved design"，是使用归化的策略，暂时省略簪子中"凤形"样式，先用"engraved design"简要概括簪子的制作工艺，因为结合上下文，后面的译文有对簪子的凤形样式进行具体的描绘。如果对"錾花金凤簪"直接翻译，中间加上"phoenix-shaped"，译文反而会变得重复累赘，读起来也拗口。因而结合上下文删去不必要的重复，简洁处理，会使句子结构更加合理，帮助读者理解原文意思，增加译文的可读性。

7. 花瓣是用纤细如发的金丝盘绕而成，均匀厚重，中心镶嵌的黄色碧玺晶莹剔透，周围环绕着16颗红蓝宝石，三色宝石交相辉映，更展示出金簪的华丽富贵。

The petals were made by spinning gold strings as slim as hair, well-distributed and thick, with a crystal-like yellow tourmaline in the center surrounded by 16 red and blue gemstones. The gems of three colors complemented each other and show the gorgeous luxury of the gold hairpin.

【注释】汉语中常常使用充满意蕴的四字词语，来增添行文的气势或表达效果，比如例句中的"均匀厚重""晶莹剔透""交相辉映""华丽富贵"等。但在英语表述中，字对字地搬用汉字的四字结构，可能会造成行文的累赘感。因此，在汉译英时应该抓住原文的重点信息进行概括化翻译，省去中文表达里有重复效果的词汇，选用目标语言中相关意义的词即可，使译文表达更为通顺和精炼。比如，译者将以上提及的四字结构分别译为"well-distributed and thick""crystal-like""complemented each other""the gorgeous luxury"，语言简洁明了，却充分传达原文的修辞效果。

8．内文图中点翠凤形银簪，采用拉丝、金银丝编织、錾刻、镶嵌、焊接等技法外，凤鸟的羽毛采用的就是点翠工艺，蓝白相间，素雅相宜。

The phoenix-shaped silver hairpin of kingfisher craft in the picture of this text was made by the wire-twisting, filigreeing, engraving, engraving and welding techniques. The feather of the phoenix was made by kingfisher craft with a harmonious match of blue and white, simple and elegant.

【注释】例句原文为一句，由若干说明银簪制作技法和点翠工艺的流水分句构成。译文为符合英语语言的表意需要，将原文分译为两句，分别对制作技法和点翠工艺进行说明，以减轻英语句子在结构上的压力，有利于更清楚地传递原文的信息。同时，"蓝白相间""素雅相宜"是具有汉语表达特色，强化语言效果的说法，联系原文语句信息，这两个词要表达的是效果是点翠工艺的直观表现，译者采用"with…"这一介宾短语作后置定语，引出对点翠工艺效果的描写，使译文结构清晰，层次分明，有助于增强读者对点翠效果的理解。

六、翻译知识——语篇与翻译的增、删、重组

语篇通常指一系列连续的语段或句子构成的语言整体。在实际翻译过程中，我们面对的往往不是孤立的词句，而是一些相互关联和制约的词句。在进行翻译前，要认真分析语篇，深刻理解原文的内容。译者既是原文的接受者，又是译文的生产者。译者在接受原文的过程中要进行语篇分析，在生产译文时同样要进行语篇分析，这样才能很好地完成翻译任务。在翻译中树立明确的语篇意识非常重要。

由于中西方文化差异的存在，汉英两种语言在词法、句法和修辞等方面存在诸多不同之处。中外学者和翻译家在研究各种语言互译的过程中找出了不少忠实传递原文信息的规律，总结了各种各样的翻译方法和技巧。对于许多翻译实践中从语言结构的角度难以处理的句子，译者可以结合语篇，在上下文语境中去理解原文的含义，并且运用一定的翻译技巧，从语篇整体角度去建构合适的译文。通常来说，语篇视角下翻译常用的方法包括增译、减译和重组。下面将逐一进行说明。

（一）增译

增译就是增加一些原文中在语言形式上没有，但语篇中包含其意义的成分，使译文在语法、结构和表达等方面更为完整，符合译入语语言的表达习惯，使译文与原文在内容、形式和风格方面都实现对等。增译是翻译方法当中比较常见的一种翻译技巧，增译必须使增补后的句子更加符合译入语的结构、意义及修辞特点。适当运用增译法可以补全原文中缺失的或者不完整的信息，还可以使译者在翻译过程中更好地发挥主动性和创造性，从而赋予译文鲜明的个性和特点。

例1 原文：在古代中国，着装规范不仅是民间习俗，更是国家礼制的一部分。

译文： In ancient China, as is stated above, how one was dressed was not merely a matter of folk customs, but also an integral part of the State rules on ceremony and propriety.

【分析】 例句原文对中国古代着装规范进行了说明，句式简单，表意明确。将例句译文与原文进行比对发现，译文中增加了"as is stated above（正如上文所说）"，单从本句来看，这样翻译并不忠实于原文。但结合语篇的上下文来分析，发现前面两段文字正是对民间习俗和国家礼制中的着装规范进行介绍。本句译文增加"as is stated above"，正是为了更好地承接上文，体现语篇的连贯性。从语篇视角进行翻译时，增译是常见的现象。

例2 原文：秦汉时的男人，不分贵贱都戴帻，只不过官员的帻衬在冠下，平民无冠。

译文： Men of all social strata in Qin and Han Dynasties all wore a kerchief, the only difference being that the officials wore kerchiefs under caps or hats but the commoners wore only kerchiefs.

【分析】 如果原文行文相对比较松散，在英译过程中，译文可以通过现在分词、表语从句、并列连词等形式将译文组织得更加紧凑。仔细分析例句原文发现，"只不过官员的帻衬在冠下，平民无冠"要表达的含义是官员与平民佩戴头饰的区别，因此，译文增添"difference"一词将原文隐藏在字里行间的含义明示出来，并以"difference being that"这一独立主格结构做状语，引出秦汉时官员与平民戴帻的不同之处。这一处增译表达了原文语言形式上没有明示，但语义上实际隐含着的意义，可以看作翻译中的明晰化方法，使译文结构更紧凑，表意更明确，体现了译者在翻译过程中的主观能动性。

例3 原文：古时的冠很小，必须用双笄从左右两侧插进发髻加以固定。

译文： The ancient crown was so small that two hairpins had to be inserted into the bun from left and right sides to fasten the crown.

【分析】 汉语中经常会省略表示逻辑关系的连词成分，但在英文的表达习惯里，句子中逻辑关系应当表示出来。例3原文中的两个分句属因果关系，译文增补"so...that..."连词成分，用状语从句将两部分内容连接起来，这样更符合英语语法，有助于读者理解原文的内在逻辑。

例4 原文：清代的王室贵族妇女所戴的扁方从材质到样式都堪称精美绝伦。在扁方仅一尺长的窄面上，工匠能雕刻出栩栩如生的花草虫鸟与亭台楼阁等精美图案。

译文： The design and material of *bianfang*, the prolate hair ornament for the royal women were exquisite. Lifelike patterns such as flowers, grass, insects, birds and landscaping were engraved on a narrow plate roughly about one Chinese Foot, around 33.3 centimeters in length.

【分析】 扁方是满族妇女梳两把头（旗头中的一种）时所插饰的特殊大簪，均作扁平"一"字形，清代宫廷梳"大拉翅"所用的扁方大多长达一尺二寸，具有典型的民族特征和丰富的中国文化内涵。对于不了解中国服饰文化的英语读者而言，扁方是一个来自异文化语境的完全陌生的文化意象。译文如果直接将"扁方"音译为"*bianfang*"，英语读者可能不知其为何物，考虑到译文读者的文化背景，译者结合语篇上下文，增译了"the prolate hair ornament"来说明扁方的形状和用途。此外，例句中的"一尺"是汉语中常用的长度单位，它与英文中的"英尺"有区别，如果将其直译为"one foot"，就不能正确传递原文信息，容易使读者造成误解。译文增补"Chinese"一词，帮助读者了解汉语中的"尺"与"英尺"的区别，再增译"33.3 centimeters in length"，说明一尺的真实长度，以向处于不同文化语境中的读者准确传递原文的信息。

例5 原文：清代礼冠自成体系，与传统冠式大相径庭，如祭祀庆典用吉服冠，平日上朝戴朝冠，燕居时戴常服冠等。

译文： The court mitres in the Qing dynasty had a system of its own widely differing to the traditional one. For example, *chaoguan*(court mitre) was used for court time, *jifuguan* (lucky mitre) for sacrifice rites, and *changfuguan* (daily mitre) for leisure time.

【分析】原文中"朝冠""吉服冠""常服冠"都是中国有特色的头饰文化，采用异化策略音译为"*chaoguan*""*jifuguan*""*changfuguan*"，在一定程度上保留了其文化特色，但从语篇角度来看，这些词汇的出现很突兀，因为从上下文没法准确推断这些词汇的含义。译者在向英文读者传递原文信息的过程中，为帮助读者理解这些音译词汇，进行了增译，即用接近英文读者的文化词汇解释其含义，这样既传播了中国文化，又帮助读者更好地了解原句中的文化意象，促进对语篇的理解。

（二）减译

减译是指原文中有些词语在译文中并不出现，因为译文虽然省略了这些词语，但其含义在上下文语篇中已经表达出来了，也就是说，减译就是删去一些在目的语中可有可无的、累赘或者违背译入语习惯表达法的词语。但是，减译不可以把原文的思想内容删去，或将某些信息漏失，而是适当地删减一些可有可无的、不影响语篇内容和思想的地方，目的在于提高语篇的可读性，让读者更高效地理解语篇。

例6 原文：清王朝建立后，中国的传统服制尽数被废，但以凤凰饰首的习俗得到了保留。

译文： After the founding of Qing Dynasty, some of China's traditional costume institutions were abolished. However, the phoenix coronet retained.

【分析】汉语中倾向于加范畴词，如问题、状态、情况、工作、精神，但这些词有时本身没有实质的意义，翻译时可以省去不译。原文中"以凤凰饰首的习俗得到了保留"译成"the phoenix coronet retained"，是省略了范畴词"习俗"，但其含义在语篇上下文中可以清楚推断出来。"the phoenix coronet retained"意为凤凰冠这种传统服制得以保留，读者可以根据语篇上下文明白以凤凰饰首的习俗没有被废除，译文表达简洁，意义的传递并未受到影响。

例7 原文：钗分两股，看似分离，实则有钗头相连，两个钗脚相互依赖，缺一不可，因此古人还将钗看成是一种寄情的物件。

译文： The design of *chai* was mostly in two strands with the connection on the top. Ancient people took *chai* as a token of love, because of the interdependence of its two strands.

【分析】原文中说明了钗作为恋人或夫妻间情感连接的信物，用了"相互依赖""缺一不可"两个词语来表达，但这两个词语共同表达一个意思，即强调两个钗脚间不可分离，象征恋人间的感情不可分离，两个四字词语在语义上前后呼应，增强了表达效果。但如果将两个词语都直译出来，译文会显得累赘，缺少简洁美，从而影响语篇的流畅性。所以，译者翻译时为了避免重复，使用了减译法，只用"interdependence"一词来译，言简意赅，又突出了重点。

例8 原文：说起中国的古代官服，人们会不约而同地想到头戴乌纱帽的小丑县官——身穿圆领袍，头上的帽翅左右翘动像两枚铜钱儿，腰间一条玉带，脚蹬白底黑靴。

译文：The mention of ancient Chinese official uniforms almost always leads to the Chinese opera type of character—one dressed in a round collar gown, a black gauze cap with wings on each side, a jade belt and a pair of black boots with white soles.

【**分析**】为了使语言表达生动形象，汉语中常常采用比喻等修辞手法。原文中"帽翅左右翘动像两枚铜钱儿"是对帽翅具体形象的刻画，但铜钱儿是中国古代铜质货币，指秦汉以后的各类方孔圆钱，一直沿用到清末"民国"初年，是汉语中特有的文化意象。如果将这个比喻翻译出来，读者需要先了解"铜钱儿"这个他们所不熟知的意象，然后以其为喻体去理解帽翅的样子，显然会增加读者的认知负担，从而影响语篇的整体解读效果。因此，译文没有将这一信息具体化，而是省略了原文中的比喻，只翻译出此官服大概的装扮特征。此外，对于"小丑县官"这个文化意象，译者同样考虑到语篇认知效果的需要，省略了其中包含的英语读者所不熟悉的具有中国特色的意象，将其解释为"the Chinese opera type of character"，以达到方便读者理解的目的。

例9 原文：冕服包括冕冠，上有一块木板称"冕板"，做成前圆后方形，象征天圆地方，呈向前倾斜之势，以示帝王向臣民俯就，就是真心惦记臣民、尊重臣民的意思。

译文：*Mianfu* is a set of garments including the *mianguan*, a crown with *mianban*, a board that leans forward, which is round in the front and square in the back, as if the emperor is bowing to his subjects in full respect and concern.

【**分析**】例句主要描述冕冠的样式及设计意义。句中"天圆地方"是中国汉族传统文化的精髓，是"天人合一"思想的具体体现，涵盖的内容十分丰富。原文中这个词语出现只是重复冕板前圆后方形的设计理念，如果译文将"天圆地方"直译出来，可能会让不了解中国文化的读者感到费解，如果再增加对"天圆地方"的解释，又势必会冲淡句子的主题，因此译文结合语篇整体的认知需要，省译了"象征天圆地方"这部分。此外，"以示帝王向臣民俯就，就是真心惦记臣民、尊重臣民的意思"这句话在语义上也有重复的地方，译者从语篇整体可读性的角度出发，采取减译法，用最简洁的方式来表达，这样不仅不拘泥于原文形式，且更加符合目的语读者的阅读习惯，读起来简洁明了易懂。

例10 原文：贵族女子们戴扁方时，会将两端的花纹露出，极具装饰性；有的还会在扁方上缀挂丝线缨穗以限制脖颈的扭动，使行动有节，增添女人端庄的仪态之美。

译文：When it was worn by the royal women, they liked to expose decorative patterns on both ends. Some would hang silk tassels on the prolate hair ornament, which would restrict their heads from turning freely so as to make them look graceful.

【**分析**】结合例句原文语篇上下文，可以推断出例句要表达的观点是佩戴扁方可以塑造仪态。汉语原句描写细腻，由多个分句构成，结构复杂。如果完全遵照原文的表述顺序直译，会使译文生硬累赘。译者在翻译例句时对原句的结构和表达方式进行了灵活调整，最明显的就是采用了减译的方法，将原文中的谓语结构"极具装饰性"减译为一个词"decorative"，清楚明了，且使译文更为紧凑；将"增添女人端庄的仪态之美"减译为"make them look graceful"，"仪态"一词在译文中没有出现，但其含义却已明显体现在译文之中了；将"使行动有节"直接省略掉，因为限制脖颈扭动和呈现端庄之美之间，已经含有了省略部分的意思，毋庸赘述，这样翻译使译文表达更加通顺和精炼。

（三）重组法

英、汉语言在语篇组织方式上存在诸多差异。在翻译过程中，译者有时需要根据译入语语篇规范对译文语篇结构进行重组，以使译文在形式结构上符合译文读者的阅读习惯和心理期待。从语篇结构层面重组译文是汉英翻译中常用的方法。

例11　原文：由于鲜花容易枯萎，于是人们又想出了用罗绢或彩纸做成假花来代替的方法。

译文： People started to use silk and colored paper to make artificial flowers, considering that the flower withered easily.

【分析】原文的叙事属于汉语中常见的表达方式，先说原因，再说结果。翻译时，译者考虑到英文的表达习惯，将语序进行了适当调整，即先说结果，再以分词短语"considering that"的结构来补充说明原因，这样表达使译文主次分明，更符合译入语的表达习惯。

例12　原文：笄除了具有固定和装饰发髻的作用，还是"分贵贱，明等威"的工具。不同身份的人所用的笄的材质也有所不同，皇帝、皇后和一些有身份的皇亲国戚用玉笄，士大夫的妻子用象牙笄，平民女子只能用骨笄。

译文： The materials of *ji* (a kind of hairpin for fastening the hair in in ancient China) used by people of different identities were different. For example, the emperor, empress and royal family members wore jade *ji*; Scholar-bureaucrat's wife wore ivory *ji*; while the common people used bone *ji*. Therefore, *ji* acted as a tool to distinguish people's social status along with its basic role of decoration and hairdo.

【分析】"笄"指古代中国女子用以装饰头发的一种簪子，用来插住挽起的头发，或插住帽子。在古代，汉族女子十五岁称为"及笄"，行笄礼表示成年。"笄"富含中国传统文化信息，是一个文化负载词。就认知语境而言，英语读者对这类文化负载词既好奇又陌生。如果将其简单音译为"*ji*"，可能会使英语读者不知所云，虽译犹不译也；如果直接用英文中的"hairpin"替代，又失去了汉语文化特点，使英语读者无法体会这个文化负载词的异域情调。为了既满足英语读者的好奇心，使他们了解"笄"这一古代中国发饰，又不至于过于陌生化，拒读者于千里之外，译文采用音译加注释的方法，在"*ji*"后以括号加注的方式对这一发饰进行简要说明。这样译在充分考虑译文读者认知语境的基础上，最大限度地帮助读者获得理想的认知体验。

为实现这种认知效果，译文在结构上也进行了适当调整。例句原文主要说明笄因材质不同，在使用中划分一定的社会层级，而这与中国古代统治秩序相关，其表达顺序是先说明笄的功能，再说明笄的不同材质与不同使用者之间的对应关系。两个句子之间相互独立，其中的逻辑关系含而不露。英语语言重理性思维，句中的逻辑关系需要以词汇或语法手段明确标记出来，从英语读者的认知语境来看，他们更习惯于读到逻辑清楚，层次分明的译文。译者为顺应英语读者的认知语境，在翻译时将语序进行调整，把笄的不同材质与不同使用者之间的对应关系提到前面进行说明，然后再增译"therefore"引出对笄的功能的说明，明示两句之间的因果关系。在翻译原文第二句时，译文同样增译了"for example"，使译文表述呈现出清晰的逻辑，从而帮助读者加深对译文的理解，达到预期的认知效果。

例13　原文：这种规范和限定各个社会阶层的穿衣戴帽并以此标识官员等级和庶民地位的做法，虽然是为了维护统治秩序，但客观上却增加了中国服饰的多样性。

译文：This practice of regulating and restricting the clothing of people from different social strata and thus defining ranks of officials and the commoners was apparently intended at upholding the order of the ruling class, but contributed to the diversity of Chinese garments.

【分析】汉语的信息重心在许多情况下不体现在形式上，而是体现在语篇内在的逻辑上。因此，汉译英时应该把汉语语篇中隐形的主次关系发掘出来，译成符合英语规范的表达方式。例句原文是常见的汉语式表达，主语中心词前面有较长的定语做铺垫，谓语部分还包含一个让步状语从句，显得比较复杂。如果照原句结构直译，难免会使译文句子结构头重脚轻，甚至可能造成误解。因此，译者在翻译时遵循英语表达习惯，适当调整顺序，将主语放在句首突出位置，修饰主语的定语后置，使译文层次更加分明。

例14　原文：裕固族妇女到了成年，开始佩戴"头面"，表示到了可参与社交并准备婚嫁的年纪。"头面"在喜庆盛装中不可缺少，是裕固族最具代表性的饰物。

译文："Toumian", the most representative Yugu ornament, is an indispensable part of festive attire. Yugu women will start to wear "toumian" when they become adults, which shows that they are ready for participating in social activities and marriage.

【分析】例句原文是汉语中典型的归纳式思维结构，开头先描述头面在几个社交场合的使用，以此作铺垫，层层推进，直到最后才点明"头面是裕固族最具代表性的饰物"。译者在汉译英时，考虑到汉英语篇思维结构的差异，对原文句子结构进行了重组，转换为英语中典型的演绎式思维结构，先对"头面"进行定性说明与评价，然后再描述其使用场合，这样使译文更贴近译入语读者的心理期待。

例15　原文：其实，古代中国官员的服饰是很丰富的，各个朝代都有自己的规定，甚至在同一个朝代里也会多次变更。

译文：In reality, Chinese official uniform is by far more complex than that. Each dynasty had rules of its own, which could be changed many times even within a dynasty.

【分析】在汉英翻译时，需依据英语语篇的要求，遵循英语的语法规范，对原文句子结构进行调整与重组，构建符合译入语特点的译文。例句中，译者首先进行了分译，将原文拆为两句，分别对中国古代官服的丰富性和相关规定的多变性进行说明。在译文的第一个句子中根据语篇上下文增译了"than that"，以显示与前文的连贯性。英文第二个句子将原文中的两个并列分句译为主从复合句，将第二个分句译为由which引导的非限制性定语从句，置于所

修饰名词 "rules" 之后，主次分明，表意清楚，更符合英文的表达习惯。

在语篇翻译实践中，增译、减译、重组方法运用十分广泛。在翻译时往往无法将原文中的每一个词直接替换为译文中的另一个词。译者应根据语篇和上下文，在译入语规范允许的范围内有所增减，即有时需要在译文中增加一些原文字面上没有的词语，有时则需要在充分传递原文含义的基础上，将译文中显得多余的词省去。不管增词还是减词，目的都是使译文通顺流畅，符合目的语读者的阅读习惯，同时又能完整地表达意思。翻译中在形式和内容的各个层面上进行的重组，也是用目标语独特的结构再现原文信息，使译文更符合读者的心理期待。对初学者而言，在翻译中树立语篇意识，熟练掌握各种翻译技巧，对提高翻译能力非常重要。

七、译文赏析

段落1

首饰作为服饰整体的一部分，是对服饰文化的补充与点缀，随着服饰形式的发展完备，商代的首饰已达到了一个新的水平。这一时期主要的首饰种类有发饰、颈饰、耳饰、手饰等。妇好墓出土的499件骨笄和28件玉笄，充分展示了当时主要头饰的精美。笄刚开始是女子用来固定发髻的，后来随着制作工艺的发展，笄又成为一种装饰用品，这也就是后来比较流行的簪子。在商代，男子冠饰同样反映阶层的差异：奴隶主贵族戴弯曲高冠，上面还装饰着许多珠玉；平民、奴隶，有裹发作羊角状斜旋而上的，有垂一短辫的，也有剪发齐颈的。男子发式，通常是将头发辫成辫子，自右向左旋转，在头顶围成一团；女子则多把长发向上拢成髻，或者卷发齐肩；小孩头发梳作两个杈状丫角儿，叫作卯（guan）角。❶

As a part of the whole dress, head ornaments are the supplement and embellishment of dress culture. With the improvement of dress styles, the head ornaments of the Shang Dynasty reached a new level. The main head ornaments during this period were hair ornaments, neck ornaments, ear ornaments and hand ornaments. The 499 pieces of bone hair-pins and 28 jade hair-pins unearthed at the Fuhao Tomb showed how exquisite the main head ornaments were at that time. Hair-pins were first used by women to secure chignon. Later, with the development of technology, the use of hair-pin was expanded as an ornament. In the Shang Dynasty, the ornament in the cap of men also showed the difference between social classes: the slave owners wore twisted high crown decorated with many jades; the commoners and the slaves only wore their hair upward in the shape of spire, or just wore a short queue, or cut short hair to the neck. Men's hair was usually queued, revolved from right to left above the head and then fixed; women's hair was usually bound upward in a bun or curled above shoulders; and children's hair was combed into the shape of two forked horns, which were called "*guan*" horns. ❷

---

❶ 语料参考：《中华文明史语》编委会. 服饰史话［M］. 北京：中国大百科全书出版社，2009：9.

❷ 语料参考：《中华文明史语》编委会. 服饰史话［M］. 北京：中国大百科全书出版社，2009：86.

【赏析】段落原文介绍了首饰在商代达到了一个新的水平，并描写了当时主要头饰的精美及特点，用词严谨，脉络清晰，信息量丰富。译文将原文的一些信息进行了灵活处理，使得整个译文段落的语言表达更加流畅，可读性强，较好地传递了原文的含义。首先，在文化特色词的翻译方面，译者采用了异化策略，以保留词汇中的中国文化内涵。比如，译者将"妇好墓"和"卯角"直译为"the Fuhao Tomb"和"*guan* horns"，既保留了中国文化意象，又结合上下文表达给读者带来了异国情调。其次，为了使译文在准确传达原文意义的同时又具有可读性，译文对原文概念进行了重组与增删，同时还改变了语序。比如，在翻译原文第四句时，使用连接副词how引导的宾语从句，将原句中的两个分句合而为一，形成一个逻辑严密，符合英文表达和描述习惯的句子。此外，为了使原文表达的含义在英译文中体现得更加缜密、紧凑，在翻译最后一句"小孩头发梳作两个权状丫角儿，叫作卯（guan）角"时，译者增译了关系代词"which"，将第二个分句译为后置定语，这样译文句内结构更加清晰，也更符合英文的表达习惯。

段落2

汉族女性也讲究在发髻上插花，而且以鲜花和翠鸟羽毛为时尚。冬季时，特别是中国人的传统大年——春节期间，女性不分年龄都爱戴红色或粉红色的绒绢花，这些花一般都被做成一定的图案，并寓意吉祥。北方妇女喜欢在髻上插一两支银簪，冬天喜欢戴皮毛头饰，兼有御寒和装饰作用；南方妇女则爱在头上横插一把精致的有图案的木梳，遮阳或挡风时常裹戴头巾，天寒时也戴帽箍———一种只围头部一圈的黑绒、黑缎子头饰，以带子结于脑后。❶

The Han women also liked to decorate their hair bun with flowers, with fresh flowers and kingfisher's feather being the most fashionable. In wintertime, especially during the traditional Chinese New Year—the Spring Festival, women of different ages all liked to wear red or pink silk flowers. These flowers were usually made into a certain pattern with lucky meaning. Women in the north liked to plug one or two silver hairpins in the hair buns and wear hair decorations made by fur in winter, which had the double functions of keeping out the cold and decorating the hair. Women from the South liked to stick a delicate patterned wooden comb into hair horizontally. They often wore headscarves to shelter them from sunshine and wind, and head hoops in cold weather—a kind of head ornaments made by black velour or satin that was wrapped around the head and tied at the back with bands.❷

【赏析】段落原文描写了汉族女性在发髻上的装饰习惯及其在南北方的差异，文中用通俗的语言，展示出非常清新生动的形象，文章脉络清晰、结构完整。在句子结构方面，原文最后一句中描写南北方妇女头饰装饰时，使用了较多的流水短句，各部分之间的关系比较隐蔽，是典型的意合语言的特征。在英译时如果按照汉语原文的行文方式就无法清楚表达原文意思，不仅不合乎逻辑，而且与上下文无法衔接起来。因此，译者重新组合了句子结构，对

❶ 语料参考：华梅. 中国服饰［M］. 北京：五洲传播出版社，2004：49.
❷ 语料参考：Hua Mei. *Chinese Clothing*［M］. 北京：五洲传播出版社，2004：53-54.

原句进行了拆分，将其分译为三句，分别说明北方妇女偏爱银簪、南方妇女偏爱木梳、南方妇女也爱头巾和帽箍三层含义。拆分之后，译文句子减轻了结构容量方面的压力，表意更为清楚。同时，译文增译了关系代词与并列连词，使译文句子间衔接紧密，这样既忠实于原文含义，又符合英文的表达习惯。比如，为方便读者理解，译文的最后一句在翻译"以带子结于脑后"时，也根据英语表达习惯，增译定语从句的引导词"that"，将其变为"a kind of head ornaments"的后置定语，使译文结构更清晰，焦点更突出。

八、翻译练习

（一）请将下列汉语句子译成英文

1. 那人头戴方巾，身穿宝蓝缎直缀，脚下粉底皂靴。

2. 花钿是古代女子的一种时尚妆饰，在唐朝时最为风行，有各种形状，颜色以红色、黄色、绿色为主。

3. 每逢春回大地时，唐都长安的仕女们便纷纷加入斗花竞赛，以插戴的花卉最多、最奇异多姿者为花中之魁。

4. 所谓朝冠、吉服冠和常服冠的区别，主要在于冠上的顶饰：朝冠之顶多用三层，上为尖形宝石，中为圆形顶珠，下为金属底座。吉服冠顶则比较简便，只有顶珠和金属底座。至于常服冠，则用红绒线编织成一个圆珠附缀于顶。

5. 由于鲜花容易枯萎，于是人们想出了用罗绢或彩纸做成假花来代替的方法。假花精巧逼真，经久耐用，深受女子喜爱。

6. 金簪的变化主要表现在簪首，常见的簪首形状有球形、花卉形、凤形、鱼形、蝴蝶形、如意形等。

7. 花翎在清代是一种高品级的标志，非一般官员所能戴用，一般被罚拔去花翎的官员一定是犯了重罪。

8. 顾姑冠是宋元时期蒙古族贵妇所戴的一种礼冠。一般以铁丝、桦木或柳枝为骨，形体较长，女子戴着它出入营帐或乘坐车辇时，必须将顶饰取下。冠外裱皮纸绒绢，冠顶插有若干细枝。

9. 忠靖冠是明代的官员退朝闲居时所戴的帽子。制作时用铁丝围成框架，用乌纱、乌绒包裹表面。冠的形状略呈方形，中间微突，前面部分装饰有冠梁，并且压有金线。后部的形状像两个小山峰。

10. 东汉末年，由于统治者的使用和士人的不拘于礼法，轻便的头巾大为流行，王公贵族也用头巾来束发，以附庸风雅，称为"幅巾"。

（二）请将下列汉语段落译成英文

冕服包括冕冠，上有一块木板称"冕板"，呈向前倾斜之势，以示帝王向臣民俯就，就是真心惦记臣民、尊重臣民的意思。前后有成串的垂珠，一般为前后各十二串，根据礼仪的轻重、等级差异，也有九串、七串、五串、三串之分。每串穿五彩玉珠九颗或十二颗，冕冠两旁垂有两根彩色丝带，丝带下端各悬系一枚玉石，提醒君王不要轻信谗言，这同冕板向前

低就的戴法一样，都有政治含义。❶

九、译笔自测

发饰

笄是中国最古老的一种发簪，用来插住挽起的长发，或插在帽子上。商周时期的男女都用笄，女子用来固定发髻，男子用来固定冠帽。古时的冠很小，必须用双笄从左右两侧插进发髻加以固定，故称为"衡笄"。笄除了具有固定和装饰发髻的作用，还是"分贵贱，明等威"的工具。不同身份的人所用的笄的材质也有所不同，皇帝、皇后和一些有身份的皇亲国戚用玉笄，士大夫的妻子用象牙笄，平民女子只能用骨笄。

"笄"是商周时期的称谓，战国以后多称为"簪"。簪是中国古代男女最常用的首饰之一，簪的最初用途仅仅是绾束头发，进入阶级社会以后，则逐渐演变成炫耀财富、昭明身份的一种标志。上古时期的石笄、蚌笄、竹笄、木笄、骨笄等相继被淘汰后，取而代之的是玉簪、金簪、银簪、翠羽簪、玳瑁簪或金镶宝石等。在各类的发簪中，玉簪和金簪均是非常昂贵的饰品，普通人家很少用得起，而银簪多为中下层妇女使用。

古代女子的首饰中还有一种发饰，称为"钗"。钗和簪都用来插发，只是簪做成一股，而钗则做成双股，形状像树枝，也有少数钗做成多股。钗的安插有多种方法，有的横插，有的竖插，有的斜插，还有自下而上倒插的。所插数量也不尽一致，视发髻需要可插两支或数支，最多的在两鬓各插六支，合为十二支。钗的造型丰富多样，名称也十分形象，多以钗首的造型命名，如燕钗、花钗、凤钗等。制钗的原料也很多，如金、银、铜、玉、骨、象牙、珊瑚、琉璃等。古代家境贫寒的女子，无钱购买名贵材质的发钗，便只能以荆枝插发，称为"荆钗"。

步摇是缀有活动的坠饰的簪或钗，由簪、钗发展而来，其坠饰会随着女子步履的走动不停地摇曳，故得名。步摇大多以金、银、玉等材质制成，其形制与质地都是等级与身份的象征。步摇又被人称为"禁步"，意在约束女子的行为。戴步摇的女子行动要从容不迫，使步摇发出有节奏的声响。

步摇始见于汉代，最初只流行于宫廷与贵族之中。魏晋南北朝时期，步摇最为盛行，花式繁多，民间女子也开始佩戴。到了唐代以后，步摇多为金玉制成、凤鸟口衔的串珠，称为"凤头钗"。除了金质的步摇以外，这一时期还出现了玉、珊瑚、琉璃、琥珀、松石、水晶等珍贵材料制作的步摇。明清时期，步摇再次流行起来，称为"步摇簪"或"步摇钗"。❷

十、知识拓展

中国古代的头饰种类繁多、造型各异、工艺精湛，是衣着服饰文化重要的组成部分，头饰在古代，不只是衬托女子容颜的单品，也成为社会身份地位、权力制度的外观表现，是时代变化的重要标志。总的来说，从古代丰富的头饰文化中，我们不仅能赏析到头饰制

---

❶ 该段落语料参考：华梅. 中国服饰［M］. 北京：五洲传播出版社，2004：53-54.
❷ 语料参考：戚琳琳. 古代佩饰［M］. 合肥：黄山书社，2013：97-117.

作工艺的美和惊艳，还能感受到深厚的民族文化内涵，感受每个时代留下的独特魅力（图3-2）。

Head ornaments in ancient China is an important part of the clothing culture with various types, different shapes and exquisite craftsmanship. Head ornaments in ancient China are not only a single piece to beautify women's appearance, but also a manifestation of social status and rights system, symbolizing the changes of the times. Generally speaking, from the rich head ornaments culture in ancient times, we can not only appreciate the stunning beauty of ornaments making process, but also feel the profound national culture and experience the unique charm left by each era.

图3-2　王蜀宫妓图轴❶

第二节　肚兜

学习目标

1. 了解中国传统肚兜的历史及文化内涵
2. 熟悉中国传统肚兜的基本语汇及其汉英表达方式
3. 掌握语篇翻译中照应、连接、省略方法的相关知识与技能
4. 通过实践训练提高中国传统肚兜文化汉英翻译能力

一、经典译言

（翻译）正如用琵琶、秦筝、方响、觱篥奏雅乐，节拍虽同，而音韵乖矣。

—— 钱钟书

二、汉语原文

肚兜的历史文化

据传说天地混沌初开之时，伏羲和女娲兄妹二人在漫天洪水之后成婚，生儿育女，并创造了人类最初的服饰——肚兜，其目的是用来遮掩人体之羞。肚兜上最初的图案是"蛤蟆蛙"，是女娲氏部落的图腾标志。

古代内衣比较早的称谓是"亵衣"。"亵"字的意思是"轻薄、轻佻、不庄重"，通过对此字的解释，可以窥见古人对内衣的心态。

中国内衣的历史源远流长。内衣在中国历史上各个时代有不同的称谓：北齐时期称内衣为"心衣"，隋唐时期则称作"宝抹"，汉代称为"抱腹""心衣"，魏晋时期称为"两

图3-3 天蓝缎地平针绣
五子夺魁肚兜❶

当"，唐代称为"诃子"，宋代称为"抹胸"，元代称为"合欢襟"，明代称为"主腰"，清代称为"肚兜""抹胸""兜兜"等。到了近代，由于近代社会的演变，西方机器织品的引进，肚兜多改为衬衫、背心。目前，偏僻地区幼儿还有穿肚兜的遗习，成人则已不多见。但是随着时代的变迁，神秘古老的肚兜已融进"服饰"之中，成为新潮前卫的时尚。这些不同的称谓展示了历代中国女性贴身服饰的变化。

肚兜是中国传统服饰中的贴身内衣。古人仅用以遮体避寒，而后才成为暖腹护胸的内衣。肚兜做工讲究，图案题材广泛，以吉祥纳福、祛灾避邪、生殖崇拜、祈求多子多福为主要内容，并逐渐成为特殊节令节日的人情礼物。肚兜形状多为正方形、长方形、椭圆形和菱形等。肚兜通常是对角设计，上角裁去，成凹状浅半圆形，下角有的呈尖形，有的呈圆弧形。肚兜的面上常有图案，大多是趋吉避凶、吉祥幸福的主题（图3-3）。红色为"肚兜"常见的颜色。

菱形肚兜最为普遍，适用于所有年龄段。正方形肚兜常用于孩童，长方形肚兜常用于孩童和成年男女，如意形、葫芦形、元宝形等异形肚兜则常常蕴含着祈求与美好的愿望。穿着肚兜时，将细带系在颈间与腰际，围裹住胸部和腹部。下面呈倒三角形，遮过小腹。材质以棉和丝绸居多。肚兜四季皆用，不仅遮胸掩腹，也可以在夹层中放入草药、棉絮等强身御寒，同时防虫叮咬肚脐。肚兜因为有兜，因此也可以纳物藏物，如香料、药品等。肚兜不仅为女性专有的服饰品，男性和长者、幼者均可穿着肚兜。总之，肚兜的材质、样式和颜色等都与穿着者的年龄、身份地位、经济能力和习俗等有着密切的联系。❷

译前提示：中国传统服饰——肚兜，是服饰文化的外在表现，从图案到款式，每针每线都记录着中华历史的变迁与民族文化的核心精神。本文从肚兜的起源说起，对中华传统服饰中肚兜文化进行介绍，从叙述肚兜的发展历史到肚兜的图案样式及其用途，简要地描述了肚兜在民族历史中所代表的文化色彩。

三、英语译文

History and Culture of Bellybands

The legend has it that, at the beginning of the world, Fu Xi and Nv Wa, the brother and sister are related by marriage after great flood, and give birth to sons and daughters. It is they who have created the original costume of human being—bellyband to hide the body against shame. The original bellyband pattern is toad, which is the totem token of Nv Wa tribal group.

The earliest ancient underwear in China was called *xieyi*. Chinese character *xie*（褻）means frivolousness, which helps understand the attitude of ancient people towards underwear.

The Chinese underwear enjoys a long history. There were different appellations for underwear in

---

❶ 图片来源：清华大学艺术博物馆网。
❷ 语料参考：訾韦力. 中国传统肚兜服饰文化（汉英对照）［M］. 北京：中国轻工业出版社，2016：24-29.（略有改动）。

various periods of Chinese history: In North Qi Dynasty, the underwear was called *xinyi,* in Sui and Tang Period, it was called *baomo*, but it was called *baofu, xinyi* in Han Dynasty, *liangdang* in WeiJin Period, *hezi* in Tang Dynasty, *moxiong* in Song Dynasty, *hehuan Jin* in Yuan Dynasty, *zhuyao* in Ming Dynasty and *dudou* (bellybands), *moxiong, doudou*, etc. in Qing Dynasty. In modern times, the bellybands were thrown aside due to the evolution of modern society and the introduction of foreign woven products, and were replaced by shirts and vests despite the remaining customs of preparing bellybands for kids in remote regions. It is rare to find adults wearing bellybands today. With the changing times, the mysterious ancient bellybands melted into costumes come to be the trendy avant-garde fashion. The various appellations in different times reflect the diversities of Chinese female underwear in application.

Bellybands are Chinese traditional lingerie. The ancient people originally wore them to cover the body against the coldness, and later, bellybands became the lingerie for warming the abdomen and protecting the chest. Bellybands of exquisite workmanship come to be accepted as the gift of returning favor on the special or seasonal occasions with wide-ranged themes of patterns, including yearning for auspicious and happy life, warding off disasters and evils, upholding reproduction and praying for many children and much bliss. Bellybands are mostly square, rectangular, oval-shaped and diamond-shaped. The common bellybands are diagonally designed, with shallow semicircle of concave shape by cutting top corner, the bottom corner of some bellybands are pointed or circular. There are various patterns in bellybands with the theme of pursuing good fortune, avoiding disaster and seeking for auspiciousness and happiness. Red is the color most commonly found in bellybands.

Diamond-shaped bellybands are quite common, which are suitable for all ages. Square bellybands are often designed for kids, while rectangular bellybands are usually for both kids and adults. Specially shaped bellybands, including *ruyi* shape, gourd shape and gold ingot shape etc., tend to be designed with connotations of pray, pursuit and best wishes.

While wearing the bellyband, two strips of cloth are tied around the neck and another two strips are fastened behind around waist to cover the belly part. Cotton and silk are the major material for bellybands. Bellybands are popular throughout four seasons, not only for covering the chest and belly, but also for strengthening the body, protecting the body against coldness by putting herb medicine or cotton fiber in the interlayer and against bites from various insects as well. The pocket in the bellybands can hold things, including spices and Chinese traditional medicine. Bellybands are not exclusively designed for female, the male and even the younger or older would wear bellybands in their daily life. In a word, the material, the style and the color of bellybands are closely related to the ages, the social status, the financial level, and custom as well. ❶

四、词汇对译
肚兜　bellyband　　　　　　　　做工讲究　exquisite workmanship

---

❶ 语料参考：訾韦力. 中国传统肚兜服饰文化（汉英对照）［M］. 北京：中国轻工业出版社，2016：24-29.

图腾标志	totem token	祛灾避邪	warding off disasters and evils
部落	tribe	椭圆形	oval-shaped
轻薄	frivolousness	菱形	diamond-shaped
称谓	appellation	对角设计	diagonally designed
遗习	remaining customs	如意形	*ruyi* shape
背心	vest	葫芦形	gourd shape
偏僻地区	remote regions	元宝形	gold ingot shape
前卫时尚	avant-garde fashion	棉絮	cotton fiber
贴身服饰	underwear	夹层	interlayer

五、译文注释

1. 肚兜上最初的图案是"蛤蟆蛙"，是女娲氏部落的图腾标志。

The original bellyband pattern is toad, which is the totem token of Nv Wa tribal group.

【注释】文中出现的"伏羲和女娲"是我国文献记载最早的创世神，是华夏民族人文先始。女娲，又称娲皇、女阴，史记女娲氏，她不但是补天救世的英雄和抟土造人的女神，还是一个创造万物的自然之神，神通广大化生万物。例句译文将"女娲"译成"Nv Wa"，是采用了音译的方法，保留了原文文化意象所具有的异国情调，是典型的异化翻译。

2. 据传说天地混沌初开之时，伏羲和女娲兄妹二人在漫天洪水之后成婚，生儿育女，并创造了人类最初的服饰肚兜，其目的是用来遮掩人体之羞。

The legend has it that, at the beginning of the world, Fu Xi and Nv Wa, the brother and sister are related by marriage after great flood, and give birth to sons and daughters. It is they who have created the original costume of human being—bellyband to hide the body against shame.

【注释】例句讲述肚兜的神话起源，由五个流水分句构成，语意连贯，一气呵成。译文根据英语语言表达的需要，采用分译法，将原文拆为两句，将伏羲和女娲的结合与创造肚兜分别进行说明，结构清楚，表意明确。此外，译文还借助特殊的英文句式增加表达的效果。比如，"The legend has it that…"是引用神话传说的地道英文表达；"It is they who…"是强调句型，强调了二人在创造肚兜方面所发挥的重要作用，赋予肚兜更强烈的民族文化色彩。

3. 古代内衣比较早的称谓是"亵衣"。"亵"字的意思是"轻薄、轻佻、不庄重"，通过对此字的解释，可以窥见古人对内衣的心态。

The earliest ancient underwear in China was called *xieyi*. Chinese character *xie*（亵）means frivolousness, which helps understand the attitude of ancient people towards underwear.

【注释】例句原文对古代为什么称内衣为"亵衣"做出了解释，遵循汉语的行文习惯，语义环环相扣。译文根据表达的需要，摆脱汉语字面的束缚，遵从英文的表达习惯对句子进行了合理调整，用定语从句引导词"which"来代替作状语的分句"通过对此字的解释"，使译文第二句结构紧凑，层次分明，逻辑清楚，充分体现出英语形合语言的特点。

4. 中国内衣的历史源远流长。内衣在中国历史上各个时代有不同的称谓。

The Chinese underwear enjoys a long history. There were different appellations for underwear in the various periods of Chinese history.

【注释】例4原文中"源远流长"是汉语成语，是一个隐喻的表达，意思是河流的源头很远，水流很长。常比喻历史悠久，根底深厚。译文将"历史源远流长"译为"enjoy a long history"是采取意译的方法传达原文的词义，简洁凝练。

5. 到了近代，由于近代社会的演变，西方机器织品的引进，肚兜多改为衬衫、背心。目前，偏僻地区幼儿还有穿肚兜的遗习，成人则已不多见。

In modern times, the bellybands were thrown aside due to the evolution of modern society and the introduction of foreign woven products, and were replaced by shirts and vests despite the remaining customs of preparing bellybands for kids in remote regions. It is rare to find adults wearing bellybands today.

【注释】例句为两句，说明近代肚兜及其穿着人群所发生的变化。如果按照原文的行文习惯直接翻译，则无法很好地体现出两句之间的连贯性和逻辑关系。译文采用合译的方法，将两句合并为一句，并使用"due to""and""despite"等连词来明示句中各成分之间的逻辑关系，使译文结构清楚，焦点突出，层次分明。译文把主语"the bellybands"提前，谓语动词使用被动语态，按照英语句子语义逻辑关系调整句子，同时译文增译了"throw"一词，传达了原文中肚兜被衬衫、背心取代的含义，使上下文更好地连接起来，句内各成分之间的关系也更加清晰。

6. 古人仅用以遮体避寒，而后才成为暖腹护胸的内衣。

The ancient people originally wore them to cover the body against the coldness, and later, bellybands became the lingerie for warming the abdomen and protecting the chest.

【注释】在汉译英翻译时，译者应该把握好主语，因为在英语中，主语是句子与句子、分句与分句之间的纽带，只有把握好主语才能保证句子的完整性，英语行文和逻辑才会更清晰。对照原文不难发现，原文的两个分句中都有省略的成分，第一个分句在动词"用"字后面省略了"肚兜"，第二个分句省略了主语"肚兜"。由于汉语重意合，这样的表达在汉语中不会造成误解。但在译成英文需要根据英语的表达习惯，确定清晰的主语及其谓语。例句译文将原文进行灵活处理，将原文中的两个分句以"and"连接起来构成英文中的并列分句，其中第一个分句根据原文以"古人（the ancient people）"作主语，第二个分句则增译"肚兜（bellybands）"作主语，使译文符合英语语言习惯，句子完整、逻辑清晰。

7. 肚兜做工讲究，题材广泛，以吉祥纳福、祛灾避邪、生殖崇拜、祈求多子多福为主要内容，逐渐成为特殊节令节日的人情礼物。

Bellybands of exquisite workmanship come to be accepted as the gift of returning favor on the special or seasonal occasions with wide-ranged themes of patterns, including yearning for auspicious and happy life, warding off disasters and evils, upholding reproduction and praying for many children and much bliss.

【注释】例句原文说明了中国传统服饰肚兜的做工、题材及用途，由四个流水分句构成，中间没有任何连词或其他的语言手段来衔接，分句间的逻辑关系隐含在字里行间，鲜明体现了意合语言的特点。此外，例句原文大量使用四字格短语，如"做工讲究""题材广泛""吉祥纳福""祛灾避邪""生殖崇拜""多子多福"等，结构工整，读起来朗朗上口，体现了汉语声调语言的特点。但英语作为形合语言和语调语言，具有完全不同的特点，在行文习惯和表达方式等方面存在明显差异。英文表达注重语言的形式逻辑，所以例句英译文舍弃了原文的流水句式和韵律，采用归化的翻译策略，改变原文的词法、句法结构。译文首先通过句式转换将原文的主动结构译为被动结构，构建起以"bellybands come to be accepted as the gift of…"为主干的复杂句；"做工讲究"译为"of exquisite workmanship"做后置定语；原文中位于中心词前面的定语"特殊节令节日的"译为"on the special or seasonal occasions"也作为后置定语；原文的四字格短语在译文中变成了能清楚传递原文信息但长短不一的短语，不再体现原文的节奏和音韵。经过译者的灵活处理，英译文层次清楚，焦点突出，体现出形合语言的特点，顺应英语语言语境，符合英文的表达习惯。

六、翻译知识——语篇与翻译的照应、连接、省略

语篇是一个完整的意义单位，是交际功能上相对完整的独立语言片段。它不是一连串的句子和段落的简单集合，而是一个结构完整、功能明确的语义统一体。语篇意义的完整性要求翻译时着重把握语篇整体，做到句子与句子之间衔接紧密，逻辑连贯，既忠实于原文的主题、功能，又尽可能体现原文的风格。语篇衔接手段主要有照应、连接、省略。

（一）照应

在语篇中，如果对于一个词语的解释不能从词语本身获得，而必须从该词语所指的对象中寻求答案，这就产生了照应关系。照应是语篇中的指代成分与所指对象之间相互解释的关系，比如用代词等语法手段来表示的语义关系。照应可以使发话者运用简短的指代形式来表达上下文中已经或即将提到的内容，使语篇不仅在修辞上具有言简意赅的效果，而且在结构上更加紧凑，使之成为前后衔接的整体。

例1 原文：孝是儒家伦理思想的核心，是千百年来中国社会维系家庭关系的道德准则，是中华民族的传统美德。基于孝的图案在传统服饰肚兜图案中比比皆是，它们以不同的内容展示百姓的精神世界以及孝的重要性。

译文：Filial piety（*xiao*）is not only the core of the Confucianism, but also the code of ethics for maintaining the Chinese family harmony for thousands of years, which is adhered to as the traditional virtues. There are many kinds of bellyband patterns promoting the filial piety. Those various patterns present the mental world of people as well as the significance of adherence to the filial piety.

【分析】例句说明了肚兜图案对中国传统美德——孝的展示。原文用"它们"回指前面提及的肚兜图案中比比皆是的孝的图案，通过代词照应前文，使原文连贯通顺。翻译中应该忠实再现原文的连贯性。译者在这里没有直接搬用原文的照应方式，而是使用了指示代词"those"，以"those various patterns"回指上文中提到的"many kinds of bellyband patterns"，清楚再现了原文照应关系，使上下文成为一个前后衔接的整体，呈现出了原文的连贯性。

例2 原文：在肚兜图案中，人们经常将各种花卉、虫鸟、动物、人物等组合成某种图

形，以表达某种象征意义和表达对美好生活的期盼。借助肚兜上的图案纹饰表达某种象征意义、文化内涵以及核心精神，是一种文化语言的呈现。

译文：In bellyband patterns, the plants, the insects and birds, the animals and the characters are often combined into a pattern, expressing some symbolic significance of expecting the happy life. This is a cultural language to express the significance, the cultural implication and the core spirit by making use of the bellyband patterns.

【分析】此例中，汉语的照应关系体现在句子的内在逻辑上。译文为确保原文信息的完整性，使之具有连贯性，使用指示代词"this"照应前文所描述的肚兜图案，通过各种图形组成的合成意象来表达象征意义，并将原文第二句的表述顺序进行调整，使本句译文结构上更平衡，同时更好地照应上一句内容，使两句在表意上更为连贯。

例3 原文：与其他服饰相比，肚兜服饰体现的文化价值比其本身价值更重要，因为其中肚兜图案蕴含的意义不仅是民族文化精神的体现，更是对情感的阐释和表现。

译文：Compared to other garment accessories, the cultural value of bellybands is more important than bellyband itself in that what is endowed in the patterns of bellybands not only embodies the national cultural spirit, but also interprets and expresses the feelings.

【分析】对照应这一手段而言，英汉语篇最显著的差别是：英语有关系代词，而且使用频率很高；而汉语则没有关系代词。例句原文中四个流水分句间没有明显的照应标志词，其连贯性主要通过语义来体现。英译文采用关系代词"what"将肚兜图案的意义与其在民族文化精神和情感方面的价值紧密结合起来，使句内各成分之间衔接自然，更加符合英语表达习惯。

例4 原文：满族女子不缠足不穿裙子，上衣的外面有坎肩，长衫的里面还穿有小衣，与汉族女子的肚兜很相似，作用相当于今日女性的胸衣。

译文：Women of the Manchu nationality didn't bind their feet, nor did they wear skirts. They wore a waistcoat over the jacket, and a small piece of clothes inside the coat, which was quite similar to the bellyband worn by women of the Han nationality, which boasted similar functions with modern bra.

【分析】例句原文为五个流水分句，分句间照应关系通过语义来体现，语言上没有明显的标记。英译文由于其形合语言的特征，每个独立的句子都需要有完整的主谓结构，译者翻译时采用分译的方法将原文的一句译为两句，同时第二句增补"they"作主语，照应上句的"满族女子"，使两句译文前后连贯。英语行文要求避免重复，使用人称代词"they"也使译文避免累赘感，显得简洁。

例5 原文：五子登科最初来源于民间故事。五代后周时期，燕山府有个叫窦禹钧的人，他的五个儿子都品学兼优，先后登科及第，故称"五子登科"。

译文：This is originally the Chinese folk story. During the later Zhou Period of the Five Dynasties, a man called Dou Yujun in *yanshan* mansion had five sons who were academic achievers. They all passed the imperial examination one after another, and that was where *Wu Zi Deng Ke* came from.

【分析】例句汉语原文中的对"窦禹钧"的回指照应通过人称代词"他"来实现，这也是汉语原文中唯一的一个照应标记，句中其他的照应关系都是通过语义来暗示。由于英语形合语言的特征要求每个句子有明确的主语，英译文采用了"they"来照应"五个儿子"，此外，还在第一个句子中采用引导定语从句的关系代词"who"来充当汉语中人称代词的角色，并使英语句子成为主从复合句。从这一点来讲，英语在表达上似乎比汉语更为简洁，句中各部分之间的关系也更为明了。

（二）连接

语篇中的连接关系是通过连接词以及一些副词或词组实现的。语篇中的连接成分是具有明确含义的词语。通过这类连接性词语，人们可以了解句子之间的语义联系，甚至可以通过前句从逻辑上预见后句的语义。语言学家韩礼德将英语的连接词语按其功能分为四种类型，即："添加""转折""因果""时序"。这四种连接词的类型可分别由"and、but、so、then"这四个简单连词来表达，它们以简单的形态代表这四种关系。

例6　原文：图案的形象性特征比服装的结构、材质更为直观。所以，它在服装中最易被着装主体用来传达信息，成为表达服装整体精神性因素最为重要的部分。

译文：The intuitiveness and vividness of patterns outweigh the structure of garment and the fabric as well. Therefore, it becomes the most important part in demonstrating garments' integral spirituality, and quite easily served to convey information by the main dress group.

【分析】汉语原文中"所以"体现了句间的因果关系。这一连接词作为语篇中的衔接手段，在英语和汉语的表达上有相同的特点，译文中使用"therefore"来表达因果关系，将前后句子紧密连接起来，使读者对句中的语义联系和逻辑关系一目了然。

例7　原文：茶花是"花中娇客"，四季常青，是吉祥、长寿和繁殖的象征。茶花又名曼陀罗花，是佛教中的吉祥花。茶花也象征着宁静安详、吉祥如意。茶花与鸟的图案常用于祝寿。

译文：It is the delicate one of all flowers, remaining green throughout the year, so it is the symbol of auspiciousness, longevity and fertility. Camellia is also called *mantuoluo* (mandragora), which is the auspicious flower in Buddhism. Camellia also symbolizes peace and tranquility, good luck and happiness. The pattern of camellia and birds tends to serve as birthday congratulations .

【注释】例句主要讲述的是肚兜上的插画图案。汉语原文简洁流畅，句间关系隐含在语义中，不言自明，不需要使用连词来明示。英语中则常常需要通过增加连词来说明句子之间或句中各成分之间的关系，例句英译文就很好地借助连词使译文表意更为清楚。比如，英译文在第一句两个分句间增加"so"明示两者之间的因果关系，使前后语义更为连贯、易懂。

例8　原文：刘海戏金蟾是古老的汉族民间传说故事，来源于道家的典故。刘海少年时上山打柴，看见路旁一只三足蟾蜍受伤，便赶快上前为之包扎伤口，蟾变成了美丽的姑娘，并与刘海成婚生子，妻子能口吐金钱和元宝。

译文：*Liu Hai Xi Jin Chan*, an old folktale of Han Nationality, comes from the Taoist allusions. Once, a teenager named Liu Hai went up hill to gather firewood, seeing a wounded three-feet toad, he quickened his step to bandage the wound, then the toad became a beautiful girl, who was married to Liu Hai and gave birth to babies. Later, his wife could vomit money and ingot (the ingot is shoe-shaped gold one , it is one kind of currency form in ancient China).

【分析】从汉语原文的表述看，没有出现连接词语，显得相对松散，但句内的含义是连贯的，读者也很容易发现句与句之间的逻辑关系，依靠语句的先后顺序，这就是汉语语篇的隐性连接，但英语更多的是依靠显性的连接。例句英译文通过时序性连接词语"then""later"等标示篇章的事件发生的时间关系，使语篇中句子连接更为紧密。

例9　原文：图案为一路连科，由白鹭、莲花、芦苇等构成。封建社会科举考试，连续考中称之"连科"。"鹭"与"路"同音。"莲"与"连"同音。

译文：The bellyband pattern of the *Yi Lu Lian Ke* (Continuous success in imperial examinations). This pattern is composed of *bailu* (egrets in English), *lianhua* (lotus) and *luwei* (reeds) etc. *Lian Ke* refers to passing the feudal imperial examinations successfully each time. 鹭 *(lu)* and 路 *(lu)* (Which means road) are homonymic, so the same case is with莲 *(lian)* and 连 *(lian)* (Which means continuousness).

【注释】在语篇层面的连接方式上，汉语呈隐性，而英语是显性的，或者说在语篇层面英语也重形合，需要用连接词作为手段把各句连起来，此句译文通过添加表示因果关系的连接词"so"，把原文各语句的逻辑关系完整体现出来了。

例10　原文：北齐时期称内衣为"心衣"，隋唐时期则称作"宝抹"，汉代称为"抱腹""心衣"，魏晋时期称为"两当"，唐代称为"诃子"，宋代称为"抹胸"，元代称为"合欢襟"，明代称为"主腰"，清代称为"肚兜""抹胸""兜兜"等。

译文：In North Qi Dynasty，the underwear is called x*inyi*，in Sui and Tang Period，it is called *baomo*，but it is called *baofu*, *xinyi* in Han Dynasty, *liangdang* in WeiJin Period, *hezi* in Tang Dynasty, *moxiong* in Song Dynasty, *hehuan jin* in Yuan Dynasty, *zhuyao* in Ming Dynasty and *dudou* （bellybands），*moxiong, doudou*, etc. in Qing Dynasty.

【注释】在汉语语篇中对偶与排比经常作为衔接手段来使用，这是汉语所独有的，而英语的平行结构之间的连接，在很多情况下则需要使用连接词语。例句原文是由相互平行的结构组成的复合句，通顺连贯，一气呵成。英译文根据英语语言的特点，增加了连接词"but"和"and"来连接句中各成分，使译文表意清楚而连贯。

（三）省略

省略是指将语言结构中不言而喻的成分省去，或基于语法规则，或源于用法习惯，其目的在于避免不必要的重复，使语言更为简洁明快，突出代表新信息和重要信息的关键词语。有时恰当运用省略策略，可以使句子或语篇前后衔接连贯、结构紧凑，表达更为生动有力。

例11　原文：肚兜不仅具有美学价值，如各种款式、装饰等，而且具有文化价值，也包含民俗礼节与纲常礼教的内涵。

译文：Bellybands not only take on the esthetic value, such as various styles and decorative functions, but also cultural value, including folk custom and etiquette, feudal ethical codes.

【分析】从汉语原文看，谓语动词"具有"在句子中重复使用，以表达语篇的完整意义。但英语是一种形合语言，在表达中尽量避免重复。因此，例句中谓语动词"具有"所对应的"take on"只出现一次，在并列结构中第二次出现时得以省略，译文中不仅语义表达上不受影响，而且增加了与前文的衔接效果。

例12 原文：因此，肚兜不只是一种内衣的形式，更是文化的传承与积淀，是对自然生命的爱护和珍视，是一种女性对服饰美的追求。

译文： Therefore, bellybands, in a sense, are the cultural heritage and accumulation, symbolizing the love for life, the pursuit for beauty of costume from women's perspective, rather than a form of lingerie.

【分析】"是"在例句原文出现三次，对原文语篇的衔接和连贯方面起着重要的作用。英文表达以简洁为美，避免重复使用相同的词汇。由此，例句译文只译出了一个"是"，其他"是"则以变通的方法省略掉了。具体方法是借助现在分词"symbolizing"作状语，将原文中的"是…"结构调整为"symbolizing"的宾语，使译文在避免重复的同时，在结构上层次更鲜明，表意也更明确。

例13 原文：这件肚兜纹饰题材采用狮子滚绣球。所谓"狮子滚绣球，好事不到头"，故而民间常广泛应用此图案取其吉祥之意。

译文： This bellyband is decorated with design of lions playing balls. As is the Chinese saying goes, "the lion plays balls, and the good thing never ends." The motif means good fortune and blessings, which is popular in the folk decorations.

【分析】原文在表现句子间的逻辑关系时，运用连接词语"故而"来说明民间谚语和肚兜常采用狮子滚绣球图案的关系。译文为更简洁而清楚地表达原文信息，进行了断句处理。译者在翻译中省略了"故而"不译，在对原文表达顺序进行调整的基础上，直接用"which"引导的非限制性定语从句，使译文更加简练、紧凑，表意清楚。

例14 原文：肚兜的系带不局限于细绳，富贵人家用金链，殷实人家用银链或铜链，小家碧玉则用缎带。

译文： The strings of the bellyband can be made of different materials, such as gold chains for wealthy families, silver and brass for the well-off, and satin ribbons for the ordinary.

【分析】例句原文说明了不同类型的人家在肚兜系带材质上的不同，四个分句中使用了三个谓语动词"用"字。英语是形合语言，在表达上力求避免重复，所以译文省略了多个重复使用的谓语动词，变换原文的表达方式，以"such as"引出原文关于系带材质的并列结构，并借助介词"for"引出不同背景的家庭，从而避免了谓语动词的重复使用，使译文句子结构主次分明，表意清楚。

七、译文赏析
段落1

在中华民族传统服饰肚兜文化中，图案的形状常常以某个形象的纹饰来表达。无论是动物、植物，还是其他图案都会成为祭祀或者表达心愿、寄托情感的载体，体现了传统文化精神。肚兜常借助图案纹饰将具体的实物形象与某一个抽象的概念相对应，借此表达广大群众追求生存的愿望和追求福、禄、寿、喜的未来目标。在肚兜图案中，人们经常将各种花卉、虫鸟、动物、人物等组合成某种图形，以表达某种象征意义和表达对美好生活的期盼。借助肚兜上的图案纹饰表达某种象征意义、文化内涵以及核心精神，是一种文化语言的呈现。❶

The shape of the bellyband patterns in Chinese traditional culture is usually decided by the image of ornamentation. The patterns of animals, plants or other things would be the carriers of practicing sacrifices, expressing the expectations or conveying the emotions, which signifies a traditional culture spirit. In Chinese traditional bellyband patterns, a concrete object image is often corresponding to an abstract concept through decoration designs to achieve people's goal of the pursuit of survival, the hope for luck, wealth, longevity, happiness, and the good future as well. In bellyband patterns, the plants, the insects and birds, the animals and the characters are often combined into a pattern, expressing some symbolic significance of expecting the happy life. This is a cultural language to express the significance, the cultural implication and the core spirit by making use of the bellyband patterns.

【赏析】段落原文描写了中华民族服饰肚兜文化中的图案纹样背后包含的文化寓意及精神。总体而言，译文简洁明快，用词严谨，层次分明、行文流畅。由于汉英语言的差异，译者采用关系连接词，对语言内容适当地进行了调整，原文第二句是由三个流水分句组成的长句，分句之间呈递进关系，没有使用衔接词，其连贯性主要靠语义实现。首先，译者在翻译时使用关系代词"which"引导定语从句，将第三个分句译为后置定语，从而使译文结构紧凑，逻辑清楚，充分体现出形合语言的特点。其次，对待原文出现的文化负载词，译者采用直译法向译文读者传递其文化含义，比如，译者联系上下文将"福、禄、寿、喜"意译为"luck, wealth, longevity, happiness"，简洁易懂。最后一句"借助肚兜上的图案纹饰表达某种象征意义、文化内涵以及核心精神，是一种文化语言的呈现"是汉语常用的表达形式，注重意义连贯，先叙述后评价。译文根据英语表达习惯，先评价后叙述，并使用指示代词"this"与上文建立联系，使译文句子自然流畅，衔接紧密。

段落2

肚兜不仅具有美学价值，如各种款式、装饰等，而且具有文化价值，也包含民俗礼节与纲常礼教的内涵。与其他服饰相比，肚兜服饰体现的文化价值比其本身价值更重要，因为其中肚兜图案蕴含的意义不仅是民族文化精神的体现，更是对情感的阐释和表现。人们通过肚兜，将山水、花鸟、云气、吉祥物展示之上，主张天、地、人同源同根、平等和谐的文化观念。著名学者潘建华教授在《肚兜寄情文化史》中提到："如果说款式是肚兜之父，美学理念是肚兜之母，那么文化意义就是肚兜之灵魂。"因此，肚兜不只是一种内衣的形式，更是文化的传承与积淀，是对自然生命的爱护和珍视，是一种女性对服饰美的追求。❷

Bellybands not only take on the esthetic value, such as various styles and decorative functions, but also cultural value, including folk custom and etiquette, feudal ethical codes. Compared to

❶ 訾韦力. 中国传统肚兜服饰文化（汉英对照）［M］. 北京：中国轻工业出版社，2016：前言.
❷ 訾韦力. 中国传统肚兜服饰文化（汉英对照）［M］. 北京：中国轻工业出版社，2016：32.

other garment accessories, the value of bellybands is more important than bellyband itself in that what is endowed in the patterns of bellybands not only embodies the national cultural spirit, but also interprets and expresses the feelings. Mountains rivers, flowers, birds, clouds even mascots are displayed through bellybands, advocating heaven, earth and human are of the same origin with the cultural value of equality and harmony. The famous scholar, professor Pan Jianhua has ever mentioned in his book *A Cultural and Metaphorical History of Traditional Chinese Bellybands*: If the style is father of bellybands, the aesthetic idea is mother of bellybands ,then cultural meaning is the soul. Therefore, bellybands, in a sense, are the cultural heritage and accumulation, symbolizing the love for life, the pursuit for beauty of costume from women's perspective, rather than a form of lingerie.

【赏析】段落原文重点描述肚兜极大的文化价值，它体现了我们民族的精神文化。整段文字逻辑清晰，表达连贯，富有感染力。原文多层次论述肚兜的文化价值，不仅涉及美学价值，也涉及具体的民族文化图案和精神文化层面，表述具有较强的逻辑层次感，因此较多使用了关系代词和逻辑连接词。译者为了明确表达原文的思想，在翻译时照应原文的表达逻辑，参照原文的句式结构，连用两次"not only...but also..."并列结构，使汉语原文中的逻辑关系在译文中同样一目了然。其次，汉语原文中"其""其中""其他"照应着上下文"肚兜服饰""肚兜图案""美学价值"，是汉语的内在逻辑表达。译文为确保原文信息的完整性，使之具有连贯性，使用了关系代词"what"，指示代词"itself"照应前文所描述的肚兜及图案意象，使两句在表意上更为连贯，结构更完整，帮助读者更好地理解原文。此外，在汉语最后两句中，"那么""因此"体现了原文的因果关系，译文中使用其英文对应词"then""therefore"来呈现这一因果关系，准确再现了原文的句子结构及所传达的意义。

八、翻译练习
（一）请将下列汉语句子译成英文
1. 尽管传统服饰图案与中华传统思想密不可分，但是随着时代的变迁，神秘古老的肚兜服饰通过传承与发展，也融进了现代服饰之中，促进了现代时尚的发展。

2. 长辈通常在孩童肚兜上绣有虎头像、"五毒"图案，以期盼孩子健康成长。儿童多用红色镶边的绣花肚兜。男孩肚兜常绣以"蟾宫折桂""五子登科"等图案，寄予盼其成才的愿望；女孩肚兜常绣"佛手莲花""牡丹蝴蝶"，希望其人生幸福。

3. 佛教莲花纹样、宝相花纹样大量出现在肚兜服饰图案中，这也正是中华民族向善的心理乃至追求永生的理想的体现。

4. 吉祥文化是中国博大精深的传统文化中不可缺少的一部分，已经渗透到生活的方方面面，传统吉祥文化思想在民族服饰肚兜图案中更是表现得淋漓尽致。

5. 和合之像多在婚礼时陈列悬挂，或者常年悬挂在堂中，与肚兜和合二仙图案一样，取谐好吉利之意。

6. 年轻人的肚兜主要表达对爱情和美好生活的向往；年轻女人常把肚兜作为传情与恩爱的信物赠予情人或丈夫；儿童肚兜主要是辟邪和祈福；老年人的肚兜则多是对安康与长寿

的祈盼。

7. 时至今日，设计师们重新推出了时尚性感的现代肚兜，并成为中式礼服的一部分而广为流传，而柔媚的传统色彩也在变化中时尚化了很多。

8. 这件肚兜以黄色缎为地，绣人物、树木和花卉，构图主次分明，疏密有致，人物造型准确生动，是现实生活场景的反映。

9. 民间百衲衣是为祈望孩童平安成长而专门做的一件衣服，百纳肚兜同样象征长辈对孩子美好未来的祈祷，同时也体现了乡村邻里的美好祝福。

10. 清代的肚兜一般做成菱形，上有带，穿时套在颈间，腰部另有两条带子束在背后，下面呈倒三角形，遮过肚脐，达到小腹。材质以棉、丝绸居多，红色则是"肚兜"常见的颜色。

（二）请将下列汉语段落译成英文

清代，女子用的肚兜也称抹胸，用绸缎或软布做成，只有前片，没有后片。以系带悬挂在脖子上，两侧各有两根带子束于腰后，遮住了肚脐。肚兜的系带不局限于线绳，富贵人家用金链，殷实人家用银链或铜链，小家碧玉则用缎带。肚兜一般常常认为是为了保护肚脐，其实，它的最主要功能还是束胸。束胸使之不显露人体曲线，是中国封建礼教的要求。但是，不暴露在外，却在贴身处表现出无比的浪漫，这可能就是中华民族的内敛之处。大红色、水粉色光亮的绸缎，精美并具有吉祥含义的绣花，菱形上下倒圆或倒三角的造型，使得抹胸具有神秘的魅力。这种浪漫，这种神秘，是我们当代人的一种感觉，也许清朝妇女怎么也想不到，百年后的今天，红肚兜又成为一种时尚，在爱美女子间悄然流行。❶

九、译笔自测

肚兜的图案文化

肚兜服饰中的图案经历数千年的锤炼之后，许多图案符号已经成为一个"民族精神象征意义的高度概括和集中体现"。图案在穿着的服装中起着标识作用，具有一定的象征意义。图案的形象性特征比服装的结构、材质更为直观。所以，它在服装中最易被着装主体用来传达信息，成为表达服装整体精神性因素最为重要的部分。

传统服饰肚兜图案题材非常丰富。花鸟、动物、神仙、神话故事、戏曲人物、生活人物等以图案的方式展示于服饰之上，与美学思想相融合，倡导天、地、人同源同根，平等和谐的中华传统文化观念，表达民间百姓的愿望以及对美好生活的祈盼，同时也反映出中华传统思想对服饰图案设计的影响。传统文化中的儒、道两家长期以来支配和影响着中华民族的精神生活，他们融合相济，和而不同，彼此共存共荣。

孝是儒家伦理思想的核心，是千百年来中国社会维系家庭关系的道德准则，是中华民族的传统美德。基于孝的图案在传统服饰肚兜图案中比比皆是，它们以不同的内容展示百姓的精神世界以及孝的重要性。

此外，受儒家思想影响的封建伦理和价值观念也渗透到了肚兜服饰图案之中。儒学思想

❶ 该段落语料参考：薛雁. 时尚百年［M］. 杭州：中国美术学院出版社，2004：44.

重视"德"，因此图案的内容多与圣君贤臣、烈女孝子、三纲五常以及伦理故事有关。儒家思想强调"修身"。松、竹、梅岁寒三友吉祥图案，以及梅、兰、竹、菊四君子组合吉祥图案都是推崇修身养性的儒家思想的体现，这些图案符合儒家道德规范而被广泛运用。儒家文化强调君臣、父子、夫妻等礼制制度，并常用一些动物来表达吉祥的寓意。

儒家思想非常重视子嗣的传承和家庭的延续，推崇多子多福的观念。因此，"麒麟送子""连（莲）生贵子"等题材被广泛用于肚兜图案之中，以求吉祥如意。符合儒家"天人合一"观点的花、鸟、虫、兽的吉祥纹样，体现着人与大自然的和谐。当然，在儒学影响下我国许多诸如科举高中、仕途顺畅和俸禄富贵等封建思想在吉祥图案中都有反映。道教集合了中国古代巫术、五行、黄老之道等各种思想，有许多"吉祥图案"都产生于道家对吉祥如意、长生不老的追求。长生不老是道教永恒的追求。道教十分重视个体生命，认为个体生命的修炼是为了成为仙人。长生不老既是历代帝王的追求，又是广大人民对生活的渴望，是道教的第一要旨。因此有关长寿的图案成为吉祥图案的一个重要部分。巫术纳入道教后，趋吉避凶成为道教的重要组成部分。例如道教的太极八卦经常出现在服饰图案中，因此古人常以此作为消灾的吉祥图案，进而成为驱邪避凶、趋利向善的象征符号。❶

十、知识拓展

"肚兜"是中国传统文化中的一种贴身内衣，是民族服饰特殊的存在。它伴随历朝历代的服饰文化不断演变，从繁到简，从遮掩到张扬，始终蕴含着丰厚的文化内涵、深刻的艺术审美、独特的民族风情。小小的贴身之物，其款式、图案、刺绣工艺等都反映着古人生活的情境与智慧（图3-4）。到现代社会，"肚兜"成为我们研究民族历史文化的载体，也影响着现代服饰设计。

图3-4 红缎地打子绣狮子滚绣球肚兜❷

"Bellyband" is a kind of intimate underwear in Chinese traditional culture, which is a special existence of national costume. It is accompanied by the continuous evolution of costume culture in the past dynasties, from complex to simple, from concealment to exposure, always containing rich cultural connotation, profound artistic aesthetics, and unique national customs. The styles, patterns and embroidery techniques of small intimate objects all reflect the situation and wisdom of ancient people's life. In modern society, "bellyband" has become the carrier of our study of national history and culture, and also affects the design of modern clothing.

❶ 语料参考：訾韦力. 中国传统肚兜服饰文化（汉英对照）[M]. 北京：中国轻工业出版社，2016：130-137.
❷ 图片来源：清华大学艺术博物馆官网。

第四章 丝绸汉译英

学习目标

1. 了解中国丝绸的历史及文化内涵
2. 熟悉中国丝绸的基本语汇及其汉英表达方式
3. 掌握与异化和归化相关的翻译知识与技能
4. 通过实践训练提高中国丝绸文化汉英翻译能力

一、经典译言

翻译是按社会认知需要，在具有不同规则的符号系统之间传递信息的语言文化活动。

——方梦之

二、汉语原文

丝

众所周知，丝是中国独特的发明，在相当长的一段时间内，中国是世界上唯一出产和使用丝的国家。

在中国的神话传说中，中华民族的祖先轩辕黄帝的元妃"嫘祖"，是公认的养蚕取丝的始祖。古代皇帝供奉她为"蚕神"。据考古资料，中国利用蚕丝的时代比传说中嫘祖生活的年代更早。战国时的荀子（约前313—前238）所作的《蚕赋》，记述了"马头娘"的传说：一个女孩的父亲被邻人劫走，只留下了她父亲的座驾——一匹马。女孩的母亲说，谁能将女孩的父亲找回，就将女儿许配给谁。结果那匹马闻言脱缰而去，真的将女孩的父亲接回来了。女孩的母亲却忘了自己的许诺。马整日嘶鸣，不肯饮食。女孩的父亲知道原委后非常愤怒，认为马不该有此妄想，一怒之下将马杀了，晒马皮于自家庭院。有一天，女孩出现在庭院里，马皮卷上女孩飞上桑树，变作了蚕，从此这个女孩就被民间奉为蚕神。蚕神的影响波及东南亚和日本等地，那里至今都供奉"马头娘"（图4-1）。

图4-1　马头娘的传说❶

除神话传说之外，关于丝的早期应用，还有更为准确的

资料。1958年，在新石器时代良渚文化遗址（位于浙江余杭）中，出土了一批4700年前的丝织品，它们是装在筐中的丝线、丝带、丝绳、绢片等，经鉴定，认为是家蚕丝制品。尽管这些文物已经炭化，但仍然能够分辨出丝帛的经纬度，这表明当时的丝织技术已经达到一定水平。

3000年前，商代甲骨文上已有蚕、桑、丝、帛等文字，可见桑蚕业已经在生产中担当重要角色。作为儒家经典，汇集了中国古代语言、文字、文学、哲学、文化思想、神话、社会生活的重要史料的《尚书》，也有关于丝的记载。

春秋战国时期，农业比之以前更发达。男耕女织成了这一时期的重要经济特征，种植桑麻，从事纺织是一种典型的社会经济图景。由于当时的养蚕方法已经十分讲究，缫出的蚕丝质量也很高，其纤维之细之匀，可与近代的相媲美。至于汉代，纺纱和织布艺术得到了进一步发展。从1972年湖南长沙马王堆西汉墓出土的织锦来看，每根纱由四五根丝线组成，而每根丝线又由十四五根丝纤维组成，也就是说每根纱竟由54根丝纤维捻成。如此高的丝纺水平，同时也推动着染、绣的发展，使它的成品更加美观也更富表现力。

在深厚的文化积淀中，丝独特的质感已渐渐成为一种风格化的象征，象征着东方美学的精神气质。或许我们可以这样说，因为有了丝，中国服饰才呈现出风神飘逸的灵动之美；因为有了丝，中国画中的人物形象才呈现出一种春蚕吐丝般的线条之美。❶

译前提示：丝绸发源于我国，并在历史的长河之中留下了耀眼的光芒，它是中华文明灿烂的一页，为世界文化贡献了辉煌篇章。本文从"马头娘"的传说入手，对中国几千年养蚕制丝的文化进行介绍，包括历史传说、考古发现、典籍记载等诸方面，并简要分析了丝对中国传统文化艺术（如中国服饰和中国画）的重要影响。

三、英语译文

Silk

As we all know, silk is the invention of China, and for a long period of time, China was the only country producing and using silk.

In Chinese legend, Leizu, the royal concubine of the Yellow Emperor, was known to be the first one to raise silkworms and make silk. The ancient Chinese emperors all worshipped her as the silkworm goddess. Archaeological data shows that the Chinese started using silk from silkworm even earlier than the days of Leizu. In the Warring States Period, Xun Zi (circa. 313–238 BC) already wrote *Praise to the Silkworm*, which told the story of the "Horse-head Girl". One day a girl's father was abducted by his neighbor, and only his horse remained. The girl's mother promised that whoever took the father back home would get to marry the young girl. Hearing this promise, the horse ran away, and returned with the man of the household. The mother, however, forgot about her promise. The sad horse refused to eat, crying all day. Finding out about the cause of all this, the angry father killed the horse and left the skin under the sun in their own courtyard. One day the girl was walking

---

❶ 语料参考：华梅. 中国服饰［M］. 北京：五洲传播出版社，2004：39–41.

by, the horse skin wrapped her up and brought her up the mulberry tree. They turned into a silkworm, and ever since that day the girl had been worshipped as the silkworm goddess. The influence of the silkworm goddess went as far as Southeast Asia and Japan, where the "Horse-head Girl" is still worshiped until the present day.

In addition to anecdotes and fairy tales, there are even more accurate proof on the early use of silk. In 1958, at the Liangzhu historical site located in what is now Yuhang of Zhejiang Province, some silk textiles were excavated made 4700 years ago, including silk threads, silk ribbons, silk strings, and pieces of silk, all held in the basket. These were made from silk of home raised silkworm according to expert opinion. Although these historical relics had been carbonized, the warps and wefts were still quite clear, which indicated that the silk textile of that time had reached some degree of sophistication.

As early as 3000 years ago, the Shang Dynasty oracle bone inscriptions already had characters meaning silkworm, mulberry tree, silk, and gauze. It is apparent that silk was already playing an important role in the production of that time. There were also clear records of silk in *Shangshu* (Book of Historical Records), a Confucian classic that recorded important information on the language, writing, literature, philosophy, cultural thoughts, mythology and social life of ancient China.

In the Spring and Autumn and the Warring States Periods, agricultural development reached a new height. One important feature of the economy at that time was the division of labor between men and women, who were engaged separately in farming and weaving. Planting of mulberry trees and weaving of textiles were typical scenes of economic activity of that time. The methods of sericulture were very exquisite at that time, and technology of silk reeling was already very developed, as the silk threads spun from this silk were as even and refined as in the modern day. By Han Dynasty, the art of spinning and weaving moved further forward. Brocade excavated from the Western Han Dynasty Tomb of Mawangdui in Changsha of Hunan Province in 1972 had yarns made from 4 or 5 strands of thread, and each thread was spun from 14 or 15 pieces of fibers. That is to say, each yarn is made up of 54 pieces of fiber. High development in silk spinning pushed further forward the art of dyeing and embroidery, giving the finished product added beauty and vivid expression.

In the profound cultural heritage, the unique beauty of silk has gradually become a symbol of eastern aesthetics. It can be said that because of silk, Chinese garments had the graceful flow, and the figures in classic Chinese paintings presented the beauty of silk-like lines. [1]

四、词汇对译

黄帝　the Yellow Emperor　　　甲骨文　oracle bone inscriptions
嫘祖　Leizu　　　　　　　　　儒家经典　Confucian classics
蚕神　silkworm goddess　　　　商代　the Shang Dynasty

[1] 语料参考：Hua Mei. *Chinese Clothing* [M]. 北京：五洲传播出版社，2004：39-42.

马头娘　Horse-head Girl

供奉　worship

丝线　silk thread

丝带　silk ribbon

绢片　piece of silk

炭（碳）化　carbonized

经纬　warps and wefts

春秋时期　the Spring and Autumn Period

战国时期　the Warring States Period

缫丝　silk reeling

纱　yarn

养蚕方法　method of sericulture

东方美学　eastern aesthetics

中国画　Chinese painting

五、译文注释

1. 在中国的神话传说中，中华民族的祖先轩辕黄帝的元妃"嫘祖"，是公认的养蚕取丝的始祖。古代皇帝供奉她为"蚕神"。

In Chinese legend, Leizu, the royal concubine of the Yellow Emperor, was the first one to raise silkworms and make silk. The ancient Chinese emperors all worshipped her as the silkworm goddess.

【注释】轩辕黄帝和嫘祖是中国传统文化中的两个重要人物。其中，轩辕黄帝即古华夏部落首领黄帝，居轩辕之丘，号轩辕氏。黄帝以统一华夏部落与征服东夷、九黎族而统一中华的伟绩载入史册。他在位期间，播百谷草木，大力发展生产，始制衣冠、建舟车、制音律、作《黄帝内经》等，被尊为中华民族的人文始祖。"嫘祖"，又名累祖，为中国远古时期西陵氏之女、轩辕黄帝的元妃。她发明了养蚕，并首倡婚嫁，母仪天下，福祉万民，和炎黄二帝开辟鸿蒙，告别蛮荒，被后人奉为"先蚕"圣母，系中国先祖女性的杰出代表。嫘祖与炎帝、黄帝同为中华民族的人文始祖。例1对这两个人物的名字采取了不同的翻译方法。译文将"嫘祖"译为"Leizu"，采用音译的方法保留了原文文化意象所具有的异国情调，是典型的异化翻译。"嫘祖"这个人物的文化内涵在下文的"蚕神"传说中进行了具体阐释，有助于外国读者理解"Leizu"这个英语中的音译外来词。"轩辕黄帝"的译法有所不同，译者遵循传统译法将其译为"the Yellow Emperor"，采用的是直译的方法。译文略去原文中的"轩辕"二字不译，是一个典型的归化式处理，因为上下文中轩辕黄帝只是作为介绍嫘祖的背景信息，这样可以减少黄帝这个人物名字上负载的文化信息，将读者的注意力导向嫘祖以及与丝有关的历史文化。

2. 据考古资料，中国利用蚕丝的时代比传说中嫘祖生活的年代更早。

Archaeological data shows that the Chinese started using silk from silkworm even earlier than the days of Leizu.

【注释】例2的译文对原文的句式结构进行了调整。原句中"据考古资料"为状语，译文将"考古资料（archaeological data）"译作主语，增加谓语"show"，并以"that"引导的宾语从句阐释原文主句的内容。这样使得英译文句子结构更为紧凑，形式更简洁。

3. 战国时的荀子（约前313—238）所作的《蚕赋》，记述了"马头娘"的传说：一个女孩的父亲被邻人劫走，只留下了她父亲的座驾——一匹马。

In the Warring States Period, Xun Zi (circa. 313-238 B.C.) already wrote *Praise to the Silkworm*, which told the story of the "Horse-head Girl." One day a girl's father was abducted by his neighbor, and only his horse remained.

【注释】"马头娘"是中国神话中的蚕神，相传为马首人身的少女。联系上下文故事情节，"马头娘"的故事发生在一个少女身上，而且她被称作"蚕神"，是蚕丝文化的源头，影响深远。在古汉语中"娘"一般指年轻女子，多指少女，有女儿、少女或女子的意思，因而此处将"娘"译为"girl"。把"马头娘"直译为"Horse-head Girl"是将中国文化中的马头娘形象直接传递到英语中去，是典型的异化翻译。此外，译文中的"circa."表示"大约"，它在系谱学以及历史学著述中被广泛采用，一般仅用在不能精确指明的年份之前，可以缩写为"ca."。

4. 1958年，在新石器时代良渚文化遗址（位于浙江余杭）中，出土了一批4700年前的丝织品，它们是装在筐中的丝线、丝带、丝绳、绢片等，经鉴定，认为是家蚕丝制品。

In 1958 at the Liangzhu historical site located in what is now Yuhang of Zhejiang Province, some silk textiles were excavated made 4700 years ago, including silk threads, silk ribbons, silk strings, and pieces of silk, all held in the basket.

【注释】例4原文中括号中的内容在译文中译成了"良渚文化遗址"的修饰成分。其中，"located in"意为"位于"，"what is now"这个结构表示"就是现在的"，"what"在这里的用法相当于"the place that"。译文将原文括号中的信息译为后置定语"located in what is now Yuhang of Zhejiang Province"，可以使原文中的补充信息自然融入译文，从而使译文读起来更流畅，并在表意方面具有明晰化的效果。

5. 尽管这些文物已经炭化，但仍然能够分辨出丝帛的经纬度，这表明当时的丝织技术已经达到一定水平。

Although these historical relics had been carbonized, the warps and wefts were still quite clear, which indicated that the silk textile of that time had reached some degree of sophistication.

【注释】例5的翻译采用了语态转换法。原文中"但仍然能够分辨出丝帛的经纬度"是省略了主语的主动语态，其中"丝帛的经纬度"是动词"分辨"的宾语。由于英文一般要求完整的句子结构，这个分句的译文以"the warps and wefts（经纬度）"为主语，采用了被动语态来译，这样更符合英文句式的特点。

6. 作为儒家经典，汇集了中国古代语言、文字、文学、哲学、文化思想、神话、社会生活的重要史料的《尚书》，也有关于丝的记载。

There were also clear records of silk in *Shangshu* (a book of historical records), a Confucian classic that recorded important information on the language, writing, literature, philosophy, cultural thoughts, mythology and social life of ancient China,

【注释】《尚书》是我国第一部上古历史文件和部分追述古代事迹著作的汇编，它保存了商周特别是西周初期的一些重要史料，乃儒家五经之一，在儒家思想中具有极其重要的地位。为帮助读者更好地理解文章内容，译文在音译书名"*Shangshu*"，以括号方式增加了对本书的解释"a book of historical records（一本记载历史的书）"，体现了译者以读者为中心的思路和传播中国经典文化的努力。

7. 男耕女织成了这一时期的重要经济特征，种植桑麻，从事纺织是一种典型的社会经济图景。

One important feature of the economy at that time was the division of labor between men and women, who were engaged separately in farming and weaving. Planting of mulberry trees and weaving of textiles were typical scenes of economic activity of that time.

【注释】例7采用了分译的方法，根据英文表意的需要，将原文的一句汉语分译成了英文中的两个句子，分别对当时的经济特征和经济图景进行说明。此外，"男耕女织"没有直译为"men plough and the women weave"，而在翻译时用who作为关系代词引导非限制性定语从句，对男女之间的分工劳动做进一步说明。封建社会的小农经济，男子耕田，女子织布，是普遍的社会分工，译文把原文隐含之意展现在读者面前，具有明晰化的效果。

8．或许我们可以这样说，因为有了丝，中国服饰才呈现出风神飘逸的灵动之美；因为有了丝，中国画中的人物形象才呈现出一种春蚕吐丝般的线条之美。

It can be said that because of silk, Chinese garments had the graceful flow, and the figures in classic Chinese paintings presented the beauty of silk-like lines.

【注释】汉语讲究均衡美，在用词造句上倾向于重复，运用相同或相似的词语、句式来强调所要表达的意思，增添文采，给人以深刻印象，是一种积极的修辞方式。但英语中，除非有意强调或出于特殊的修辞需要，总是倾向于尽量避免重复。例8中汉语原文重复使用"因为有了丝"，以强调丝对中国服饰和中国画的重要影响。英译文中避免重复，只使用了一次"Because of silk（因为有了丝）"，使行文简洁、有力，更符合英语民族的语言心理和表达习惯。

六、翻译知识——异化与归化

异化与归化是翻译研究中常见的一对术语，由意大利裔美籍翻译学者劳伦斯·韦努蒂提出，首见于其1995年出版的专著《译者的隐身：一部翻译史》。这对翻译学术语反映的是翻译中的策略问题，即译者在多大程度上使某个文本符合译入语的文化规范。韦努蒂的想法源自德国哲学家施莱尔马赫的观点。施莱尔马赫在《论翻译的方法》中提出："翻译的途径只有两种，一种是尽可能让读者安居不动，而引导作者接近读者；另一种是尽可能让作者安居不动，而引导读者去接近作者。"前者强调目的语或译文读者，后者强调源语或原文作者。施莱尔马赫认为翻译要帮助目标语读者在不脱离目标语境的情况下正确而完整地读懂原文，并由此演绎出两种翻译策略，即让读者靠近作者，或让作者靠近读者，也就是韦努蒂所指的"异化"与"归化"。根据韦努蒂的观点，"归化"是指在翻译中迫使外语文本符合目的语文化的价值观和表达习惯；"异化"则指在翻译中显示外语文本的语言及文化异质性，把读者推到国外。

（一）异化

劳伦斯·韦努蒂在著作《译者的隐身》中提出："异化翻译是一种通过减少本民族文化对异域文化的压力，进而来表达外文文本的语言与文化差异的翻译策略，它让读者直接接触异域文本。"异化这种翻译策略使用不同于目标语中所盛行的文化符码，来表明异域文本的独特性，刻意打破目标语的行文规范，保留源语文化，给读者带来一种异域感觉。异化翻译策略保留了原文的意象以及语言模式，有助于丰富目标语文化和目标语言表达方式。就汉译英而言，异化翻译是译者尽量保持汉语中的语言与文化特色，并将其直接转化到英语译本中，让目标语读者直接体验到来自汉语的异域风情。下面举例说明。

例1　原文：西兰卡普被称作"土家之花"，在土家族人民生活中有着实用的、礼俗的和审美的三方面意义，不仅以经久耐用著称，而且是土家族婚俗中的主要嫁妆。

译文：*Xilankapu* brocade, the most famous handicraft of Tujia ethnic group that has practical, ceremonial and aesthetic values in people's life, is not only famous for its outstanding durability, but also the principal dowry in Tujia wedding customs.

【分析】"西兰卡普"是一种土家织锦，在土家族语言里，"西兰"是铺盖的意思，"卡普"是花的意思，它是土家族人民智慧和技艺的结晶，是中华民族文化中一种独特的存在。为了将这一中国特色文化体现出来，并满足英美读者了解中国文化的愿望，译者保留其"异味"，用异化的策略将其音译为"*xilankapu*"。同时，考虑到英语读者的文化背景，为帮助他们理解这一具有文化特色的术语，译文附以brocade对"*xilankapu*"进行补充说明。译文对原文中的"土家族"一词也采用近似的翻译策略，用音译法保留了"Tujia"所带给英语读者的异国情调，然后以"ethnic group"来对其进行补充说明，使英语读者在领略译文中的异国风情的同时，也能理解原文所传递的信息。

例2　原文：迢迢牵牛星，皎皎河汉女。纤纤擢素手，札札弄机杼。

译文：Far, far away resides the Cowherd Star,

Fair, fair the Weaver Maiden Star.

Slim and soft are her tender hands,

Click and clack sounds the loom that stands.

【分析】这是《古诗十九首》中的《迢迢牵牛星》的前两句诗句，诗人借神话传说中牛郎、织女被银河相隔而不得相见的故事，抒发了因爱情遭受挫折而痛苦忧伤的心情。第一句"迢迢""皎皎"两个叠词互文见义，表现了情人眼里的咫尺天涯，第二句用"纤纤""札札"两个叠词生动描写了一个饱含离愁的少妇形象。这些叠音词的使用让这首诗音节和谐，富有节奏，自然而贴切地表达了物性与情思。我们可以看到，译者按照原文的顺序进行了翻译，保持了形式与韵味相对应的原则，很明显使用了异化的翻译策略，使原文的字面形式及节奏得到了很好地表达。此外，原文中的"牵牛星"和"河汉女"在中国文化中分别代表牛郎和织女，译为"the Cowherd Star"和"the Weaver Maiden Star"，同样以异化的翻译策略，在译文中保留了中国传统文化意象。

例3　原文：楼兰出土的多块云气动物纹锦上都可以找到麒麟的形象，头上都有明显竖起的一只肉角，非常独特。

译文：The images of *kylin* were found in several pieces of *jin* silk unearthed in Loulan, all with a distinctive horn on their heads.

【分析】本句中"麒麟"是中国传说中的一种神兽，饱含中国文化特色，因此"麒麟"也就进入了英语词汇，英文"*kylin*"专门表达"麒麟"。所以译文用异化翻译策略，充分保留其文化特色，向英文读者展现了中国文化。

例4 原文：缂丝被认为是古代丝织品的奇葩，这种丝织品织出的花纹具有特殊的效果，它的纹样边饰清晰，织物的图案具有立体感，像镶嵌在绸面上似的。

译文： *Kesi* was considered one of the most wonderful silk species in ancient China, boasting three-dimensional patterns with clear-cut rims, as if inlaid onto the textile.

【分析】 "缂丝" 是中国传统丝绸艺术品中的精华，在原文中是负载有中国文化特色内涵的词汇，译者在翻译时为了传播中华文化，使用异化翻译策略进行音译，保留了其民族文化特色，虽然目标语读者对 "kesi" 这个词汇感到陌生，但译文可以在一定程度上满足目标语读者了解中国文化的愿望。

例5 原文：作为万民景仰的天子，龙纹和带有特定内涵的 "十二章纹" 成为皇帝服饰的专用图案。

译文： As son of the heaven, emperors had the privilege to wear costumes with dragon patterns and the "twelve emblems".

【分析】 天子是中国古代皇帝的称谓，在这里翻译成 "son of the heaven"，体现了译者忠实于中国传统文化，选用 "heaven" 一词而非 "God"，表明天子是建立在宇宙天命理论基础上的神所选择的天下共主，与基督教里的上帝（God）无关。译者在这里使用异化的翻译策略，为英语引入了一个具有中国传统文化内涵的新的文化意象。

（二）归化

劳伦斯·韦努蒂认为归化在英美翻译文化中占据主导地位，是西方译者经常采用的一种翻译策略。它要求译者在翻译中采用透明、流畅的风格，减少译文中的异域因素，最大限度地减弱译文读者对外语语篇的生疏感。作为一种翻译策略，归化翻译尽可能地使源语文本所反映的世界接近译文读者所熟悉的世界，减少外文文本中所包含的民族特色，使译文更加符合目标语的文化价值理念。归化策略采取目标语的读者所习惯的表达方式，使译文流畅、自然，不留翻译的痕迹，减少给读者带来的陌生感。就汉译英而言，归化翻译就是译者要将汉语原文中语义、形式及文化层面上的中国特色转化为目标语英语中相对应层面的内容，以增强英文的流畅性和可读性，从而使目标语读者更容易理解原文。

例6 原文：锦和绣两大工艺是当时丝绸生产技术中的最高代表，后世常将锦绣两字合用以表示事物的美好，如锦绣河山、锦绣前程等。

译文： *Jin* (brocade) and *xiu* (embroidery) were the two most classic silk species of the time, and they are often applied together in modern Chinese to describe something positive and splendid, such as a land of splendors and a glorious future.

【分析】 锦绣是一个汉语词汇，指精美鲜艳的丝织品。原文中 "锦绣" 这个词并不单指丝织品种类，这里它是属性词，用来比喻美丽或美好的事物。所以译者将 "锦绣河山、锦绣前程" 译成 "a land of splendors and a glorious future"，没有采取直译的形式，避免引起读者的误解，而是在上文对锦和绣二字进行解释铺垫的基础上，使用归化策略，舍弃锦绣作为丝织品的意象，只译出这个词的衍生意义，使译文表达流畅，可读性强。

例7 原文：隋唐时期的丝绸图案一方面继承了中国传统的风格，另一方面是从西域的装饰艺术中吸收了大量的营养，创造了融合中西方艺术特色的丝绸纹样，达到了中西合璧的效果，诞生了新的混血儿，其中不得不提的就是"陵阳公样"和"红地太阳神锦"。

译文：Silk patterns of the Sui and Tang period not only took on traditional Chinese styles, but also borrowed many inspirations from decorative art of the West. As a result, many patterns integrating both Chinese and western characters were produced, among which the "pattern of Duke Lingyang" and "samite with image of Apollo on red ground" were the most representative ones.

【分析】这句话描述的是唐代服饰纹样的特点。唐代服饰纹样既有鲜明的本土特色，同时又吸收了外来优秀文化元素，从而形成了精美的民族图式纹样。"陵阳公样"和"红地太阳神锦"是典型的代表，在中华纹样史上有高度的美学价值。"陵阳公样"是窦师伦主持设计的纹样，大都以团窠为主体，围以联珠纹，团窠中央饰以各种动植物纹样。窦师伦因设计方面的贡献被封"陵阳公"高爵，这些纹样因此被称为"陵阳公样"。译文为了方便读者理解，采用归化的翻译方法，将这里的"公"字译为目标语读者常见的英文词"Duke"，最大限度地淡化了原文的陌生感。同时，使用"Duke"一词也迎合了读者文化认同感。同样，鉴于中英文化的差异，译者运用归化策略，将"红地太阳神锦"中"太阳神"译为"Apollo"，用西方文化中人们熟悉的意象替换了源语文化意象，有助于读者迅速理解原文内容，增加译文的可读性。

例8 原文：明晚期趋于华丽的民间丝绸图案迎合趋吉纳福的心理需求，具有较浓的民俗色彩，寓意吉祥的纹样十分流行。

译文：Towards the later half of the Ming Dynasty, the patterns of increasingly gorgeous silk for civilian use were adorned with symbols of good luck and happiness to suit the spiritual needs of the common people, thus having a strong folk taste. The patterns that had a connotation of good luck were very popular.

【分析】"趋吉纳福"四个字具有中国文化的意蕴，表达了人们向往吉祥安康的幸福生活。吉祥在《说文解字》中解释为"善""福"，意思是"吉利""祥和"，即凡事如意、美满。人们自古以来对美好生活的愿望与追求产生了吉祥文化，吉祥文化成为中国博大精深的传统文化中必不可少的一部分，它已经渗透在我们生活的方方面面。在译文中若采用异化策略，原封不动地翻译过去，就显得累赘冗长，甚至词不达意。译文依据源语的表达内容进行改写，没有直译"趋吉纳福"，而用"symbols of good luck and happiness"来解释这四个字的含义，这样更容易被译文读者理解和接受，是归化策略的具体体现。

例9 原文：每逢过年，人们常用"年年有余（鱼）""岁岁平（瓶）安"等年画、剪纸来表示吉祥祈福，这里的鱼和瓶就是所谓的吉祥图案。

译文：With the approach of every new year, people would produce new year paintings and paper cuttings with auspicious patterns such as fish and bottle, fish (yu) as a pun for surplus, and bottle (ping) for peace, to express good wishes towards the coming new year.

【分析】"年年有余""岁岁平安"是中国传统吉祥祈福最具代表的语言之一，众所周知，在中国文化中"鱼"与"余"谐音，"瓶"与"平"谐音，所以人们经常用这种传统吉祥符号，来祈求生活富足、安定，这也是汉语的语音特色。另外，原文中用"年年""岁岁"叠词的形式形容时光，是汉语的表达习惯，英语中极少使用类似的叠词，如果在翻译时直接翻译，能保留源语的词汇特色，但会影响译文的可读性以及读者对原文的理解。译者在翻译"年年有余（鱼）""岁岁平（瓶）安"时采用的是归化策略，利用英语介宾短语"with auspicious patterns such as fish and bottle"直接点明两个词语要表达的含义，从而使译文更加符合目标语的表达方式。

例10 原文：有人称赞湘绣，"绣花能生香，绣鸟能听声，绣虎能奔跑，绣人能传神"。

译文： As praised in a poem, "Xiang embroidery is able to produce fragrant flowers, chirping birds, running tigers and lively figures".

【分析】原文"绣花能生香，绣鸟能听声，绣虎能奔跑，绣人能传神"通过拟人化的手法表现了湘绣形象生动、逼真、质感强烈的艺术特色，并且通过四个排列工整的分句，使语言具有韵律感，朗朗上口，中间没有任何连词或其他的语言手段来衔接，分句间的逻辑关系隐含在字里行间，鲜明体现了意合语言的特点。但英语作为形合语言，具有完全不同的特点，在行文习惯和表达方式等方面存在明显差异。英文表达注重语言的形式逻辑，所以例句英译文舍弃了原文的排比句式和韵律，采用归化的翻译策略，改变原文的词法、句法结构，译为以"Xiang embroidery is able to produce..."为核心结构的句式，并将原文中的谓语动词"生香""听声""奔跑""传神"转变为修饰名词的定语"fragrant""chirping""running""lively"，使译文层次清楚，焦点突出，体现出形合语言的特点，顺应英语语言语境，符合英文的行文习惯。

归化与异化作为翻译的两种策略，在汉英翻译的过程中，彼此相辅相成，融合与互补。在汉英翻译的过程中，语言形式方面的归化翻译可较好地传达原文的中心思想，有利于西方读者理解原文的真实含义，然而原文的形式会受到破坏。文化层面的归化翻译，会给西方读者带来"熟悉感"，便于读者接受，然而原文中所含有的中国文化以及作者的意图会在某种程度上受到破坏。相反，语言形式方面的异化翻译可以较好地传递语言形式以及原作意境，并且有利于汉语更好地根植到英语语言中，增加汉语的传播与普及性，但会让西方读者产生"异域"与"奇怪"之感，从而失去一些读者。文化层面的异化翻译，可以较好地传播中国传统文化。在中国文化走出去的大背景下，涉及文字与句法结构的翻译，我们可以尽量采取归化的翻译策略；而涉及文化层面的内容，可以尽量采用异化的翻译策略。

七、段落赏析

段落1

包括清代宫廷服饰在内的中国古代宫廷服饰，最具有表现魅力的是制作服饰的丝绸。丝绸自古即以其优良的服用性能和华丽的装饰效果而备受人们的青睐。中国古代帝王无不以其为奢侈生活的珍贵之物。大量的田野考古资料已证明，中国是丝绸的发源地，它与具有五千年岁月的中国古代文明几乎同时产生并同步发展。公元前5世纪，中国丝绸已开始远播海外。到汉唐时期，举世闻名的"丝绸之路"将中国丝绸源源不断地传到了世界各地。它犹如一条蜿蜒万里的绚丽丝带，把欧亚大陆和东西方文明紧紧地联系起来，促进了东西方政治、经济和文化的广泛交流，对人类的进步和繁荣做出了巨大贡献。因此，丝绸也赢得了中国古

代的"第五大发明"的美誉。❶

　　The salient feature of Chinese court costume, including Qing court costume, lies in their fabric——silk, which has been favored since immemorial time for its excellent quality and magnificent decorative effect. It was cherished by Chinese emperors as part of their luxurious lifestyles. Abundant field archeology has demonstrated that silk originated in China almost at the same time as the birth of her civilization 5000 years ago and developed alongside the latter. By the 5th century BCE, it had spread overseas. From Han to Tang dynasties, the world-famous Silk Road provided a continuous supply of silk to various parts of the world, connecting Eurasia with Western and Eastern civilizations like a splendid silk ribbon winding thousands of miles. It helped promote extensive political, economic and cultural exchanges between East and West, making a tremendous contribution to mankind's progress and prosperity. As a result, silk was acclaimed as "the fifth invention" of ancient China.

　　【赏析】本段原文介绍了丝绸在古代中国的尊贵地位及其巨大的文化价值，语言严谨，逻辑清晰，信息量丰富。译文在充分传递原文信息的同时，做到了语言流畅，再现了原文的文体风格。首先，在文化特色词的翻译方面，译者采用了异化的方法，以保留词汇中的中国文化内涵。比如，译者将"清代宫廷服饰"和"丝绸之路"直译为"Qing court costume"和"Silk Road"，中国文化意象得到了保留，为英语读者提供的是原汁原味的中国式表达。这种异化的处理方式，在充分考虑译文读者接受能力和审美期待的情况下，如果加以合理使用，将有助于中国文化的对外传播与交流。其次，为使译文更符合英语的表达习惯，译者对部分句式进行了调整，如将被动语态转换为主动语态。原文第三句中"中国古代帝王无不以其为奢侈生活的珍贵之物"，以及最后一句"丝绸也赢得了中国古代的'第五大发明'的美誉"都是主动语态，主动语态在汉语中的广泛使用与中国人的传统思维方式注重内因和主观体验有关。英语重理性思维和语言表达的客观性，被动语态使用广泛。译者将两句分别译为被动语态"it was cherished by Chinese emperors as part of their luxurious lifestyles"和"silk was acclaimed as 'the fifth invention' of ancient China"，使英译文更符合英语读者的阅读习惯，增加了译文的可读性。再次，本段翻译中译者没有拘泥于原文的句子结构，而是根据译文的需要，对原文进行了灵活的重组，尤其体现在对分译法和合译法的使用方面。比如，译者将原文的前两句合译为一句，将原文的第二句译为以"which"引导的定语从句修饰前一句的"silk"，使译文在结构上更紧凑，也使原文两句之间的逻辑关系在译文中表达得更清楚。再如，作者根据表达需要，将原文的倒数第二句拆分为两部分，第一部分"它犹如一条蜿蜒万里的绚丽丝带，把欧亚大陆和东西方文明紧紧地联系起来"译为"connecting Eurasia with Western and Eastern civilizations like a splendid silk ribbon winding thousands of miles"，以状语形式并入上一句，这样的调整使译文从地理角度对丝绸之路的丝绸传播及文化联系功能表述得更清楚透彻，为下一句对丝绸之路更为宏观的价值描写做好了铺垫。结构变了，但译文依然逻辑清楚，表达流畅。最后，为增加译文的连贯性，译者还采用了替代的衔接方式，比如

---

❶ 语料参考：故宫博物院编. 天朝衣冠［M］. 北京：紫禁城出版社，2008：序言.

译文第三句使用"the latter"指代前面提及的"中国古代文明",避免了重复,使表达简洁凝练,同时译文前后呼应,增强了语内连贯性。

段落2

由于丝绸之路的通畅和文化交流的发达,隋唐时期的丝绸图案远比前代丰富,流行变化之快也超越前代。在艺术风格上既继承了中国传统,又从中亚、西亚的装饰艺术中吸收营养,有西域风格的联珠纹样大量出现在中原丝绸织物上。图案题材从早期的动物纹样转向动、植物纹样并重,动物纹样则从兽类纹样转向以飞禽类纹样为主,这在艺术史上是一次重大转折。❶

The continuous flow of cultural exchanges along the Silk Road brought a huge variety of design patterns to Tang silks. There were much more to choose from than any other previous periods and the styles also changed more frequently. Artistically, the Chinese tradition was inherited along with new influences from decorative arts of Central Asia and West Asia. The Western regions of beaded roundels appeared overwhelmingly on textiles produced in the heartland of China. The subjects turned from the early animal motifs to the combination of both animals and plants. Among the animal motifs birds became a dominant element as compared to beasts only found in earlier textiles. This marked a turning point in art history.

【赏析】原文描写了隋唐时期的丝绸文化艺术的特点及风格,用词精确,语言通畅,结构严谨。译文忠实于原文表达方式,不仅完整阐释了原文信息,也符合英文读者的阅读习惯,具有可读性。首先,汉语原文中"中原"是独具中国特色地域的习惯表达方式,其与英文中的表达不是很贴近,此时,译者采用归化策略,将其意译为"in the heartland of China",比较清楚地传达出了汉语文化所表示的信息,避免了英文读者对原文信息理解错误。但在句子结构上,译者没有遵循英文结构中"主句在前,偏句在后"的行文习惯,而是遵循原文语序进行翻译,保留了汉语表达中偏正复句的特征。比如,最后一句总结句"This marked a turning point in art history"并未与前一句合译,而是仍然放在译文末尾,单独成句,这样的表述方式,在一定程度上保留了原文的结构特征,忠实传达了原文主旨。

八、翻译练习

(一)请将下列汉语句子译成英文

1. 提花织造技术不断提高,更多精美的丝织品被织造出来,如闻名天下的漳绒、漳缎、云锦,并形成了许多各具特色的地方名绣。

2. 重锦是宋锦中最贵重的品种,常用于宫殿中各式龙椅、宝座的铺垫、靠背等,所以其纹样风格与宫殿室内外环境相协调,多为龙纹、云纹、蝙蝠纹,以及各种大型的宝花纹等。

3. 龙作为帝王的化身,以多种姿态出现在皇帝的丝绸服饰上:正面的龙头,腾云上升的升龙,从天而降的降龙,行走于云彩之间的行龙,以圆形呈现的团龙,等等。

---

❶ 语料参考:赵丰. 中国丝绸通史［M］. 苏州:苏州大学出版社,2005:187.

4．事实上，宝花并不是某一种花卉专有的名称，而是一种想象性的图案，它把各种植物花卉进行变形，并且重新组合，宝花图案的发展经历了从简单到丰满，从想象到写实的过程。

5．两宋时期的丝绸技术最显著的一点就是与唐代丝绸艳丽、丰满的风格相比，更注重寻求自然轻淡的天然之美，这点可能是由于温和儒雅的宋王朝面对北方强悍的异族，只能寄情于世外桃源的隐逸生活，陶醉于山水花鸟的闲情逸趣。

6．明代丝绸产品种类繁多，其中妆花和绒类产品是明代丝织技术的重要进步，缂丝和刺绣工艺也在继承传统的基础上进一步发展。

7．丝绸产品的丰富为以服装为代表的礼仪制度提供了不可或缺的物质基础，为中国礼乐文化体系的形成与发展做出了贡献。

8．在新疆尼雅遗址出土东汉丝织物和毛织物上，既有西域的植物纹和葡萄纹，又有希腊神话中人首马身的"堪陀儿"，还有中原的武士形象，可以说，织物记载着服饰文化的交流。

9．明代丝绸技术由于专业化生产而有较大发展。在养蚕和缫丝方面有"出口干"和"出水干"工艺，丝织络并捻技术和提花织造技术已相当完备。

10．早在春秋战国时期，就有以丝绸为材料的帛画出现，湖南长沙楚墓出土的《人物龙凤》《人物驭龙》帛画是中国迄今发现最早的完整的独幅绘画实物。

（二）请将下列汉语段落译成英文

丝绸在华夏文化中处于特殊的地位，这种地位，首先可以从它对汉字的影响说起。远在殷商时期，神秘的甲骨文中就刻写了"桑蚕丝"等字。这些象形字正是来源于植桑养蚕、丝绸织造。今天我们习以为常地以"纟"为偏旁的字，正是由蚕桑丝绸业衍生出来的文字，如"经、纬、组、织、线"等。此外，我们常用的一些中文词汇都和丝绸业有关系。如"机构"一词泛指机关、团体等工作单位或内部组织，而最早是指织机的结构。❶

九、译笔自测

丝绸与礼仪

从文学到艺术，都能看到丝绸对这些文明做出的特殊贡献，而丝绸作为集优良实用性与高度艺术性于一体的优美产品，本身就是高贵和身份的象征。普通百姓只能以粗布麻衣为服饰用品，这也是以"布衣"指称平民的由来。丝绸在产生之初，就和宗教礼仪密切相关。

早期，丝绸被认为是"通天"的载体。因为先民们在观察蚕的生长时，发现蚕的一生由卵而蚕，作茧自缚成蛹，最后破茧而出羽化成蛾的过程，万分神奇。他们开始把蚕的变化和天地生死联想到一起，认为蚕的一生正是人的一生的写照。所以，如果想要死后灵魂到达天国，就必须像蚕一样破茧而出。于是人们在死后，用丝织物把身体包裹起来，形成一个用丝质材料做成的人工"茧子"，以求灵魂"升天"。

而丝绸礼仪制度的最极致的表现就是"天子"的服饰——龙袍，以及表明官员品级高

---

❶ 该段落语料参考：《中华文明史话》编委会. 丝绸史话［M］. 北京：中国大百科全书出版社，2010：4-5.

低的补服。作为万民景仰的天子，龙纹和带有特定内涵的"十二章纹"成为皇帝服饰的专用图案。龙纹意味着皇帝乃真龙天子，十二章纹则是"日、月、星辰、山、龙、华虫"和"宗彝、藻、火、粉米、黼、黻"。日、月、星辰代表三光照耀，象征着帝王皇恩浩荡；山，代表帝王性格稳重；龙，是一种神兽，象征帝王们善于审时度势；华虫，通常为一只雄鸡，象征王者"文采昭著"；宗彝，是古代祭祀的一种器物，通常是一对，里面分别有虎纹和长尾猴，虎纹，象征威武，长尾猴遇雨以尾塞鼻，象征智慧；藻，则象征皇帝的品行冰清玉洁；火，象征帝王处理政务光明磊落；粉米，就是白米，象征着皇帝给养着人民，安邦治国，重视农桑；黼，为斧头形状，象征皇帝做事干练果敢；黻，为两个弓形相背，寓意君臣离合及善恶相背。

至于文武百官的补服，我们今天在电视上也经常能看到，即前胸后背各缀有一块方形补子的官服，上面的图案要么是禽鸟，要么是猛兽。一般文官用禽鸟，武官则是猛兽，从一品大员到九品芝麻官都有相对应的图案，如仙鹤，代表一品文官。可以说，龙袍和补服成为"分贵贱，别等威"的工具，也很好地阐述了丝绸这种特殊材料在"礼仪之邦"所起的重大作用。❶

十、知识拓展

中国一直以来都有"丝绸之国"的雅誉，五千多年前，我们的祖先就开始学习栽桑养蚕，缫丝织造，后经过历朝历代技术的不断革新，大量轻柔光亮、色彩绚丽的丝织品被制造出来，为中华文明留下无数宝贵的财富。男耕女织，也成为中国古代先民基本的生产生活方式。但丝绸的魅力远不仅此，穿越欧亚大陆的丝绸之路、海上丝绸之路，以及南方丝绸之路，都留下了中国丝织文化对外传播交流的身影（图4-2）。

图4-2　捣练图❷

China has always been known as "the country of silk". More than 5000 years ago, our ancestors began to learn how to plant mulberries, raise silkworms, reel silk and weave cloth. After continuous technological innovation in successive dynasties, a large number of soft, bright and colorful silk fabrics were made, leaving countless precious wealth for Chinese civilization. Men's farming and women's weaving also became the basic way of production and life of ancient Chinese ancestors. But the charm of silk is far beyond that. The Silk Road, the Maritime Silk Road and the Southern Silk Road across Eurasia have witnessed the external communication of Chinese silk culture.

❶ 语料参考：《中华文明史话》编委会. 丝绸史话［M］. 北京：中国大百科全书出版社，2010：8-10.
❷ 图片来源：搜狗图片网。

第五章　刺绣汉译英

学习目标

1. 了解中国刺绣的历史及文化内涵
2. 熟悉中国刺绣的基本语汇及其汉英表达方式
3. 掌握习语及文化负载词翻译的相关知识与技能
4. 通过实践训练提高中国刺绣文化汉英翻译能力

一、经典译言

两国文字词类的不同，句法构造的不同，文法与习惯的不同，修辞格律的不同，俗语的不同，即反映民族思想方式的不同，感觉深浅的不同，观点角度的不同，风俗传统信仰的不同，社会背景的不同，表现方法的不同。以甲国文字传达乙国文字所包含的那些特点，必须像伯乐相马，要"得其精而忘其粗，在其内而忘其外"。

——傅雷

二、汉语原文

苗绣

苗族服饰中无一支系不以刺绣、织锦或蜡染进行装饰，而且其装饰的精美所反映出的经济水平，大大高于整体各方面的生活现状，特别是节日集会时的盛装，五彩缤纷，灿烂夺目，与吃、住之简陋，形成非常鲜明、强烈的反差。很难想象，在这样奇丽的服饰背后，许多人家竟连床铺被盖尚不齐全，食品常年以酸汤菜为主。

纺织绣染于苗族妇女来说，是能力的体现和审美的体现。妇女的这两种价值体现，与其社会声誉和择偶、婚姻缔结产生直接而重大的影响。故苗族流行一句俗话，叫"人比人，花比花"。所谓"人比人"是指比身材；"花比花"是指比服饰上的织绣染等技艺。其含义是，人与人的比较，既要看外在的美，又要看心灵手巧、勤劳等内在的美。

苗族是以家庭为生产单位的自给自足小农经济，吃的、用的都基本上由家庭自己生产，妇女除参加农业劳动和家务劳动之外，纺织、刺绣或蜡染是她们一生的主要劳作。因此，纺织绣染成为人们评价妇女能力高下的一个主要标准，而这种能力关乎生计，故男青年择偶把它作为一个基本条件。故对不好好学习纺织绣染的姑娘，老人常以"小心嫁不出去"来责骂。苗族姑娘从小就必须认真学习两项技能：一是纺织绣染，二是唱歌。

苗族一百多种服饰类型，其中一部分类型只采用刺绣或辅以织锦作装饰，另有一部分类型则以刺绣、织锦与蜡染并用。从汉文献的记载和服饰发展史的规律来考察，苗族古代服饰应是编织、染绘、刺绣并用，编织、染绘更普遍一些。这由《后汉书》对西汉时期（距今2000年前后）生活在西南地区的苗、瑶、彝等"蛮""夷"民族服饰的记述可以得到说明。书中说到长沙武陵蛮（主要是苗族、瑶族先民）"织绩木皮，染以草实，好五色衣服""衣裳斑斓""椎髻斑衣"，即指染绘装饰工艺。书中所说的"哀牢人……知五彩、文绣、罽氍、帛叠，栏干细布，织成文章如绫锦"，表明编织、染绘、刺绣并存。刺绣工艺是从染绘发展形成的，它是工艺进步和审美要求提高的结果。

图5-1　平绣❶

苗族刺绣在长期的发展过程中，形成了成熟精湛的多种技法和材质的巧妙配搭，以取得更悦目的观赏效果。有些技法非常独特，如扁带绣、梗绣、锡片绣，在其他民族中罕见。各个服饰类型都有一种至二、三种主要的绣法，它与材质、纹样的变化构成各自服饰类型的刺绣风格。有些类型的刺绣，往往由几种技法配合，具有起伏的、变化的丰满表现力。绣法的大致归类，有平绣（图5-1）、锁丝绣、扁带绣、梗边绣、打籽绣、布贴绣、堆绣、锡片绣、挑绣。

纹样可归为三大类：一类是具象性纹样，即接近于写实的纹样；第二类是半抽象半具象纹样，这一类即由几何形线条组成的实物具象；第三类是纯几何形纹样。苗绣的色彩强调对比强烈，不像湘绣、苏绣、蜀绣等汉族名绣那样以色彩的明暗过渡去反映立体影调，其基本上是一色一块面。❷

译前提示：苗绣是苗族人民智慧的结晶，是记载着苗族先民生产生活的无字史书，更是中华文明光彩夺目的瑰宝，为中国乃至世界文化多样性做出了杰出的贡献。本文介绍了苗绣之于苗族人民的重要性，不同地区苗族人服饰的差异，苗绣主要技巧以及基本纹样，帮助读者搭建起有关苗绣知识的基本框架，实现对苗族文化的初步了解与掌握。

三、英语译文

Miao Embroidery

An overview of Miao costumes shows that every branch decorates itself with embroidery, brocade or batik and that the economic level suggested by the exquisiteness of decoration goes far beyond the general living conditions in any other aspect. In particular, the finery worn at festivals or gatherings is so splendid and iridescent as to form a striking contrast with the crudity of their food and shelter. It is hard to imagine that despite such brilliant dresses, many families suffer a shortage of

---

❶ 图片来源：曾宪阳，曾丽. 苗绣：一本关于苗绣收藏与鉴赏的书［M］. 贵阳：贵州人民出版社，2009：39.
❷ 语料来源：钟涛，宛志贤. 苗绣苗锦［M］. 贵阳：贵州民族出版社，2003：3-4.

bedding and eat "sour soup dishes" as their staple.

For Miao women, embroidery, brocade and batik embody their ability and aesthetic taste. Such embodiment has direct and grave impact on their reputation, choice of mates and marriage. For this reason, a popular saying among the Miao people goes, "compare the persons and compare the designs they make". The comparison of persons refers to figure, and the comparison of the designs they make concerns skills in embroidery, brocade and batik. The meaning is that in comparing one woman with another, one should pay as much attention to ingenuity and diligence as to outward beauty.

The Miao people practice a family-based self-contained peasant economy, in which the family produces almost all the food and articles for everyday use on its own. Apart from farming and housework, women spend most of their time weaving, embroidering or making batik. Thus embroidery, brocade and batik are a major criterion for judging a woman's ability. Since it concerns livelihood, young men considers it a basic requirement when choosing their wives. For this reason, girls who will not apply themselves to these skills are harshly reminded of their bleak prospect of getting married by their elders. Since their childhood, Miao girls must learn two kinds of skills well; one is weaving, embroidery and batik, and the other is singing.

The Miao people have over a hundred types of dresses. Some of them are only decorated with embroidery, sometimes supplemented with brocade, and some are decorated with a combination of embroidery, brocade and batik. Han historical records and the pattern of the development of dresses suggest that weaving, dyeing and embroidery were used in combination for ancient Miao dresses, the former two techniques being more widely used. This can be seen in the descriptions in *Book of the Later Han Dynasty* of the ethnic costumes of such "barbarian" peoples as Miao, Yao and Yi in Southwest China in the Western Han Dynasty (around 2,000 years ago). According to the book, "barbarians" in Wuling, Changsha (mostly ancestors of Miao and Yao) "wove with bark fibers and dye the fabric with grass fluids, preferring variegated clothing". The phrases "gorgeous clothing" and "motley dresses" refer to the decorative technique of dyeing. The descriptions also suggest the coexistence of weaving, dyeing and embroidery. Embroidery developed from dyeing as a result of technical progress and higher aesthetic demands.

Over the past centuries, Miao embroidery has developed ingenious combinations of mature and exquisite techniques and a variety of materials for more pleasant effects. Such special techniques as flat band embroidery, stalk embroidery, and tin piece embroidery are rarely used in other ethnic groups. Each dress type has one, two or three primary embroidery methods, which, together with variations in material and design, form the embroidery style of that particular type. Some types of embroidery are highly expressive, with undulations and variations, thanks to the combination of several techniques. Embroidery methods can be generally classified into flat embroidery, lock-silk embroidery, flat band embroidery, stalk edge embroidery, granular embroidery, appliqué embroidery, barbola, tin piece embroidery and couching.

Designs fall into three broad categories. The first is the concrete design, which is somewhat realistic; the second is the half abstract and half concrete design, in which geometric lines are combined to form concrete images; the third is the purely geometric design. Unlike famous Han embroidery types, such as Hunan embroidery, Suzhou embroidery and Sichuan embroidery, in which gradations of colors are used for three-dimensional effects, Miao embroidery features sharp contrast between colors, with single-colored patches.❶

四、词汇对译

华丽的服饰　finery

刺绣　embroidery

织锦　brocade

蜡染　batik

纺织　weaving

《后汉书》　*Book of the Later Han Dynasty*

西汉（时期）　the Western Han Dynasty

审美要求　aesthetic demands

扁带绣　flat band embroidery

梗绣　stalk embroidery

锡片绣　tin piece embroidery

平绣　flat embroidery

锁丝绣　lock-silk embroidery

打籽绣　granular embroidery

布贴绣　appliqué embroidery

堆绣　barbola

挑绣　couching

几何纹　geometric design

半抽象半具象纹样　half concrete and half abstract designs

立体感　three-dimensional effect

五、译文注释

1. 故苗族流行一句俗话，叫"人比人，花比花"。

For this reason, a popular saying among the Miao people goes, "compare the persons and compare the designs they make".

【注释】该例句中涉及苗族流传的一句俗语："人比人，花比花"，是苗家人心目中评价一个姑娘是否优秀的重要指标，"人比人"比的是身材，而"花比花"比的则是织染刺绣等传统技艺。作者在文中引用本句，意在突出苗绣在苗族人生活中的重要地位。译者在翻译本句时并未停留在原句的字面意思，而是将深层次的理解展现在了译文之中。在格式上，译者注意到了原文前后两个小短句之间对称的格式，在译文中用"compare"引领两句，对应起来，保留了原句的语言风格和形式。

2. 书中说到长沙武陵蛮（主要是苗族、瑶族先民）"织绩木皮，染以草实，好五色衣服"。

According to the book, "barbarians" in Wuling, Changsha (mostly ancestors of Miao and Yao) "wove with bark fibers and dye the fabric with grass fluids, preferring variegated clothing".

【注释】本句译者在翻译时基本保留了原文的语序结构，并没有进行大的调整。尤其是在《后汉书》句子的翻译中体现得尤为明显。"织绩木皮，染以草实，好五色衣服"，作者在抓住语句核心意思的基础上，用通俗易懂的语言表达出来，化解了译入语读者对于汉语古文理解的障碍。同时作者也兼顾了语言形式的问题，保留了原文的表达顺序及特色。

3. 有些技法非常独特，如扁带绣、梗绣、锡片绣，在其他民族中罕见。

Such special techniques as flat band embroidery, stalk embroidery, and tin piece embroidery are rarely used in other ethnic groups.

【注释】本例句涉及有关苗族独特刺绣技法的几个文化负载词，分别是"扁带绣、梗绣、锡片绣"。在翻译这些文化负载词时，译者采用了直译的翻译方法。使译文在意义和形式上都贴近原文，使读者有机会了解到苗族文化的原貌。

4. 很难想象，在这样奇丽的服饰背后，许多人家竟连床铺被盖尚不齐全，食品常年以酸汤菜为主。

It is hard to imagine that despite such brilliant dresses, many families suffer a shortage of bedding and eat "sour soup dishes" as their staple.

【注释】本句翻译的难点在于找句子的主干，中文的表达方式"形散神聚"，重语义的内在逻辑，而西方语言却是由严密的语言框架结构组织起来的。原文想表达的是，在绮丽的服饰背后，苗族贫穷的生存状况是难以想象的。翻译到英文中，苗家生活拮据自然而然成为句子的逻辑主语，同时，译者采用"it"作为本句的形式主语，避免了头重脚轻的问题。本句中还涉及一个文化负载词"酸汤菜"的翻译。译者采用了直译的翻译方式，完整地向译入语读者传递了苗族饮食文化信息。

5. 这由《后汉书》对西汉时期（距今2000年前后）生活在西南地区的苗、瑶、彝等"蛮""夷"民族服饰的记述可以得到说明。

This can be seen in the descriptions in *Book of the Later Han Dynasty* of the ethnic costumes of such "barbarian" peoples as Miao, Yao and Yi in Southwest China in the Western Han Dynasty (around 2,000 years ago).

【注释】本例涉及文化负载词《后汉书》的翻译。《后汉书》，作为"二十四史"之一，是一部记载东汉时期历史的纪传体断代史，由中国南朝宋时期的历史学家范晔编撰，与《史记》《汉书》《三国志》合称"前四史"，是中国历史上具有重要地位的一部著作。译者在翻译"后汉书"时，遵循了常规译法，即用直译的方式进行翻译，书名意义清楚明了。译文中用"barbarian"一词如实翻译出了"蛮""夷"等字面意义，且加以双引号表示对历史称谓的再现，并不含任何鄙夷的色彩，忠实再现了原文的含义。如果从民族团结以及文化外宣的出发点考虑，也可以将"蛮"，"夷"进行省译，或用"Miao, Yao and Yi minority group"来代替。

6. 苗绣的色彩强调对比强烈，不像湘绣、苏绣、蜀绣等汉族名绣那样以色彩的明暗过渡去反映立体影调，其基本上是一色一块面。

Unlike famous Han embroidery types, such as Hunan embroidery, Suzhou embroidery and Sichuan embroidery, in which gradations of colors are used for three-dimensional effects, Miao embroidery features sharp contrast between colors, with single-colored patches.

【注释】"湘绣、苏绣、蜀绣"同属中国四大名绣，享誉世界，它们分别是湖南、苏州、四川刺绣产品的总称。在翻译时，译者将地域简称译为人们所常用的全称"Hunan""Suzhou""Sichuan"，这样的翻译使读者对三种刺绣的地域属性一目了然。但如果要保留原语文化特色，使读者更为深入地了解中国的刺绣文化，可以将其分别译为"Xiang embroidery""Su embroidery"或是"Shu embroidery"，再加注进行解释。以湘绣为例，译者可以将其译为"Xiang embroidery"，再在其后进行补充说明，如"Hunan embroidery"或是"embroidery in Hunan province"。

7. 苗族是以家庭为生产单位的自给自足小农经济，吃的、用的都基本上由家庭自己生产。

The Miao people practice a family-based self-contained peasant economy, in which the family produces almost all the food and articles for everyday use on its own.

【注释】原文中，本句由两个分句组成，有各自的主语，前半部分的主语为"苗族"，后半句主语为"吃的，用的"，译者在翻译时用一个复合句处理本句，将"The Miao people"作为主语，将"peasant economy"作为先行词，由"in which"引导非限制性定语从句，以此理清原文内在的逻辑关系，即在小农经济的条件下，人们生产自己吃穿所用。

8. 故对不好好学习纺织绣染的姑娘，老人常以"小心嫁不出去"来责骂。

For this reason, girls who will not apply themselves to these skills are harshly reminded of their bleak prospect of getting married by their elders.

【注释】例8翻译的优秀之处在于它缓和了原文的语气，适度美化了苗族长者的形象。在翻译时，译者并没有直接将"责骂"翻译出来，而是采用了"harshly reminded"这样比较委婉的说法，既体现出了长者的忧虑心焦，又不至于让外国读者觉得苗族家长专断可怖。同样译文在选用主语时，也没有延续原文的"老人"，而是换作"girls"，用被动的方式进行表达，弱化了原文语气。"小心嫁不出去"一句并非特定俗语，并没有词对词照实翻译的必要，在中文写作中，作者许是想幽默或是引起中文读者的共鸣，但是在外宣过程中，可能会引起西方读者的误解和反感，译者将其进行意译，既表达了原文的含义，又柔和了语气，无形中化解了可能存在的问题。

六、翻译知识——文化负载词的翻译

语言作为人类表达和交流思想、情感的工具，是文化中不可或缺的重要组成部分，是文化的重要载体，具有某一语言社会所遗留下来的难以磨灭的具有地域与时代特色的烙印。由于不同语言社会、不同文化群体之间所可能存在的种种差异，某一特定语言中就不可避免地会存在一些标志其背后文化中特有事物的词、词组和习语。这些词、词组或是习语往往"反映了特定民族在漫长的历史进程中逐渐积累的有别于其他民族的独特的活动方式"（蒋继彪，2015），在翻译界被称为"文化负载词（culture-loaded terms）"。

文化负载词具有两个显著特征：其一，文化负载词是某一特定民族所特有的，其二，文化负载词极难在译入语文化和语境中找到能够与其完全对应的存在。王东风曾撰文论及作用于翻译过程的跨文化因素，其中包括以下四类，即词语的文化身份、审美干涉、政治干扰和伦理干预。文化身份是指不同文化中某一特定词汇在附加意义和语用意义上的差异。审美干

涉则涉及特定语言使用群体的审美标准、规范以及个人的审美偏好。政治干扰和伦理干预，则顾名思义，分别指意识形态以及道德观念对于翻译的影响。这些作用于翻译过程的跨文化因素也影响着译者对于文化负载词的翻译。在处理文化负载词的翻译上，译者往往会根据具体情况，灵活运用音译、直译以及综合翻译的方式对其来进行翻译。下面将举例进行具体分析。

（一）音译

原文：

剖帕生最早？剖帕算最老？

剖帕生的晚，剖帕不算老。

哪个生最早？哪个算最老？

修狃生最早，修狃算最老。

修狃生最早？修狃算最老？

修狃生的晚，修狃不算老。

哪个生最早？哪个算最老？

译文：

Is Poupa the earliest and oldest?

Poupa was born later and he is not old enough.

Then who was born the first and the oldest?

It is Xiuniu.

Is Xiuniu the earliest and oldest? He is not.

Then who is the earliest and oldest?

【分析】该小段选自《苗族古歌》中的开天辟地歌，讲述了苗家神兽修狃相关的内容。修狃在苗族文化中，在苗家人心目当中占有十分重要的地位。在苗族神话中更是有"水汽变云雾，云雾生修狃，修狃蛋生盘古"的说法。而这位盘古又是开天辟地的存在，他死后化为了世界万物。所以说，在苗族的神话世界中，倘若没有修狃，就没有盘古，而没有盘古，世界也不会存在，可见修狃在苗族文化中不可或缺、至关重要的地位。在苗家人的观念中，修狃具有与汉文化中麒麟相似的寓意，因而，象征祥瑞的修狃也常常作为纹样出现在苗族刺绣当中。修狃属于苗族文化的特殊符号、特殊意向，其背后蕴藏着独属于苗族的特殊文化传统，因此"修狃"一词也应该理所当然被视为具有特殊含义的文化负载词。在本文段中，译者选用了音译的翻译方法来翻译"修狃"，在译文中极大程度上保留了苗族特有的文化特色，帮助译入语读者最大程度上贴近了原语文化。

例2　原文：蜀绣，又称川绣，发源于四川成都一带。蜀绣讲究针法均匀，色彩细腻，有地方风味。

译文： *Shu xiu*, also called *chuan xiu*, originates from areas around Chengdu, Sichuan. *Shu xiu* emphasizes even stitching, delicate color and local flavor.

【分析】蜀绣又名"川绣"，与苏绣、湘绣、粤绣齐名，为中国四大名绣之一，是利用桑蚕丝在丝绸或其他织物上绣出花纹图案的传统工艺，色彩明丽清秀、针法精湛细腻，具有独特韵味。译者在翻译"蜀绣""川绣"这两个饱含中国独特文化的文化负载词时，采用了直接音译的方式。由于整篇文章都是围绕着中国刺绣展开，且前文已经进行过铺垫，所以译者在这里采用音译的方式并不会使读者难以接受，通过对后半句的阅读，读者可以清楚地了解到，这是一种来自四川成都的刺绣。这样的翻译非但没有给读者造成难以克服的阅读障碍，反而在译入语环境中重现了原语文化特色。

（二）直译

例3 原文：锡绣技法讲究图案的整体布局是否整齐、对称，锡粒制作是否细致，钉绣是否均匀整齐、细密。

译文：Tin embroidery cares about the symmetry of the pattern and the exquisiteness of the tin dots, and whether tin dot lines are in order.

【分析】"锡绣"是贵州剑河地区苗家人独有的刺绣技法，采用彩丝线、金属锡等材料，依据底布的经纬线数纱对称布局，制作诸如"万"字纹或"寿"字纹等相对固定的几种几何图案，做工十分复杂。刺绣时，制作者需要从薄锡片上剪下宽约2毫米的锡片条，并将条头剪成尖角，卷边形成钩状，再根据图案布局用丝线在底布上按定式钉成一个个线套，用小锡钩钩住，然后将这些锡条卷合、剪断、压实，使之成为一个按定式排列，用于拼组图案的小锡粒。锡绣是苗族特有的刺绣技法，是苗族人民智慧具体的、艺术的体现。"锡绣"作为特色文化词汇，在译入语语境中难以找到与之完全对应的存在。译者在处理这样一个文化负载词的翻译时，选择了直译的翻译方法，在未影响读者理解的情况下，极大程度上保留了源语词汇特色，向译入语读者展现了苗族文化。

例4 原文：最著名的可能还要数苏州的双面绣了。人们可以从双面绣作品的任意一面看到图案，甚至可以在两面看到不同的图案。

译文：Perhaps most famous of all is Suzhou's double-sided embroidery, where a single image can be viewed from either side of a piece or even different images on each side.

【分析】双面绣又称两面绣，作为一种中华民族传统工艺，始于宋代。双面绣，顾名思义是经同一绣制过程，在同一块底料上，绣出正反两面图像。双面绣后又发展出双面三异绣，即在刺绣时做到绣品正反两面对应部位图样不同，针法不同，色彩不同，使观赏者能在一幅绣品上欣赏到两个不同图案、不同针法、不同色彩的刺绣艺术形象。例句中，译者在翻译"双面绣"这样一个文化负载词时，采用了直译的方法，在保留原语语言特色和格式的同时，突显了双面绣"在两面刺绣"的特点，使读者可以在第一时间了解到双面绣的核心内涵，便于读者阅读理解。

例5 原文："网绣"有时也称织绣或铺针绣，顾名思义，就是绣品最终成型后，其效果有一种"网织"的感觉。

译文：Fishnet Stitch or Knit Stitch gets its name for the fishnet effect of the works.

【分析】网绣是苗绣中比较有特色的一种刺绣技法，以其成品效果形似网织而得名。译文中，译者采用直译的方式对"网绣"一词进行翻译，在意义和形式上都做到了贴合原文，同时也并未给读者造成阅读障碍，反而使读者产生了一种"顾名思义"轻松顺畅的阅读感受。

（三）综合翻译

1. 音译加直译

例6　原文：蜀绣源于四川，受其地理环境和当地风俗习惯的影响，风格雅致、轻快。

译文：Originated from Shu, the short name for Sichuan, Shu embroidery, influenced by its geographic environment and local customs, is characterized by a refined and brisk style.

【分析】蜀绣属中国四大名绣之一，是四川地区刺绣作品的总称。译者在翻译这个独属中国文化的词汇时采用了音译加直译的方式，形成音译"Shu"与直译"embroidery"的组合。这样的翻译在形式上贴合蜀绣在原语文化中的名字，同时让读者顾名思义，可以在第一时间抓取其内涵意义，此外，"蜀"的音译让读者了解到四川的简称，增进了他们对于四川的了解。

例7　原文：中文的"绣"意味着美丽和宏伟。例如，在广东深圳，人们用"锦绣中华"来描述中国的宏伟壮丽。"锦"指锦缎；"绣"是刺绣；"中华"是中国。"绣"也组成了其他的短语，比如说"绣楼"和"绣球"。

译文：The Chinese word "xiu" implies beautiful and magnificent. For example, name for "Splendid China" in Shenzhen, Guangdong was "jin xiu zhonghua". "*jin*" is brocade; "*xiu*" is embroidery; "zhonghua" is China. "Xiu" is also a part of phrases such as "xiu lou" (embroidery building) and "xiu qiu" (embroidered ball).

【分析】本段中说，由于中国刺绣织锦的华丽，人们也常常用"锦"和"绣"组成其他的词语，来形容美好的事物，比如说"锦绣中华"，当然本身就与刺绣相关的词语中也自然少不了"绣"的影子，比如说"绣楼"和"绣球"。译者在翻译"绣楼"和"绣球"时采用了音译加直译的方式，在帮助读者了解原语文化的同时，起到了辅助读者理解的作用。

2. 音译加意译

例8　原文："饕餮"是夏商周时期青铜器、古玉器、古石器上著名的纹饰。

译文：Ferocious Glutton (Taotie) is an image on very ancient bronze, jade and stone carvings as Chinese call it Taotie.

【分析】饕餮是中国传统工艺中的常见纹样，在苗绣中也有出现。例子中，译者在翻译处理时采用了意译加音译的方式，意译出"饕餮"的内涵，再通过音译的方式传递原汁原味的原语信息，在降低读者阅读难度的同时，达到了帮助中国文化走出去的目的。

从对中国文化负载词的英译分析来看，为传播中国文化，译者往往倾向于使用音译、直译以及综合翻译的方法。音译和直译的方法有时显得生硬，从而影响目标语读者的接受，翻译中加注解释能起到一定的作用。然而，复杂的文化词语、风俗习惯和风土人情并非译者加注解释一下，目标语读者就能完全明白。目标语读者对文化负载词的理解需要结合亲身体验才能更为透彻。尽管如此，音译、直译以及综合翻译等翻译方法一直在中国文化负载词的翻译实践中得到使用。中国文化对外传播的障碍会随着众多译者的努力以及国际跨文化交流的发展逐渐减少。

七、译文赏析

段落1

关于苗绣起源的详细情形我们今天已经很难考证了。不过现在大家比较认同的一个说法是：5000年前苗族人文始祖蚩尤时代的苗族服饰大多是战服模样，处于萌芽状况的刺绣服饰并不普遍存在。关于史书中记载的对苗族"鸟章卉服"特征的描述，应该是到了苗人建"三苗国"之后。至春秋战国时的楚国，苗族的刺绣工艺才有了较大的发展。❶

The exact origin of the Miao Embroidery is extremely difficult to trace today. And the consensus the people have now is that four or five thousand years ago during the Chieftain Chiyou times the people wore military clothing and their embroidery was still in infant period and the current Miao embroidery costumes were non existent then. Historical literature does have a description of the Miao clothing as "bird and flower patterns" and that is pinpointed to Sanmiao Kingdom times about 3000 years ago when the Miao tribes established tribal alliance. Later on in Chu Kingdom time at Spring and Autumn Period (BC 770–476), Miao embroidery skills had developed significantly.

【赏析】段落原文介绍了苗绣的起源。该段落原文篇幅较短，但信息量丰富，语言简洁明了，三言两语便将苗绣的起源交代得清清楚楚。译者在翻译过程中基本遵从了原文的行文顺序，并适当进行调整，整体上流畅、达意。首先，在文化特色词汇的英译上，译者基本采用了异化的英译策略。例如译者在翻译"三苗国"时，采用了音译与直译相结合的方式，将其译为"Sanmiao Kingdom"，将最为原汁原味的表达和其所蕴含的文化以最为直接的方式展现在读者面前。其次，为更好地适应读者的阅读习惯，译者将句式结构进行了相应调整。例如在翻译"……应该是到了苗人建'三苗国'之后"时，译者将最为关键的信息，也就是"三苗国"这个历史节点，借助"be pinpointed to"的短语结构，放在了这个分句的开头；同时为了方便读者更好地理顺时间脉络，译者在"Sanmiao Kingdom times"后面增译了"about 3000 years ago"来进行补充说明；而苗人建立"三苗国"的内容，则通过由when引导的定语从句的形式译出，置于句末。经过灵活的调整，译文结构更紧凑，表意更清楚。最后，从篇章结构上来看，译者基本遵从了原文的行文顺序，从5000年前的蚩尤时代，到3000年前的"三苗国"时期，再到春秋战国时期，由远到近，基本按照历史发展的顺序来进行内容上的呈现。

段落2

服饰是苗族人文化的一个载体，苗绣又在其中占有重要的地位，苗装的制作和穿用也具有宗教活动的性质。正是这样的精神支撑，使得苗族先民宁愿耗费4~5年的时间制作一件服饰。也正是这种虔诚的心态，促成了苗族先民对刺绣工艺的极致追求。苗绣无论在工艺、内容还是内涵上，都远非其他民族可比，这也使得这些古老的苗装苗绣具有不可复制的特质。❷

As one way of expression of the totem on clothing, Miao embroidery occupies an important

---

❶ 语料参考：曾宪阳. 苗绣：一本关于苗绣收藏与鉴赏的书［M］. 贵阳：贵州人民出版社，2009：14.

❷ 语料参考：曾宪阳. 苗绣：一本关于苗绣收藏与鉴赏的书［M］. 贵阳：贵州人民出版社，2009：18-19.

place in Miao's daily and cultural life and the making and wearing of the embroidery costume has a religious implication. So that is the reason the Miao people are happy to spend four to five years to make one costume. With this piety, the Miao people have been in search of excellence for embroidery skills and techniques. As a result, the Miao embroidery excels other embroideries in skills and technique, in contents and meanings. These characters make the old Miao embroidery beyond any possibility in today's world for replicating.

【赏析】段落原文介绍了苗绣的重要地位，它是苗族文化的重要组成部分，是苗族人民的精神寄托，是苗族妇女对于艺术的极致追求，无论是技艺还是内涵方面，苗绣都是独一无二的重要存在。译文简明扼要，通顺畅达，内容丰富，充分传递了原文信息。译者在具体词句的翻译上，没有生搬硬套原文，而是在充分理解原文内容并考虑到中西语言差异的基础上，适当发挥主观能动性进行翻译。例如，译者将"虔诚的心态"译为"piety"。中文经常使用一种复合式的表达（如由定语与中心语构成的偏正词组），来体现结构之工整，内容之正式，但是这种复合式表达往往不够凝练。而英文又是一个崇尚简洁的语言，一些凝练的名词常常可以取代汉语中的复合式结构。译者这样译，更符合英语表达规范和西方读者的阅读习惯。又如，在翻译第一句时，译者对原文结构进行了调整。原文由三个短句组成，主语分别是"服饰"、"苗绣"和"苗装的制作与穿用"，译者在翻译时，首先将句子划分为两个部分，前两个分句为一组与第三个分句用连词"and"相连。第一个分句被译为由介词"as"引出的定语，修饰和突出句子主语"Miao embroidery"。随后将"苗装的制作与穿用"中的"苗装"译为"the embroidery costume"，与前文"苗绣"相对应，从而将该句重点聚焦于"苗绣"之上。内容上，译者也进行了适当调整。如将文中第一句的"文化载体"译为"expression of the totem"即"图腾表达"，体现了译者的个人思考与理解，对原文内容进行了更为具体化的阐释。

八、翻译练习
（一）请将下列汉语句子译成英文

1. 如果把这些苗乡艺人的刺绣作品和现代艺术大师们的旷世杰作相比较，你会惊异地发现，"得意忘形"是他们共同的东西，他们表达的是一种宇宙观，一种认识论，简言之，是一种哲学。

2. 苗绣苗锦特殊的生存环境和传承方式，使一些支系的纹饰保存着许多远古文化和原始艺术的特征。

3. 辫绣不直接用绣线刺之于布面作绣，而是将绣线编成粗于单支线的辫子状带子，盘锁成图案。

4. 刺绣、织锦在色彩的丰富性、鲜亮性和材质、工艺变化表现的质感、立体感等视觉效果及经久耐用性等诸多方面均胜于蜡染的单调、平板，这可能是各型苗族服饰普遍喜欢使用刺绣、织锦的原因。

5. 花溪型现代挑花称为新花，底布用多种色布，绣线色彩也鲜艳且多变化，不像其他类型的挑花，底布色彩和纹样都比较固定、单一。

6. 苗族服饰中的诸多奇幻刺绣、织锦纹样，蕴含着人类童年时代形成的神秘观念，具有若干鲜明的原始文化特征。

7. 由于底布和纹饰主调均采用相近的暖色，显出一派热烈的喜气。

8. 我们不得不叹服每一个苗绣的创造者都是艺术大师、工艺大师，她们没有受过专业的美学美术训练，她们对美的领悟力和创造力是与生俱来的!

9. 苗族没有文字，当你凝视一张绣片、当你试图去了解这个在大山皱褶里生活的民族的时候，当你试图去弄懂苗绣里那些神秘的纹饰的时候……你的思想已经闯入苗的世界。

10. 没有文字的苗族，用刺绣表达了她们对宇宙起源的认识，它厚重的文化内涵，也让它长存千年，世代相传，永不灭变。

（二）请将下列汉语段落译成英文

"辫绣"的技法主要是在贵州雷山地区生活的苗族使用，刺绣时先将剪纸纹样粘贴在绣布上，将所用的彩线（一般是8根、9根或13根）用手工的方式辫织成3毫米宽左右的辫带，再根据图案轮廓要求，按照一定的纹理，由外向内将辫成的辫带平盘绕织盖在剪纸上，用同色彩线将辫带固定，图样铺满即得成品。❶

九、译笔自测

挑花工艺

挑花即挑绣，又称数纱绣。因其工艺简便，易于操作，只需一块平纹底布和绣线就能制作，可卷放在怀里，随身携带，走到哪里就做到哪里，几乎所有苗区都采用，而且绝大部分支系的服饰以其为主要装饰。

挑花必须以平纹布为底布，以便利用其经纬纱交叉的十字点作为坐标施针布线。老的绣件，均为自织棉布或细麻布。细麻布流行在西支，因其旧时的布料为麻纺。现今的绣件有一部分系化纤平纹布。挑花底布不使用绸缎，因其太软而织路又不便数纱。挑花的底布不用垫裱布壳或纸壳，用单层布，从背面朝上施针线，即"反面挑，正面看"，背着针的一面才是正面。有些挑花正反两面区别很小，仅是光洁度不同。

挑花针法用得普遍的是十字针，即绣线以十字交叉组成一个小点，由十字小点构成线条或块面，故这种针法又称十字挑花。十字线的长度可长可短，一般1~2毫米，短的则重叠为一颗点子，看不见十字。除此之外，有齐针、长针技法。齐针即不作十字交叉，针脚的起止点占底布一个纱格，像坐标纸上的格点。长针有点类似平绣，起针至落针把绣线拉得较长。全用长针做出的挑花，与织花很接近，有时难以区分。

挑花多是纯几何纹，或几何纹组成的半抽象半具象纹样。后者包括植物花果和少量动物、人物及用具。由于几何纹组成的客体形象是半抽象半具象的，即使同一个花纹，各地赋予它的名称并不一样，而且就是同一地，各人称呼法也不尽相同。比如回纹、螺旋纹，有的称为蕨菜花，有的称为牛旋，有的称为水旋（即水波纹）。挑花纹样中出现的动物变体主要以蝴蝶、蛙、鸟等为多，尤其蝴蝶，各种几何线变形都能构成，人物出现极少。

---

❶ 该段落语料参考：曾宪阳. 苗绣：一本关于苗绣收藏与鉴赏的书［M］. 贵阳：贵州人民出版社，2009：45.

挑花类型中，贵阳市郊花溪区苗族花型最多。此型服饰的挑花用于围腰、上衣、背带，幅面较大，老辈人的绣件以暗淡花纹的蜡染为底，新一代人的绣件以各种色布、白布为底。针法以十字针为主，纹样中点线较多，空露底面的地方较多。根据当地苗族的称法，此纹样的花型有：猪蹄叉、狗牙、牛牙、蕨菜花、狼鸡草、荞子花、马齿苋、鸡冠花、蝴蝶花、地米花、阳雀花、麦穗花、鱼刺花、蜜蜂、飞蛾、燕子、山雀、鱼、龙、蛙、灯笼、银钗、雪花等等。

黔西北多个型类苗装的挑花所用的十字针针脚间距极短，看不出纵横两针的交叉十字，几乎成一个小点；有的则用齐针，不用纵横十字交叉，也成小点子，每个点如同坐标纸的一个小方格，即1毫米正方大小。满幅绣件以这样密挨的小点组成粗细不同的线条，再由线条构成形状。这类挑花，色彩对比强烈、跳跃，线条极为分明，而且有一定的厚重感，与织花工艺非常相似。

黔东南黄平型挑花在工艺上与众不同。其以青黑或深紫黑自织土布为底面；纹样以红色为主调，辅以一点蓝色，采用长针法，类似平绣，每根线从一轮廓边平直拉到另一轮廓边，平针挑绣完之后，用与施洞型相同的锁绣针法锁一条轮廓边，锁线多为橘红色。一幅绣面由众多细小的几何形组成多视觉的花纹，即把不同的几何纹分成不同的组合观察，得出不同的相似物像。比如合起来看是蝴蝶形状，另合几个几何块又是别的物像。由于每个小几何块锁上边线，图案显得丰满，给人很精细的质感。

都柳江下游岜沙型挑绣工艺也独具一格，而且是苗族挑花中仅有的一种技法。其以青黑土布为底，采用长针针法为主，可以斜拉、横拉、竖拉，不是一律的固定拉法，能形成折曲线、波纹线。绣线以白色为主调，间以水红色、绿色、黄色等点缀，色彩素净而又不呆滞。可作的纹样主要有太阳花、鱼篓、龙纹，三角齿、钉耙（劳动工具）等。

此外还有很多不同风格，篇幅有限，不一一介绍了。❶

十、知识拓展

蜀绣源于四川，受其地理环境和当地风俗习惯的影响，风格雅致、轻快。最早的有关蜀绣的记载出现在西汉时期。在当时，刺绣产品是一种奢侈品，只有皇家才能享受，它的使用被政府严格控制。在汉代和三国时期，蜀绣和蜀锦则被用来交换马匹和清偿债务。

到了清代，蜀绣进入市场，形成产业。工作坊和政府部门全力投入，促进了行业的发展。蜀绣因而变得更优雅，题材的覆盖范围更加广泛。从大师的绘画作品，到专人设计的图案，再到山水、花鸟、龙凤、瓦片、古钱币，似乎都可以成为刺绣的主题。民间故事，如八仙过海、麒麟提子，以及其他吉祥图案，如喜（喜鹊）上眉（梅）梢和鸳鸯戏水也是受人们喜爱的刺绣主题。带有浓厚地方特色的图案在当时很受外国人的喜爱。这些地方特色图案包括莲花和鲤鱼，竹林和熊猫（图5-2）。

---

❶ 语料参考：钟涛. 苗绣苗锦［M］. 贵阳：贵州民族出版社，2003：9.

图5-2　蜀绣[1]

Originated from Shu, the short name for Sichuan, Shu embroidery, influenced by its geographic environment and local customs, is characterized by a refined and brisk style. The earliest record of Shu embroidery was during the Western Han Dynasty. At that time, embroidered products was a luxury enjoyed only by the royal family and was strictly controlled by the government. During the Han Dynasty and the Three Kingdoms, Shu embroidery and Shu brocade were exchanged for horses and used to settle debts.

In the Qing Dynasty, Shu embroidery entered the market and an industry was formed. Workshops and governmental bureaus were fully devoted to it, promoting the development of the industry. It became more elegant and covered a wider range. From the paintings by masters, to patterns by designers, to landscape, flowers and birds, dragons and phoenix, tiles and ancient coins, it seemed all could be the topic of embroidery. Folk stories like *the Eight Immortals Crossing the Sea*, *Kylin Presenting a Son* and other auspicious patterns such as *Magpie on Plum* and *Mandarin Ducks Playing on the Water* were also favorite topics. Patterns with strong local features were very popular among foreigners at that time. These local features included lotus and carp, bamboo forest and pandas. [2]

[1] 图片来源：腾讯网。
[2] 内容参考：*Chinese Embroidery*，来源：中国旅游指南（travelchinaguide）网。

第六章 新时代中国服饰汉译英

学习目标

1. 了解新时代中国服饰发展现状及其文化体现
2. 熟悉新时代中国服饰的基本语汇及其汉英表达方式
3. 掌握语境翻译的相关知识与技能
4. 通过实践训练提高新时代中国服饰文化汉英翻译能力

一、经典译言

翻译是以符号转换为手段、意义再生为任务的一项跨文化交际活动。

——许钧

二、汉语原文

20世纪的"中国风"时尚

20世纪末，国际时装界青睐起东方风格来，东方的典雅与恬静，东方的纯朴与神秘，开始成为全球性的时尚元素。随着中国在世界地位的提高，穿上华服已经成为海内外华人自豪的象征。中国女性自然而然地穿起了中式袄，很多男性也以一袭中式棉袄为时尚。如今的华服，并不完全是纯正的中式袄裙，很多女式华服已经时装化——上身是一件印花或艳色棉布镶边立领袄，下身配牛仔裤和一双最新流行款式的皮鞋，既现代又复古。

2001年，我国香港电影《花样年华》在海内外上映，剧中的女主人公在幽暗的灯光下，不断变换着旗袍的颜色和款式（有二十几种之多）时，人们看到了东方美人的古典气质。剧中人身着旗袍，美丽、优雅而略带忧伤，许多人第一次发现中国传统的服装穿起来竟有如此的神韵。借着电影的魔力，旗袍热再度升温。

也许没有人会想到，在中国举行的APEC会议——一次颇具影响的国际性区域合作的经济和政治活动，掀起了新一轮华服热。2001年秋天的上海，当与会各国首脑身穿蓝缎、红缎、绿缎面料的中式罩衫（图6-1）亮相时，全世界都轰动了。国际媒体纷纷登载了元首们身着华服的合影，并撰文作有关服装的评论。政治家们为华服做了一次最成功的广告，与其说中式对襟袄迷人，不如说是布什、普京等身着华服所带来的巨大效应，商场里就有顾客对着服装导购人员直言要买一件"普京穿的对襟袄"。而APEC引起华服热，还有一个潜在的基础就是蓬勃发展的中国经济。华服热所表现的是中华民族在国际舞台上发挥着日益重要的影响力。

图6-1　APEC领导人礼服 "新中装" ❶

也正是从20世纪90年代开始，国外著名时装品牌纷纷瞄准了中国的消费市场，在北京、上海、深圳、广州等大城市开设专卖店，中国本土的时装品牌和时装模特也逐渐引起了人们的兴趣。而随着1988年中国第一本引进国外版权的时装杂志的诞生，越来越多的报纸、杂志、广播、电视、网络等媒体进入传播时尚的领域，世界最新的流行信息可以在最短的时间内传到中国来。来自法国、意大利、英国、日本、韩国的时装、发式、彩妆潮流直接影响着中国的流行风，"时尚"所代表的生活方式和着装风格已为越来越多的中国人所接受和追逐。

改革开放以后，中国服装设计经历快速发展，出现了真正的时尚产业。高销售量的服装、配饰、化妆品市场与日益强大的传媒业的发展，使越来越多的人得以走近时装、欣赏时装、以时装为美。时装已构成了大众理解并乐于投资的一种生活方式。❷

译前提示：中国近现代服饰风格，在历史的变革中不断演绎出新的变化，本文从社会文化、经济、政治等多角度回顾20世纪的中国服饰，新时代中国服饰在与西方时尚文化接轨后，出现了不同的风格，无论是端庄优美的旗袍，还是标致儒雅的中式华服等，新时代的中国服饰已成为当今国际时尚界不容忽视的时尚元素。

三、英语译文

"Chinese Style" Fashion in the 20th Century

At the end of 20th century, international fashion world began to favor oriental style. Oriental elegance, tranquility, simplicity and mystery became the global fashion elements. With the rising of China's position in the world, overseas Chinese started to feel proud to wear Chinese costumes, with women in Mainland China naturally put on Chinese jackets and many Chinese men consider

---

❶ 图片来源：新浪博客。
❷ 语料参考：华梅. 中国服饰［M］. 北京：五洲传播出版社，2004：150-153。

Chinese cotton jackets as fashionable. Chinese dresses of nowadays are not like those classical traditional Chinese coats or jackets. Many female Chinese costumes adopt the fashionable elements. The costume arrangement looks rather interesting when girls put on print or flamboyant cotton cloth coats with edgings and stand collar, jeans and leather shoes in the latest fashion, with both modern and vintage feel.

In 2001, Hong Kong movie *In the Mood for Love* was played in China as well as many other countries. The actress in the movie changed *qipao* of different colors and styles (more than 20) under the dark lights. Audiences were amazed by the classical charm of the oriental beauty. The actress looked beautiful, elegant and sentimental with *qipao*. People for the first time found that Chinese traditional dresses had a kind of special charm. Due to the magic of the movie, the *qipao* once again was the rage.

Nobody would expect that APEC conference held in China—a very influential political activity, created another round of Chinese dress fashion. In autumn 2001 in Shanghai, the whole world was stirred when leaders from all countries who put on Chinese blue, red or green satin jackets appeared in the public. International media published the photos of the leaders wearing Chinese dresses and wrote articles with comments. The charm of Chinese "button in the middle" style jacket coupling with the huge effect created by Bush and Putin made a very successful advertisement for the Chinese costume. Some customers asked the shop assistant to give them a "Putin" style Chinese jacket. Another background behind this trend is the more and more important influence that China has on the world stage. The Chinese dress rage also signifies the constantly increased confidence and cohesion of the Chinese nation.

It was from the 1990s that many oversea famous brand costumes one after another aimed at Chinese consumer market and opened the monopolization stores in big cities such as Beijing, Shanghai, Shenzhen and Guangzhou. Domestic Chinese brand clothes and fashion models gradually attracted people's interests. Along with the first fashion magazine that used foreign copyrights coming into being in China in 1988, more and more newspaper, magazines, radio stations, television stations and networks entered the fashion promotion field. World latest fashion information could be introduced to China very quickly. The trend of garments in fashion, hairstyles and make-up styles from France, Italy, U.K and South Korea directly influence the fashion trend in China. Life styles and dressing styles represented by "fashion" are accepted and followed by more and more Chinese.

After reform and opening up, China's fashion design has gone through rapid development, which brought about the emergence of a real fashion industry. The markets of garments, accessories and cosmetics that are sold in large scale, together with the development of the increasingly powerful media encourage more and more people to approach and appreciate the beauty of fashionable garments. Fashion has become a lifestyle that people understand and love to invest in.❶

❶ 语料参考：Hua Mei. *Chinese Clothing* [M]. 北京：五洲传播出版社，2004：168-173.

四、词汇对译

东方风格　oriental style

恬静　tranquility

时尚元素　fashion elements

中式袄　Chinese jacket

中式棉袄　Chinese cotton jacket

印花　print

艳色　flamboyant

皮鞋　leather shoes

棉布镶边立领袄　cotton cloth coats with edgings

《花样年华》　*In the Mood for Love*

旗袍　*qipao*

古典气质　the classical charm

神韵　special charm

中式对襟袄　"button in the middle" style jacket

专卖店　the monopolization stores

中国本土　domestic Chinese

时装品牌　brand clothes

国外版权　foreign copyrights

彩妆　make-up

配饰　accessories

五、译文注释

1. 20世纪末，国际时装界青睐起东方风格来，东方的典雅与恬静，东方的纯朴与神秘，开始成为全球性的时尚元素。

At the end of 20th century, international fashion world began to favor oriental style. Oriental elegance, tranquility, simplicity and mystery became the global fashion elements.

【注释】例句原文中的"青睐"是一个汉语词汇，睐：看，青：眼，指人高兴的时候正看着，黑色的眼珠在中间，用正眼相看。在英语中没有直接恰当、贴切的英文词与其对应，在翻译时译者结合汉语的原意表达，"青睐"在原文中指喜爱或重视，所以用"favor"来译既能贴切地体现汉语的意义，又符合英语的表达习惯。

2. 随着中国在世界地位的提高，穿上华服已经成为海内外华人自豪的象征。中国的女性自然而然地穿起了中式袄，很多男性也以一袭中式棉袄为时尚。

With the rising of China's position in the world, overseas Chinese started to feel proud to wear Chinese costumes, with women in Mainland China naturally putting on Chinese jackets and many Chinese men considering Chinese cotton jackets as fashionable.

【注释】汉语构句重意合，句子的线性扩展常常采用意合对接的方式。例句汉语原文句法框架简约，语法范畴模糊，是用多个动词和形容词作谓语平行铺排的句子。但英语构句重形合，译文将句子主语提到明显位置，用介宾短语"with…"、分词短语"putting on…""considering…"等形式明示译文句子各成分之间关系，这样围绕句中主谓结构形成上下递迭前后呼应的势态，逻辑结构严谨。

3. 2001年，我国香港电影《花样年华》在海内外上映，剧中的女主人公在幽暗的灯光下，不断变换着旗袍的颜色和款式（有二十几种之多）时，人们看到了东方美人的古典气质。

In 2001, Hong Kong movie *In the Mood for Love* was played in China as well as many other countries. The actress in the movie changed *qipao* of different colors and styles (more than 20) under the dark lights. Audiences were amazed by the classical charm of the oriental beauty.

【注释】例句原文描述了香港电影《花样年华》中女主人公穿着旗袍的场景，这部电影是一个关于迁徙的爱情的故事，如果将"花样年华"直译为"Life is a flower"，表达不出电影的主题，影响读者感受文章的氛围。译文采用意译策略，将其译为*In the Mood for Love*，传达出电影的内涵，帮助英文读者对电影的理解与接受。同时，译者根据英文表达的需要，在翻译中将由多个流水分句构成的原文句子，分译为三个句子，使表意更清楚。

4. 剧中人穿着旗袍，美丽、优雅而略带忧伤，许多人第一次发现中国传统的服装穿起来竟有如此的神韵。借着电影的魔力，旗袍热再度升温。

The actress looked beautiful, elegant and sentimental with *qipao*. People for the first time found that Chinese traditional dresses had a kind of special charm. Due to the magic of the movie, *qipao* once again was the rage.

【注释】汉语的谓语与英语的谓语存在着较大的差异，所以在翻译时谓语及动词的选择要有利于整个句子的行文和内容的整合。本例在这方面提供了典范。原文的第一句"剧中人穿着旗袍，美丽、优雅而略带忧伤"中除了谓语动词"穿着"，还有几个形容词在句中作谓语。但在英语中没有这种结构。为了使英语行文更为方便、达意，同时也使原文信息传达更为紧凑，例句译文采用系表结构作谓语，搭配介词短语。

5. 2001年秋天的上海，当与会各国首脑身穿蓝缎、红缎、绿缎面料的中式罩衫亮相时，全世界都轰动了。

In autumn 2001 in Shanghai, the whole world was stirred when leaders from all countries who put on Chinese blue, red or green satin jackets appeared in the public.

【注释】汉语中习惯于将详细的内容放在句子的前面进行陈述，将概括性或评注性文字放在句末，英语中则恰恰相反，往往先将该概括性或评注性文字放在句首，而将详细的内容置于句末。例句就说明了这个问题，译文用"stirred"为谓语，采用一般过去时和被动语态，将句子的信息重心传达出来，然后以when为连词，who为关系代词，引导从句。使英语译文通畅自然，既符合逻辑，又具有鲜明的层次感。

6. 也正是从20世纪90年代开始，国外的著名时装品牌纷纷瞄准了中国的消费市场，在北京、上海、深圳、广州等大城市开设专卖店。

It was from the 1990s that many oversea famous brand costumes one after another aimed at Chinese consumer market and opened the monopolization stores in big cities such as Beijing, Shanghai, Shenzhen and Guangzhou.

【注释】译文为强调时间，更好地表达出"正是"的意义，采用了强调句型。英语是形合语言，英语句子往往必须有主语，如果找不到表示事物或人的名词或代词充当主语时，则用"it"作形式主语，这是英语主语显著的重要体现。例句原文为了突出时间概念，在翻译时，以it为形式主语，构成强调句式"it + be + 强调部分 + that..."，统领主句的主语，并将并列的分句连接起来，更符合英语行文逻辑。

六、翻译知识——翻译与语境
语境从宏观来说，就是语言使用的环境，指影响言语交际者交际的各种主观因素和客观因素。语境对于语言的理解是必不可少的。语境可分为语言环境和非语言环境。所谓非语言环境是指语言使用的社会文化物质环境，也称为言外语境或隐性语境。所谓语言环境，也称为言内环境或显性语境，即研究与词、短语、句子乃至更长的语篇同现的上下文。关于语境在翻译中的作用，英国翻译理论家彼得·纽马克认为"语境在所有翻译中都是最重要的因素，其重要性大于任何法规、任何理论、任何基本词义"。从这个意义上来说，语境对于语言形式及其所要表达的意义起着制约和解释作用，同时，也制约着翻译过程中对该语言形式的理解和表达。因此，译者在翻译过程中应充分考虑到语境。

例1 原文：30年代末，一种"改良旗袍"出现了，它使旗袍更西化，采用了西式服装上的胸省和腰省，使用肩缝和装袖，甚至出现了垫肩，人性化的设计理念在旗袍上得到了充分的体现。

译文：Towards the end of the 1930s, a type called "reformed *qi-pao*" appeared, narrowing the gap between the Chinese dressing and the Western style of close fitting to the chest and the waist. decorative sleeve pleats, detachable sleeves and shoulder pads were used. More practical design concepts were demonstrated on the dressing gown.

【分析】汉语民族同英语民族因各自的社会文化背景不同，语言环境与交际情境各异，所以在翻译过程中，可根据具体语境适当转换表达视角，对原语文化信息进行整合，使译文更加符合目的语的语言习惯，以方便目的语读者理解和接受。例句原文中对改良旗袍的描述，使用了"改良旗袍""西化"这些汉语读者容易理解的词语，但在将该句翻译成英语时，则要适当改变原文形式，整合整段信息，"narrowing the gap between the Chinese dressing and the Western style of close fitting to the chest and the waist"就是经过语言结构的整合与转换，采用了更符合目的语语境的表达，以帮助英文读者更好地理解和接受原文所要表达的意义。

例2 原文：到了20世纪70年代初，虽然极"左"思潮仍然统治着服装行业，人们的穿着还受着种种限制，服饰上的清规戒律还没有清除，但是比60年代后半期要好些了，人们开始心有余悸地试探新款式。

译文：The 1970s saw testimony to the above statement. Although various restrictions on dress codes still existed and the leftist influence remained strong in the apparel industry, it was much better as compared with the late 1960s. Some people were starting to test new modes, cautiously of course.

【分析】翻译是在不同语言之间进行交流，而不同语言形式承载着不同的文化内涵。受文化缺失因素的影响，在译文中不可能都一一对应，因此译者要结合具体语境，对部分文化现象进行隐化。本文中的文化现象隐化主要表现在对部分文化现象省略不译或者将原文中的文化现象隐藏。例句原文中"清规戒律"是具有中国文化特色的词汇，它对于中国读者来说很好理解，但对于外国读者来说则难以理解，因为它负载了丰富的文化信息，并且是汉语中习以为常的表达。所以在翻译时为使译文容易理解，需要充分考虑目的语具体语境，译文只需要将"清规戒律"的意义表达出来就好，不需要直译或详细解释该成语的文化背景。

例3 原文：改革开放后的20年，是中国人时尚观念从复苏到逐渐成熟的20年。政治与世俗传统从来也无法压制人们内心对美丽的渴望，虽然这种对美的追求一开始还是不太成熟。

译文： The past 20 years of reform and open door policy has given the Chinese a lengthy period to become awakened to the need for fashion in their lives and to grow mature with concepts relating to fashion. Politics and old traditions have hardly ever succeeded in stifling the longing for beauty although that longing might be immature at the beginning.

【分析】 原文中"改革开放""复苏""世俗传统"都在传达一种新旧更替的感觉，描述中国时尚在新时代有新的发展，这样描写是为了顺应整个文章的语境，体现新旧交替的社会环境。译者结合原文语境，选择"open door policy""awake""grow mature""old traditions"等前后意义相对应的词语来传达原文信息，虽然不能完全对应，但译者有意区分，顺应译语读者的认知语境。这样的翻译保留了原文中对应关系，充分考虑译语结构和表达习惯，顺应目标语境，使译文读者感受到改革开放所带来的革新与变化。

例4　原文："日韩流"现象并不只是一个单纯的流行现象，它让我们看到中国时尚流行模式的变化，时尚不再是一呼百应的全民狂欢。

译文： The popular acceptance of the "Japanese and Korean Influence" is more than just a phenomenon of fashion. It helped to illustrate the changing landscape in China's fashion mode. No longer was fashion the kind of thing that would inaugurate a nationwide carnival upon a call from the leaders of the country.

【分析】 例句中汉语原文由对"日韩流"现象的说明，引出中国时尚流行模式的变化，包含三个流水分句，结构较松散，文化信息隐含在字里行间，并不会影响汉语读者的理解。在将该句翻译成英语时，为了更加准确地传达原文语义，消除文化差异，顺应不同的语境，"显化"翻译策略非常重要。例句译文吃透了汉语原文的精神实质，摆脱了汉语字面的束缚，按照英文表达习惯对句子进行了断句和重组，适当添加代词、连接词，采用倒装句、非限制性定语从句，对原文信息进行整合，使上下文保持连贯，清晰地将原文之意传达出来了。

例5　原文：1978年，中国改革开放之初，正值喇叭裤在欧美国家的流行接近尾声之际，中国的年轻人几乎一夜之间就穿起了喇叭裤，这股流行风尚传遍了全中国。

译文： In 1978, it was about the end of bell-bottomed pants' popularity in Europe and America when China opened its door to outside world. Bell-bottomed trousers became popular among young people in China overnight and then were quickly spread to the whole country.

【分析】 汉英翻译时，译者要善于充分捕捉原文所承载的信息，再现原文的交际情景，让英语读者充分了解原文的语境意义。例句原文用简单的时间线铺垫，描述了改革开放之初喇叭裤在中国流行的现象，句子由若干流水分句构成，是典型的意合语言特征。从情景语境角度来看，原文具有语域标志结构"正值……之际"，是较为正式的表达，比常用的"正当……的时候"的表达方式要正式得多。由此推断，例句原文是学术类、报道类等严肃文体中的句子，情景语境方面属于正式交际场景，而非聊家常似的非正式场合的语言。在汉译英过程中，译者为再现原文的情景语境，使用了同样体现正式语域特点的"it was...when..."的强调句结构，这一结构强调喇叭裤在中国开始流行的时代背景，有助于不了解中国服饰发展历史的英语读者更清楚地了解喇叭裤在中国的流行特点。同时，译文为体现表达的严谨性，用分译的方法将喇叭裤的流行与其所处的背景分译为两句，且英译文第二句将原文的后两个分句进行整合，由 "and" 连接两个谓语。这样，译文清楚再现了原文的逻辑关系，并对原文所包含的信息进行了全面而准确的传递，充分再现了原文的情景语境，有利于读者轻松解读译文。

例6 原文：1978年改革开放后，恰如一夜春风入城来，打开国门的中国人看到了外面的世界。近20年来的时尚流行变化日新月异，节奏越来越快。

译文： Commencement of the reform and open door policy in 1978, like a spring breeze, brought sweeping changes into the Chinese society and enabled the people to see the outside world. Tremendous changes have taken place during the past 20 some years with fashion trends came and gone. The pace change of has become increasingly faster.

【分析】 译文读者的认知语境是译者在翻译时需要考虑的因素之一。例句原文中，为了生动地表达改革开放对中国市场带来的积极影响，以春风比喻改革，"恰如一夜春风入城来"生动形象地表现了改革带来的活力与希望。为充分传递原文信息，译文采用与原文贴近的表达，"like a spring breeze" 既生动，又简单易懂，再现了原文的表达效果。从译文读者的认知语境来看，"spring breeze" 在英文中指一种春天里的柔和的风，结合紧接后面的文字 "brought sweeping changes into the Chinese society"，读者很容易解读出这个比喻的含义，从而理解中国的改革开放为社会生活带来的蓬勃生机与变化。可见，译文在充分考虑译文读者认知语境的基础上，保留原文的意象，形象再现了原文信息。

例7 原文：穿上这套新装，给人以娇小飘逸之感，给19世纪末服装的四平八稳沉着气象，画上了句号。

译文： This new dress does give an elfin and airy impression, putting an end to the overly-secured costumes cast on the folks at the end of the 19th century.

【分析】 例句原文描述了文明新装带给时代的变化，重点突出文明新装对原有服饰风格的改变。原文中的"四平八稳"是一个成语，出自《水浒全传》第四十四回，原指说话做事稳当，现指做事只求不出差错，缺乏进取精神。对于这个文化含义丰富的词语，如果将其文化内涵完全译入英文，对于不太熟知中国文化的读者，则会造成理解上的负担。考虑到译文读者所处的具体文化语境，译文采用了意译的方法，将该成语译为 "overly-secured"，简洁明了，却表达出了词语在本句上下文中的含义，有助于译文读者了解新装对于不合理的旧装的冲击，以体现服饰变革带来的积极影响。这一译法是在充分考虑译文读者所处的文化语境、词语所在的上下文语境，以及读者认知语境的基础上，做出的合理选择。

例8 原文：如今的华服，并不完全是纯正的中式袄褂，很多女式华服已经时装化——上身是一件印花或艳色棉布镶边立领袄，下身配牛仔裤和一双最新流行款式的皮鞋，既现代又复古。

译文： Chinese dresses of nowadays are not like those classical traditional Chinese coats or jackets. Many female Chinese costumes adopt the fashionable elements. The costume arrangement looks rather interesting when girls put on print or flamboyant cotton cloth coats with edgings and stand collar, jeans and leather shoes in the latest fashion, with both modern and vintage feel.

【分析】华服，是指具有中华民族传统服饰特征的服饰，承载着丰富的中国传统文化内涵，有些特征，尤其是具有鲜明时代气息的特征，是只有曾经亲眼见证过中国服饰变迁的中国读者才可以理解的。比如例句中上身着传统华服袄，下身配牛仔裤，脚穿皮鞋的搭配，在不了解中国服饰文化变迁的英美读者眼中可能会觉得荒诞而不可思议，但这种搭配现象在中国的特定时期却真实存在，而且体现出彼时彼刻的时髦感。为帮助读者了解中国文化语境中的这种特定的服饰搭配，译者充分考虑译文读者所处文化语境的差异性，增译了对这种服饰搭配的评价"the costume arrangement looks rather interesting"，为读者解读搭配细节做好心理铺垫，使读者在读完搭配细节之后自然产生如原文所说的"既现代又复古"的感觉，而非荒诞滑稽的感觉，使译文读者跨越文化语境的差异，最大限度地达到对新时期华服的认知。

七、译文赏析

段落1

文明新装由袄和裙组成。上衣为一种大襟紧身的短袄，衣摆呈圆弧形，后来也有较多是平摆的。摆长不盖住臀，衣袖长至肘，袖口一般大为七寸，为喇叭形，也称倒大袖。裙为黑色，裙摆较大，为穿套式，长至足踝，后来逐渐提高到小腿。张爱玲曾在《更衣记》中对这类时装有过生动形象的描写："喇叭管袖子飘飘欲仙，露出一大截玉腕，短袄腰部极为紧小。上层阶级的女人出门系裙，在家里只穿一条齐膝的短裤，丝袜也只到膝为止。裤与袜的交界处偶然也大胆地暴露了膝盖。"穿上这套新装，给人以娇小飘逸之感，给19世纪末服装的四平八稳沉着气象，画上了句号。❶

Consisting of lined jacket and skirt, the new civilized dress, *wen ming xin zhuang*, had a closely fitted top jacket with a single side opening and a curved hem (some later models also had straight cut hems). The hem line usually reached to the point where the hip was not covered. The sleeves were at elbow-length. They were shaped like a trumpet and wider at the cuffs (usually about seven Chinese inches wide). The skirts were black slip-on types with wide bottoms reaching to the ankle at first and to the calf in later models. In her novel *Changing the Dress*, Zhang Ailing had vivid description about this type of dress: "With their trumpet sleeves swinging angelically, and a large fleshy section of the arms exposed, the ladies of the upper class wear short lined jackets tight at the waists. They dress themselves in skirts when outdoors but wear only knee-length shorts at home, though with stockings but reaching up barely to the knee. And occasionally the knees are bravely exposed between the stockings and the bottom of the pants." This new dress does give an elfin and airy impression, putting an end to the overly-secured costumes cast on the folks at the end of the 19th century.

【赏析】段落原文通过精练的语言细致描述了文明新装的特点与风格，字里行间刻画了一个完整的文明新装形象。行文工整，信息量丰富。原文由多个流水短句组成，文化信息

---

❶ 语料参考：薛雁. 时尚百年［M］. 杭州：中国美术学院出版社，2004：64–65.

隐含在字里行间，易于为汉语读者所理解。但英语语言规则与表达习惯不同于汉语，如果拘泥于汉语原文的句法结构来翻译，势必影响英语读者对原文信息的理解。因此，译者在翻译时遵循英语语言规范进行了灵活调整，通过使用现在分词、过去分词、关系副词、短语、从句等形式使译文结构更为紧凑，逻辑更为清楚。比如，在译文倒数第三句中，译者进行了逻辑梳理，增译了一些原文隐含或省略的信息，用连词"when"和"though"分别引导时间状语从句和让步状语从句，使译文长短句结合，表达更紧凑，较好地传达了原文的内容，符合英文的行文习惯和英文读者的阅读习惯。另外，原文中"文明新装"是中国独特的服饰文化，蕴含中国独特的时代气息，译者在翻译时采用了音译加注释的方法，将其译为"the new civilized dress, *wen ming xin zhuang*"，以帮助英文读者更好地理解和接受这些文化信息。

段落2

喇叭裤，是一种立裆短，臀部和大腿部剪裁紧贴收身，而从膝盖以下逐渐放开裤管，使之呈喇叭状的一种长裤。这种裤型源于水手服，裤管加肥用以盖住胶靴口，免得海水和冲洗甲板的水灌入靴子。与之相配的上装则是修身的弹力上衣，呈现为A字形的着装形象。喇叭裤最初在美国塑造了"垮掉的一代"的服饰形象，20世纪60年代末至70年代末在世界范围内流行。1978年，中国改革开放之初，正值喇叭裤在欧美国家的流行接近尾声之际，中国的年轻人几乎一夜之间就穿起了喇叭裤，这股流行风尚传遍了全中国。❶

Bell pants are also called bell-bottomed pants. This kind of trousers has short crotch, thin and tight cut in hip and thigh part. From the knee part, the legs of trousers are widened and loosened, making the trousers look like bell shape. Originally it was a kind of sailor clothes. Loosening the trousers legs was to cover the rubber boots, which was to prevent seawater or deck cleaning water being poured into the boots. The slim fit elastic top takes on an "A" look. Bell-bottomed trousers originally were American decadent style clothes, popular from the end of 1960s to the end of 1970s in the world. In 1978, it was about the end of bell-bottomed pants' popularity in Europe and America when China opened its door to outside world. Bell-bottomed trousers became popular among young people in China overnight and then were quickly spread to the whole country.❷

【赏析】段落原文生动形象地描述了中国喇叭裤的历史由来及其造型。总体而言，原文表达可读性较强、语言流畅。原文第一句是典型的长句，由多个流水分句组成，信息含量大，与英文的句式结构存在较大差异。如果翻译时拘泥于原文的形式，可能会给读者阅读带来一定的困难。因此，译者根据原文语义进行了拆分，将原文第一句译为三句，表意清楚，易于理解。原文中"垮掉的一代"意在完整表达喇叭裤的历史背景，加深汉语读者对这一服饰文化的理解。但"垮掉的一代"源自英文"Beat generation"，译为失去信念的一代。英文读者对此英语文化名词十分熟悉，如果照字面意思将其原封不动的翻译出来，可能会在英美读者心中会产生较为负面的效应。因此，译者对其省略不译，避开文化差异性可能造成的困

❶ 语料参考：华梅. 中国服饰［M］. 北京：五洲传播出版社，2004：145.

❷ 语料参考：Hua Mei. *Chinese Clothing*［M］. 北京：五洲传播出版社，2004：163.

扰，方便读者理解原文的重点信息。最后，为使译文更符合目的语语境，译者按照英文表达习惯，对译文最后一句进行了拆分与重组，使译文层次更清楚，表达更符合英语语言规范。

八、翻译练习

（一）请将下列汉语句子译成英文

1. 牛仔装也是从20世纪70年代末传入中国的，穿着者的队伍不断壮大，从时髦青年扩大到各阶层和各年龄段。

2. 因此，工人、知识分子和在校学生都以一身军装为荣，头戴绿军帽，背一个打成井字格的行军背包，再斜背一个军用书包和水壶，脚穿胶鞋。

3. 采用国产面料和土布，省去复杂的绲边和盘纽，改用拉链，减去领的高度，缩短下摆长度，色调淡雅和谐，从而形成了20世纪40年代旗袍的特色。

4. 随着企业形象设计意识的加强，无论设计人员还是服饰理论界，甚至使用者，都认识到职业装的重要地位和广阔前景。

5. 事实上20世纪60~70年代，我们的服饰不仅色彩单一，而且品种、款式、面料、纹样都处于极度贫乏的状态，所谓"新三年，旧三年，缝缝补补又三年"就是这一时期的典型写照。

6. 20世纪60年代初期，顺应当时简朴实用的着装风格，青年装悄然兴起。这是一种以中山装为基础，进行简化和修改而成的新款式，其基本构造和轮廓与中山装一致，只是将中山装的四个口袋改为一上两下三个口袋，其中上袋为手巾袋，并采用了暗门襟结构，看上去更加简洁精致，受到青年人的喜爱。

7. 电视电影是人们最初十年的主要时尚的启蒙者，爱情电影唤醒了沉睡在人们心中多年的青春涌动，片中人物的一颦一笑举手投足都能引起人们一呼百应的模仿。

8. 蝙蝠袖是一种袖笼底部开得几乎和腰线一般低，像蝙蝠翅膀形状的袖型。当年蝙蝠衫流行时，多半通过港台影视这个媒介传入神州大地，国人的时尚观仅处于被动地接受、模仿、跟风阶段，新奇感和从众心理是蝙蝠衫在中国流行的原因。

9. 20世纪80年代初，随着邓丽君的《甜蜜蜜》《何日君再来》等流行歌曲的风行，男青年从电影中学起了穿喇叭裤，于是这个时期时髦的男青年形象是蓄长发、戴蛤蟆镜，上穿花格子衬衫，下穿紧裹臀部和大腿的牛仔裤。

10. 1932年旗袍从短向长发展，1935年长度达到极致，下摆近地，被称拖地旗袍。

（二）请将下列汉语段落译成英文

2001年APEC会议期间，中式服装大行其道，这种"唐装"是传统和现代的结合品，既吸取了传统服装富有文化韵味的款式和面料，同时又吸取了西式服装立体剪裁的优势。它在继承了传统"马褂"精髓的基础上，取中西合璧的形式，在基本款式、色彩、纹样、织物、盘花纽、手工绲边上都体现了浓郁的中国特色，而面料则不再局限于真丝、软缎、丝绒、织锦缎，而是广泛采用了棉、麻、化纤、莱卡、牛仔布等面料。可以说这种"唐装"是中国的设计师配合现代中国的盛世而创作的洋为中用的改良中装，它的流行含有民族自尊的文化符号，也反映了我们一个文化大国在当今全球化时代十分必要的自我敏感，反映了人们在服饰

文化上的一种新的时代性的需求和回归。❶

九、译笔自测

世界潮流与中国时尚

20世纪80年代中期以后，时装的款式越来越多，流行周期越来越短，时装的款式、面料不断推陈出新，中国已与世界潮流同步而行了。中国人的日常着装有各种T恤衫、拼色夹克、花格衬衣、针织衫，而穿西装扎领带已开始成为郑重场合的着装，且为大多数"白领阶层"所接受。下装如直筒裤、弹力裤、萝卜裤、裙裤、七分裤、裤裙、百褶裙、八片裙、西服裙、旗袍裙、太阳裙等，60年代在西方诞生的"迷你裙"也在那段时间再度风行一时。

20世纪90年代初，以往的套装秩序被打乱了。过去出门只可穿在外衣里的毛衣，因为样式普遍宽松，这时可以不罩外衣单穿，堂而皇之地出入各种场合了。"内衣外穿"的着装风格，经过两三年的时间，已经见怪不怪。过去，外面如穿夹克，里面的毛衣或T恤衫应该短于外衣，但是年轻人忽然发现，肥大的毛衣外很难再套上一件更大的外衣，就将小夹克套在长毛衣外。本来只能在夏日穿的短袖衫，也可以罩在长袖衫外。很快，服装业开始推出成套的反常规套装，如长衣长裙外加一件身短及腰的小坎肩，或是外衣袖明显短于内衣袖。

那段时间，巴黎时装中出现了身穿太阳裙、脚蹬纱制长筒黑凉鞋的形象。太阳裙过去只在海滩上穿，上半部瘦小，肩上只有两条细带；而作为时装出现时，裙身肥大而且长及脚踝。几乎与此同时，全球时装趋势先是流行缩手装，即将衣袖加长，盖过手背；后又兴起露腰装乃至露脐装，上衣短小，腰间露出一截肌肤。这类时装也在中国流行过，但款式没有东邻的日本开放大胆，日本流行的露脐装甚至引发了"美脐热"。而由露腰露脐引发的露肤装，倒是在中国较为广泛地流行开来，还有一种微妙的趋势：将以往袒露的手、小腿等部位遮起来，将原来遮挡的如腰、脐等部位露出来。

在世纪之交的几年间，中国的时装潮流顺应国际趋势，着装风格趋向严谨，特别是白领阶层女性格外注重职业女性风采，力求庄重大方。所谓"原始的野性"，如草帽不镶边、裤脚撕开线等，不再那么受青睐；袒露风开始在一些阶层、一些场合有所收敛，尽管超短裙依然流行，但为了在着装上尽力去表现女性的优雅仪态，很多年轻姑娘穿上了长及足踝的长裙。

与之相映成趣的是，一些时尚青年崇尚西方社会中的反传统意识，故意以荒诞装饰为时髦，如仿效美国电影《最后的莫西干人》的发型，两侧剃光，仅留中间一溜，染成彩色；穿"朋克装"——西方社会继嬉皮士以后，又一颓废派青年装，用发胶粘发成兽角状，黑皮夹克绣饰骷髅等；或将衣裤故意撕或烧出洞。❷

十、知识拓展

当代中国精神与时尚文化、时尚产业具有非常密切的关系。实现中华民族的伟大复兴必

---

❶ 该段落语料参考：薛雁. 时尚百年［M］. 杭州：中国美术学院出版社，2004：126.
❷ 语料参考：华梅. 中国服饰［M］. 北京：五洲传播出版社，2004：146–148.

须弘扬中国精神，时尚文化包含在中国精神之中，与物质文明的生产相伴而行，又总是以其契合人民幸福生活的方式展示中国文化的软实力。将时尚文化与中国精神结合在一起，不仅是时尚文化自身的要求，更是中国精神的具体化，对中国时尚产业朝向高附加值、先进制造业与优质服务业等多产业集群组合发展起到凝心聚力的重要作用（图6-2）。❶

图6-2　现代服饰设计中的敦煌壁画元素❷

Contemporary Chinese spirit is closely related to fashion culture and fashion industry. To realize the great rejuvenation of the Chinese nation, we must carry forward the Chinese spirit. Fashion culture is included in the Chinese spirit, accompanied by the production of material civilization, and always shows the soft power of Chinese culture in a way that suits the people's happy life. The combination of fashion culture and Chinese spirit is not only the requirement of the characteristics of fashion culture itself, but also the embodiment of Chinese spirit, which plays an important role in the development of China's fashion industry towards the combination of high value-added, advanced manufacturing and high-quality service industries.

❶ 内容参考：中国社会科学网。
❷ 图片来源：京艺网。

参考文献

［1］Baker, Mona. *In Other Words: A Coursebook on Translation*［M］. Beijing: Foreign Language Teaching and Research Press, 2000.

［2］Bassnett, Susan. *Translation Studies*. Shanghai: Shanghai Foreign Language Education Press, 2004.

［3］Benjamin, Walter. The Task of the Translator［C］. Venuti, Lawrence（ed.）. *The Translator Reader*. London: Routledge, 1998.

［4］Berman G., *Tang Elite Women and Hufu Clothing: Persian Garments and the Artistic Rendering of Power*［D］. the University of Wisconsin-Milwaukee, 2020: 6.

［5］Bowker, Lynne et al (eds.). *Unity in Diversity? Current Trends in Translation Studies*［C］. Manchester: St. Jerome, 1998.

［6］Chesterman, A. *Memes of Translation: The Spread of Ideas in Translation Theory*［C］. Amsterdam/ Philadelphia: John Benjamins Publishing Company, 1997.

［7］CHO, Dong-il. Encyclopedia of Korean Culture［J］. *Seoul Journal of Korean Studies*, 1992.

［8］Delisle, Jean. *Translation: An Interpretive Approach*［C］. Ottawa & London: University of Ottawa Press, 1988.

［9］Gentzler, Edwin. *Contemporary Translation Theories*［M］. London / New York: Routledge, 1993.

［10］Hatim, B & Mason I. *Discourse and the Translator*［M］. Shanghai: Shanghai Foreign Language Education Press, 2001.

［11］Holz-Manttari, J. *Translational Action: Theory and Method*［M］. Helsinki, Finland: Annales Academicae Scientiarum Fennicae, 1984.

［12］Hu M. The Scholar's Robe: Material Culture and Political Power in Early Modern China［J］. 中国历史学前沿，2016, 11（3）: 339-375.

［13］Hua Mei. *Chinese Clothing*［M］. 五洲传播出版社，2004.

［14］Jacobson, Roman. On Linguistic Aspects of Translation［C］. *The Translation Studies Reader*. Ed. Venuti, Lawrence. New York: Routledge, 2000: 113-118.

［15］Lefevere, A. (ed). *Translating/ History/ Culture: A Sourcebook*［M］. Shanghai: Shanghai Foreign Language Education Press, 2004.

［16］Lullo S. A . Trailing Locks and Flowing Robes: Dimensions of Beauty during China's Han dynasty (206 bc– ad 220)［J］. *Costume*, 2019, 53(2): 231-255.

［17］Munday, J. *Introducing Translation Studies: Theories and Applications*［M］. London/ New York: Routledge, 2001.

［18］Newmark, P. A *Textbook of Translation* ［M］. Shanghai: Shanghai Foreign Language Education Press, 2001.

［19］Newmark, Peter. *About Translation* ［M］. Beijing: Foreign Language Teaching and Research Press, 2006.

［20］Nida, E. A. *Approaches to Translating in the Western World* ［M］. Beijing: Foreign Language Teaching and Research. 1984 /Nida & Taber, 1982: 12

［21］Nida, E. *Language and Culture: Contexts in Translating* ［M］. Shanghai: Shanghai Foreign Language Education Press, 2001.

［22］Nida, E. *Toward a Science of Translating* ［M］. Shanghai: Shanghai Foreign Language Education Press, 2004.

［23］Nord, C. *Translating as a Purposeful Activity: Functionalist Approaches Explained* ［M］. Shanghai: Shanghai Foreign Language Education Press, 2001.

［24］Nord, Christiane. *Text Analysis in Translation* ［M］. Amsterdam: Rodopi, 1991.

［25］Richards, I. *Toward a Theory of Translating* ［C］. Arthur F. Wright (ed.). Studies in Chinese Thought. Chicago: University of Chicago Press, 1953.

［26］Robinson, Douglas. *Becoming a Translator—An Introduction to the Theory and Practice of Translation*, second edition. London/ New York: Routledge, 2003.

［27］Sheng A. The Disappearance of Silk Weaves with Weft Effects in Early China ［J］. *Balkan Journal of Medical Genetics*, 1995, 30(1): 37–45.

［28］Snell-Hornby, M. *Translation Studies: An Integrated Approach* ［M］. Shanghai: Shanghai Foreign Language Education Press, 2001.

［29］Steiner, G. *After Babel: Aspects of Language and Translation* ［M］. Shanghai: Shanghai Foreign Language Education Press, 2001.

［30］Sue, Phelps. Ethnic Dress in the United States: A Cultural Encyclopedia ［J］. *Reference Reviews*, 1997.

［31］Venuti, Lawrence. *The Translator's Invisibility: A History of Translation* ［M］. London/ New York: Routledge, 1995.

［32］Vermeer, Hans. *A Framework for a General Theory of Translating* ［M］. Heidelberg: Heidelberg University, 1978.

［33］Wang C. Conservation study of Ming dynasty silk costumes excavated in Jiangsu region, China[J]. *Studies in Conservation*, 2014, 59(Supp 1): S177–S180.

［34］Wilss, W. *The Science of Translation: Problems and Methods* ［M］. Shanghai: Shanghai Foreign Language Education Press, 2001.

［35］Zhong Tao. *Miao's Embroidery and Brocade* ［M］. Guizhou Ethnic Publishing House, 2010.

［36］《北京文物鉴赏》编委会. 明清金银首饰［M］.北京：北京美术摄影出版社，2005.

［37］《中华文明史话》编委会. 服饰史话［M］.北京：中国大百科全书出版社，2016.

［38］《中华文明史话》编委会. 丝绸史话［M］.北京：中国大百科全书出版社，2010.

［39］巴尔胡达罗夫.语言与翻译［M］.蔡毅，译.北京：中国对外翻译出版公司，1985.

［40］曾宪阳，曾丽.苗绣：一本关于苗绣收藏与鉴赏的书［M］.贵阳：贵州人民出版社，2009.

［41］陈福康.中国译学理论史稿［M］.上海：上海外语教育出版社，1992.

［42］陈宏薇.高级汉英翻译［M］.北京：外语教学与研究出版社，2009.

［43］陈毅平，秦学信.大学英语文化翻译教程［M］.北京：外语教学与研究出版社，2014.

［44］陈友勋.汉英笔译教程［M］.北京：科学出版社，2019.

［45］程永生.汉英翻译理论与实践教程［M］.北京：外语教学与研究出版社，2005.

［46］方梦之.译学词典［M］.上海：上海外语教育出版社，2004.

［47］冯庆华.实用翻译教程［M］.上海：上海外语教育出版社，2002.

［48］何刚强.笔译理论与技巧［M］.北京：外语教学与研究出版社，2009.

［49］胡伟华.新编翻译理论与实践教程［M］.北京：外语教学与研究出版社，2016.

［50］华梅.中国服饰［M］.北京：五洲传播出版社，2004.

［51］黄忠廉，方梦之，李亚舒.应用翻译学［M］.北京：国防工业出版社，2013.

［52］季羡林.季羡林论中印文化交流［M］.北京：新世界出版社，2006.

［53］姜倩，何刚强.翻译概论［M］.上海：上海外语教育出版社，2016.

［54］蒋继彪.文化翻译观下的《伤寒论》文化负载词英译研究［J］.中国中西医结合杂志，2015，35
（7）：877–881.

［55］康志洪.科技翻译［M］.北京：外语教学与研究出版社，2012.

［56］连书能.英汉对比研究［M］.北京：高等教育出版社，1993.

［57］刘敬国，何刚强.翻译通论［M］.北京：外语教学与研究出版社，2011.

［58］刘宓庆.翻译教学：实务与理论［M］.北京：中国对外翻译出版公司，2003.

［59］刘宓庆.新编当代翻译理论［M］.北京：中国对外翻译出版公司，1999.

［60］罗新璋.翻译论集.［M］.北京：商务印书馆，1984.

［61］潘惠霞.商务翻译（英译汉）［M］.北京：对外经济贸易大学出版社，2010.

［62］戚琳琳.古代佩饰［M］.合肥：黄山书社，2013.

［63］秦洪武，王克非.英汉比较与翻译［M］.北京：外语教学与研究出版社，2010.

［64］邵志洪.汉英对比翻译导论［M］.上海：华东理工大学出版社，2005.

［65］沈从文.中国古代服饰研究［M］.上海：世纪出版集团，2005.

［66］沈立新.中外文化交流史话［M］.上海：华东师范大学出版社，1991.

［67］沈周.中国红：古代服饰［M］.合肥：黄山书社，2012.

［68］思果.翻译新究［M］.北京：中国对外翻译出版公司，2001.

［69］谭载喜.翻译学［M］.武汉：湖北教育出版社，2000.

［70］王传铭.汉英服装服饰词汇［M］.北京：中国纺织出版社，2005.

［71］王传铭.现代英汉服装词汇［M］.北京：中国纺织出版社，1996.

［72］王建国.汉英翻译学基础理论与实践［M］.北京：中译出版社，2019.

［73］王克非.论翻译研究之分类［J］.中国翻译，1997（1）.

［74］王宪生.英汉句法翻译技巧［M］.北京：中国人民大学出版社，2013.

［75］王佐良.翻译：思考与试笔［M］.北京：外语教学与研究出版社，1989.

［76］卡拉希帕塔.仙童英汉双解服饰词典［M］.郭建南，译.北京：中国纺织出版社，2005.

［77］肖开容，罗天.汉英翻译教程［M］.北京：中国人民大学出版社，2018.

［78］谢天振.中西翻译简史［M］.北京：外语教学与研究出版社，2009.

［79］薛雁.时尚百年［M］.杭州：中国美术学院出版社，2004.

［80］严勇，房宏俊.天朝衣冠［M］.北京：故宫出版社，2008.

［81］杨绛.失败的经验［J］.中国翻译，1986（5）：23-29.

［82］杨自俭.翻译新论［M］.武汉：湖北教育出版社，1994.

［83］袁杰英.中国旗袍［M］.北京：中国纺织出版社，2000.

［84］臧迎春.中国传统服饰［M］.北京：五洲传播出版社，2003.

［85］张健.新闻翻译教程［M］.上海：上海外语教育出版社，2019.

［86］张今.文学翻译原理［M］.开封：河南大学出版社，1987：8.

［87］赵丰.中国丝绸通史［M］.苏州：苏州大学出版社，2005.

［88］赵刚，杜振东.突破汉英翻译二十讲［M］.上海：华东理工大学出版社，2019.

［89］钟涛.苗绣苗锦［M］.贵阳：贵州民族出版社，2003.

［90］庄译传.英汉翻译简明教程［M］.北京：外语教学与研究出版社，2002.

［91］訾韦力.中国传统肚兜服饰文化（汉英对照）［M］.北京：中国轻工业出版社，2016.

附录1　参考答案

第二章　传统服装汉译英

第一节　深衣

翻译练习

（一）请将下列汉语句子译成英文

1. According to the Confucian *Book of Rites*, the robe must be long and spacious enough to completely cover the body, which is a traditional Chinese ethic.

2. With the development of the Chinese economy, interest in national dress prompted a search for popular, authentic, and intrinsically Chinese styles.

3. *Shenyi* was a form of formal wear for the scholar-officials in Song Dynasty and Ming Dynasty.

4. The scholar-dress *shenyi* was introduced to Korea during Goryeo Dynasty, and it was worn by followers of Confucianism.

5. The *shenyi* grew in popularity during the transition from the Warring States Period to the Former Han Dynasty and its shape deviated further from its earlier description.

6. There have been gradual changes but clear distinctions in the form of the *shenyi* between the early and late period of the Western Han Dynasty.

7. In the Ming Dynasty, in line with the attempt of the Hongwu Emperor to replace all the foreign clothing used by the Mongols of Yuan, and the support of the Chinese elites, the *shenyi* rose in popularity.

8. The lower part of the clothes is made up of 12 panels of fabric sewn together, representing 12 months a year.

9. Front-piece-winding *shenyi* is the variant of *quju shenyi*, and Women in Western Han Dynasty mostly wore this style.

10. *Chanyi* is a single layer garment without lining, which is similar to *shenyi*. It includes two styles—*quju* and *zhiju* that are suitable for both men and women.

（二）请将下列汉语段落译成英文

In the Spring and Autumn-Warring States Period, the royal family of the Zhou Dynasty began to decline, the five powerful lords of the Spring and Autumn Period and the seven powerful states of Warring States Period each did things in their own way. On the one hand, they were competing in the development of production and paid attention to commodity circulation; on the other hand, they annexed the small and weak states and seized the land and wealth of other states. Due to the development and application of iron tools and a large number of skilled craftsmen captured by these states in particular, the communication and improvement between various handicraft industries were promoted. As a result of the industry competition, all the aspects ranging from textile materials to the technology of tailoring and the ornament art had been profoundly influenced. There were a lot of quality dress materials and brands appeared in the market, booming the development of dress style. Meanwhile, the atmosphere of "contention of a hundred schools of thought" among the scholars of the Warring States Period played an important role at that time in promoting cultural and academic development, making fine dress popular in the society.❶

译笔自测

Shenyi and Its Development

Shenyi, a gown with a piece of triangle cloth sewn to its left front, was popular during the pre-Qin Dynasty period. The triangle cloth is rounded to the back of the jacket and tied fast there, in order not to have any part of the wearer's body exposed. This was necessary when underwear was yet to become popular. *Shenyi* was usually made of soft materials with stiff brocade edges, giving them both a strong decorative effect and durability.

Each part of *shenyi* was given an individual symbolic meaning. In the process of making, firstly, the upper part and lower part were tailed separately and then stitched at the waist, which meant respecting the ancestors and succeeding to old traditions. Round sleeves and square collar meant faying in with a rule.There was a vertical line at the back implying the meaning of being an honest person. Horizontal line on the dress tail implied to be impartial.

Shenyi, or deep garment, literally means wrapping the body deep within the clothes.This style is deeply rooted in the traditional mainstream Chinese ethics and morals that forbid the close contact of the male and the female.At that time,even husband and wife were not allowed to share the same bathroom, the same suitcase,or even the same clothing lines. A married woman returning to her mother's home was not permitted to eat at the same table with her brothers. When going out, a woman had to keep herself fully covered. These rules and rituals were recorded in great detail in the Confucian *Book of Rites*.

Shenyi first appeared in Spring and Autumn Period and Warring States Period(770B.C. -221B. C.). It was home-worn dress for scholar-bureaucrat. Owing to its simple design for comfortable wear, it was widely spread among common people. When *shenyi* was replaced by gown-style dress, it gradually disappeared from the stage of history. However, *shenyi* was the prototype of *kuzhe* and *ruqun*.

---

❶ 译文参考：《中华文明史话》编委会. 服饰史话［M］. 北京：中国大百科全书出版社，2009：91.

Before *shenyi* appeared, clothes for ancient people consisted of the upper part and lower part. *Shenyi* was a kind of loose straight dress covering the upper and lower of the body; However, there remained a boundary line between the upper part and the lower part of the body by tailoring halves and sewing together.

By Han Dynasty, *shenyi* evolved into what is called the *quju pao* or curved gown, a long robe with triangular front piece and rounded under hem. In the mean time, *zhiju pao,* the straight gown also known as *chanyu,* was also popular. The wide-sleeved *shenyi* of the Han Dynasty, the round collar gown of the Tang Dynasty and the straight gown of the Ming Dynasty are all typical wide *changpaos*, mainly preferred by the intelligentsia and the ruling class. Time went by, and the *changpao* became a typical garment for those with leisure, as well as a traditional garment of the Han people. ❶❷❸

第二节　胡服

翻译练习

（一）请将下列汉语句子译成英文

1．The "Silk Road" was considered to be the intersection of ancient Eastern and Western civilizations connecting Asia and Europe, and silk was the most representative goods.

2．Some forms of *hanfu* worn in the Eastern Han Dynasty started to be influenced by the costumes of the *hu* people and the gown with round collar started to appear.

3．The reformed martial attire absorbed the style of *hu* dress. The reform included changing the traditional loose costume into short clothes with narrow sleeves for the convenience of shooting arrows, and changing the leggings into pants with crotch and pant legs as a whole. Strapped pants were known as *kun*.

4．Tang costume, which is characterized by cross collar, right lapel and lace up, is a kind of Chinese costume. It is a style in the Han costume system.

5．In the middle and late Tang Dynasty, the traditional aesthetic concept of China was strengthened in the clothing, and began to restore the ancient style, from showing women's figure to the elegant style of wide clothes and big sleeves in the Qin and Han Dynasties.

6．The Xianbei rulers continued to wear own distinctive Xianbei clothing in order to maintain their ethnic identity and avoid merging with the Chinese majority population.

7．In the Warring States Period, the wearing of short upper garment worn by the Chinese which is belted with a woven silk band and had a right-opening also influenced the *hu* dress; it was worn together with trousers allowing greater ease of movement.

8．Tang costume is a modern dress with traditional elements based on the Qing Dynasty jacket

❶ 译文参考：Hua Mei. *Chinese Clothing* [M]．北京：五洲传播出版社，2004：14.
❷ 沈周．中国红：古代服饰 [M]．合肥：黄山书社，2012：59.
❸ 臧迎春．中国传统服饰 [M]．北京：五洲传播出版社，2003：16.

with a stand collar and Western three-dimensional cutting.

9. Some female servants depicted on the tomb mural of Xu Xianxiu appear to be dressed in clothing which looks closer to the Xianbei style garment than the Chinese-style clothing due to the use of narrow sleeves.

10. The Tang capital at Chang'an was the eastern terminus of the Silk Road and therefore was a place where foreigners gathered to sell their goods from the West and to buy merchandise from China for the return trip.

（二）请将下列汉语段落译成英文

To a certain extent, dress may represent a social custom. Both ideology and life convention may be shown through dress culture. The emergence of the Chinese metaphysics and the prevalence of the Taoism and Buddhism not only broke through the world of Confucianism, but also promoted the diversification of the dress cultures. During the period, all the literati preferred to wear loose dress. On one hand, the society had changed, and people no longer advocated the force but began to be indulged in comforts; on the other hand, the literati's psychology was also changed, and they neither cared about the national affairs nor attached importance to rituals.❶

译笔自测

Wearing *hu* Dress and Shooting on Horse

In 307 B. C., King Wuling of Zhao State promulgated the decree of *hu* dress to implement "wearing *hu* dress and shooting on horse", which was an important reform in the history of costume.

Hu dress referred to the dress worn by the northwestern barbarian tribes in ancient China. In pre-Qin and Han Dynasties, costume of the Central Plains tended to be large and loose. The matching clothes were so complicated that it's not only time-consuming to wear but also inconvenient to dress, while the clothes worn by the northwestern barbarian tribes were in narrow shape tied with the waistband and boots. Such costumes were suitable to ride and shoot on horseback.

The state of Zhao, located in northwestern China often engaged with the northwestern barbarian tribes. Facing the rugged valley terrain, soldiers of Zhao lost their original advantage in fighting on chariots. Considering their drawback, King Wuling of Zhao State decided to implement the military reform of training cavalry. The priority was given to costume reform so as to meet the requirement of battles on horseback.

The reformed martial attire absorbed the style of *hu* dress. The reform included changing the traditional loose costume into short clothes with narrow sleeves for the convenience of shooting arrows, and changing the leggings into pants with crotch and pant legs as a whole. Strapped pants were known as *kun*. It was used to protect the thigh and buttocks muscle against much friction on

---

❶ 译文参考：《中华文明史话》编委会. 服饰史话［M］. 北京：中国大百科全书出版社，2009：109.

horseback. No additional skirt was added outside.

He also carried out a series of reform on shoes and hats. *Hu*'s hat was made of the leather or hide to prevent against sand and to keep forehead warm, which was different from the hat of the Central Plains. King Wuling of Zhao State ordered his soldiers to wind the black silk around their heads. Later it evolved into hats. In terms of shoes, *hu* troops wore leather boots which were neither too hard, nor too soft. The upper of the boots reached below the knees to protect the legs and fit for horse riding. King Wuling ordered his soldiers to wear this style of boots and have dozens even hundreds of dazzling bronze bubbles decorated on their upper surfaces to show off his mighty army.

After a series of reform, China's earliest formal martial attire appeared. It gradually developed into armor equipment. With the spreading of *hu* dress, the state of Zhao not only fought many brilliant battles but also inaugurated a new era in China's cavalry history. In the meantime, the reform of the dress weakened the symbolic representation of ranks and social status and strengthened the practicability of the costume, which had profound influence in the later development of costume.

From then on, "learning to wear *hu* dress for convenience" has become the general tendency of China's costume innovation. Han nationality has been absorbing the essence of ethnic minorities to enrich its own costume culture.❶

第三节　龙袍

翻译练习

（一）请将下列汉语句子译成英文

1. The sitting dragon pattern is generally located in the middle part of the chest of the robe, with a curved torso and seven bends, and a dynamic but steady tail, further highlighting the unshakeable supremacy of the ruler.

2. The most outstanding feature of the Chinese royal attire is the embroidered dragon. In Ming and Qing Dynasties, the robe had to have nine dragons embroidered on it—eight on the two shoulders and two sleeves symmetrically, one inside the front lapel, displaying the royal prominence bestowed by the gods.

3. In the Qing Dynasty, the use of yellow became stricter than ever, and a hierarchy of shades of yellow to differentiate ranks within the imperial clan was instituted.

4. The robes were made of rich silk, sometimes in several layers or with silk padding to add warmth.

5. Court boots were made of dark leather and black satin. Starting from Tang Dynasty, when emperor and government officials went to court, they wore court boots.

6. The image of the dragon is interpreted by the technique of silk scales and silk gold and the use of gold threads will bring out the majesty and wealth of the royal family in the overall pattern.

---

❶ 译文参考：沈周. 中国红：古代服饰［M］. 合肥：黄山书社，2012：144.

7. Horse-hoof sleeve tends to be narrow, and wrap the arm tightly. The cuff is tailored in arc shape and its upper part covers the back of the hand for the convenience of shooting, hence called "arrow sleeves".

8. After Qing troops entered *shanhaiguan*, formal attire adopted the design of "arrow sleeves", whose shape looked like a horse hoof, hence its name of horse-hoof-shaped cuff.

9. Summer cape is a kind of decoration on the collar of court dress for emperors and empresses in the Qin Dynasty. It's usually tailored in diamond shape with embroidered dragons and rims as decoration. It's made of silk and satin with dragon and python design.

10. Cape is sometimes sewed on the dress. Sometimes it's separated from the dress, draped on the shoulders and buckled on the neck end.

（二）请将下列汉语段落译成英文

By the Qing Dynasty, the rulers continued the feudal tradition for thousands of years in Chinese history. The Dragon Robe was mainly in yellow, with round neck, right lapel and arrow sleeves, and the golden or the apricot-yellow could be used as subsidiary colors. On the robe, patterns of golden Chinese dragon with five claws and five-colored auspicious clouds were embroidered. Among the auspicious clouds scattered patterns of "twelve symbolic tokens". According to the records of Chinese ancient documents, there are nine Chinese dragons embroidered in the imperial dragon robe. However, only eight Chinese dragons can be found on the imperial robe. Therefore, some people believe that such design was purposely to imply that the emperor himself was counted as the "ninth Chinese dragon". In fact, this "ninth Chinese dragon" was embroidered inside the robe which cannot be seen easily. The imperial dragon robe was the typical dress for the emperor and can better show the emperor's dignity as well as his worship to the "Chinese dragon". In addition, the rulers of the Qing Dynasty also stipulated that except the royal family all others were prohibited from using the design of the Chinese dragon.❶

译笔自测

Dragon Robe and its Origin

Dragon Robe, also named as "Yellow Robe" was the gown-style dress of the ancient Chinese emperors. Starting from the Tang Dynasty, it was for the emperor's exclusive use. Yellow-colored clothes were popular and common people were allowed to wear them before the Tang Dynasty. However, after the Tang Dynasty was established, the yellow color became the royal color, and the yellow robe became the royal costume.

Yellow robe was also known as "Dragon Robe" because there were embroidered dragons on it. Dragon, the head of all beasts, was a fictitious animal created by ancient Chinese. According to the legend, dragon had a snake-like body, fish scales, deer antlers and hawk talons. As the head of

❶ 译文参考：《中华文明史话》编委会. 服饰史话［M］. 北京：中国大百科全书出版社，2009：162–164.

all beasts, it had the capability of overturning rivers and seas, calling up the wind and invoking the rain. Dragon symbolized the emperor in ancient China and any utensils used by the emperor had the pattern of dragon.

The basic form of the Dragon Robe was simple. It was a long robe, reaching to the ankles, with long sleeves and a circular opening for the neck. But the fabric and decoration of the Dragon Robe were complex and rich. The key element of a Dragon Robe was, of course, the dragon pattern. Most Dragon Robes had a large dragon in the center of the garment, with smaller dragons on the sleeves and lower hems. The dragons swam in a sea of intricate patterns, with geometric designs, natural scenes, waves, or other brightly colored figures adorning the lower half of the garment and the sleeves. The robes were made of a lot of silk, sometimes in several layers or with silk padding to add warmth. Occasionally, the robes would include embroidery at the neck fastening or the cuffs.

Emperor Liyuan of Tang Dynasty was the first emperor who wore Yellow robe. He wore it in daily life and banned common people to wear yellow clothes. It was reiterated by Emperor *Gaozong* of Tang in his words "all are forbidden to wear yellow dress except for emperors". Traditional Chinese culture accounted for the reason why Emperor Liyuan of Tang chose yellow-colored robe. In ancient China when the first emperor of a new dynasty was crowned, he had to trace the ruling foundation for his dynasty, that is, following the "the Doctrine of Five-virtue Circulation" raised by an *yin-yang* expert, Zouyan who believed that each dynasty was ruled by five virtues in order—earth, wood, gold, fire and water following one after another in circular. Following this doctrine, Tang Dynasty was established in "earth" of five-virtue circulation which was represented by yellow, thus emperor Liyuan of Tang chose to wear Yellow Robe.

Ever since Yellow Robe was firstly worn by the emperor in Tang Dynasty, it became a symbol of imperial power. After Zhao Kuangyin was made Emperor Taizu of Song in Yellow Robe, anyone who wore Yellow Robe would be punished for treason. Emperor Renzong of Song stated: common people were not allowed to wear yellow gowns or any dress with yellow patterns. Since then, the yellow color and the Yellow Robe were the exclusive use for the royal family. This system has been followed until the demise of Qing Dynasty.❶

第四节　唐装

翻译练习

（一）请将下列汉语句子译成英文

1. The basic design idea of Tang Dynasty costume is a historical necessity, the choice of the times, which has a far-reaching influence on later development of garments.

2. An excellent design can best reflect the features of its national tradition and embrace a strong personality different from other nationalities in terms of the spirit and form.

---

❶ 译文参考：沈周. 中国红：古代服饰［M］. 合肥：黄山书社，2012：21-24.

3．In that period, political stability, developed economy, great progress in production and textile technology and frequent external exchanges prosper the clothing unprecedentedly.

4．The costumes in Tang Dynasty integrated the elaboration of the patterns in Zhou Dynasty, the relaxation in the Warring States Period, the liveliness in Han Dynasty and the grace in Wei and Jin Dynasties.

5．Materialistic abundance and a relatively relaxed social atmosphere gave Tang Dynasty the unprecedented opportunity to develop culturally, reaching its height in poetry, painting, music and dance.

6．In the Tang Dynasty, textile industry made enormous progress. With advancement in in silk reeling and dyeing techniques, the variety, quality and quantity of textile materials reached unprecedented height, and the variety of dress styles became the trend of the time.

7．Compared with the gorgeous dress of the *ruqun*, a full set of men's riding attire on women had its own unique flavor.

8．Although in Confucian teachings long ago it was said that men and women should never cross- dress, women in men's dress were frequently seen in Tang paintings and Dunhuang Grottos.

9．Aristocrats or commoners, at home or going out, many women dressed like this in those days. It is not hard to imagine that Tang was a rather open society as far as women are concerned.

10．The typical men's wear in the Tang Dynasty included the *futou* or turban, round collar jacket and gown, belt on the waist and dark leather boots.

（二）请将下列汉语段落译成英文

During Sui and Tang Dynasties (581–907), *futou* was the main men's wear. After padded with infilling (known as *jinzi*) to make it strong, *futou* didn't look as soft as it used to be. *Futou* underwent great changes in the Five Dynasties(907–960). It looked like a square-shaped hat with two layers on the top tilting in front. In Sui and Tang Dynasties(581–907), hats followed the styles in Northern and Southern Dynasties(420–581).Gauze hats were till popular. *Li mao*, woven by bamboo, palm bark, grass and felt, was commonly used at that time. The round shape *li mao* was very common, which had a big and wide brim to keep away from sunshine and protect people from getting wet when they worked in the field. Costume of Han ethnic group was deeply influenced by Western ethnic minorities in Tang Dynasty (618–907). *hu mao* which was worn by the ethnic minorities was introduced into the Central Plains and prevailed at that time.❶

译笔自测

The Tang Dynasty Women's Dresses

The most outstanding garments in this great period of prosperity were women's dresses, complimented by elaborate hairstyles, ornaments and face makeup. The Tang women dressed in sets

---

❶ 译文参考：沈周. 中国红：古代服饰［M］. 合肥：黄山书社，2012：124.

of garments, each set a unique image in itself. People no longer dressed by their whims, but played up the full beauty of their garment based on their social background. Each matching set of garments had its own unique character, as well as a deep cultural grounding. In general, the Tang women's dresses can be classified into three categories: the *hufu* (or alien dress that came from the Silk Road), the traditional *ruqun*, as well as the full set of male garments that broke the tradition of the Confucian formalities.

The Tang women favored the *peizi* or cape, or as an alternative, a large piece of silk draped over the arms. The difference lies in that the cape was wide and short, and draped over one shoulder of the wearer. The cape can be seen on many clay burial maid figurines unearthed from Tang tombs. There was a story that the Imperial Concubine Yang Yuhuan had her cape blown away onto someone's hat during an open royal banquet. Judging from this story the cape must have been light and thin, although we cannot exclude the possibility that heavier capes made from wool were used in winter to shield the body from the cold wind. The *pibo*, however, is much longer and narrower. Draped over the shoulder from back to front with both ends hanging naturally, it is what we normally call the "ribbon" —a beautiful piece rarely forgotten in classic Chinese paintings.

Footwear to go with *ruqun* includes brocade shoes with tipped-up "phoenix head" toes, and shoes made with flax or cattail stems, all very light and delicate. In addition to images in classic paintings, we are able to see real pieces unearthed in Xinjiang and other places. When wearing the *ruqun*, the Tang women rarely wore hats. Sometimes they wore decorative flower crowns, but when out, they often covered their faces with a veil. This kind of veil hat became the trend in the early Tang Dynasty, but by mid- Tang Dynasty many no longer bothered to wear it, but chose to show their hair buns when they were out riding. There was a large variety of hairstyles at that time, all competing for opulence and extravagance, including over 30 kinds of tall buns, double buns, and downward buns, most of which were named after their shapes. Some of the names of these hairstyles came from ethnic minority groups. As is seen from the beauty paintings of the Tang Dynasty, a full range of ornamental objects was used on the buns, including gold hairpins, jade ornaments, as well as fresh or silk flowers.

Such is the extravagance of the Tang garments. Nowadays people call any front closure Chinese jacket the "Tang costume" as a general term for addressing all traditional Chinese garments. However, the term is used only because people today take pride in those prosperous days. In reality, the modern "Tang costumes" have far less of the luster, extravagance and vitality compared to those of the Tang Dynasty. The grandeur of the metropolis where all nations came to admire made the Tang Dynasty nothing short of the kingdom of garments.❶

---

❶ 译文参考: Hua Mei. *Chinese Clothing*［M］. 北京: 五洲传播出版社, 2004: 30.

第五节　传统裙装

翻译练习

（一）请将下列汉语句子译成英文

1．Tulip skirt was also colored with vegetable dye. The skirt had the beauty of the tulips as well as the fragrance that came out of tulips.

2．A feather skirt worn by a princess in the mid Tang Dynasty was woven with feathers from a hundred birds. As an outstanding piece of work in the history of Chinese costume and textiles, the skirt had varying colors in daytime and at night, under sunlight and under night light, held upright and upside-down. Moreover, images of birds were woven all over the skirt, coming to life in the play of light.

3．Pomegranate had close relations with Chinese costume. Women of ancient times mostly wore skirts in pomegranate-red color, and the dye was extracted from the pomegranate at that time.

4．Emperor felt the wrongs she suffered and released an order: once meeting concubine Yang, all the civil and military officials must bow to her. The one who didn't bow down to her shall be given a severe punishment.

5．The Tang aesthetics was that of suppleness and opulence, like peonies in flowers, men and women with short necks and shoulders, and horses with small head, thick neck and a large backside.

6．Barrel-shaped dress is made of light and thin gauze. The dress shapes a barrel pulling over the head. It appeared in southwestern ethnic minorities, then it was introduced to the Central Plains during Sui and Tang Dynasties.

7．Pleated skirt is a women's dress with many pleats on it. The number of the pleats ranges from several dozens to more than a hundred. The dress is made of several pieces of cotton and silk with same width of pleat and fastened on the waist. It was mostly used among dancers in the court of Sui and Tang Dynasties. It became popular in Song Dynasty.

8．The horse face skirt is the most common and popular dress in Qing Dynasty. There are pleats on both flanks with a glossy satin in the middle, known as "horse face".

9．*Luoqun* was made of a kind of silk gauze and prevailed in Tang Dynasty. In a painting, a lady of Tang Dynasty, with her hair in high buns, wore a round collar dress that exposed the cleavage and a red cappa. She also wore *luoqun* in two intervening colors on lower part of the body. She was dancing stretching her arms.

10．Phoenix tail skirt is made of strips of cloth. Colorful satins are cut into pieces among which two are wide others are narrow. Each strip has embroidery patterns and both sides are trimmed with rolling gold thread or sewn with lace.

（二）请将下列汉语段落译成英文

Zhang Xuan and Zhou Fang, women painters of the Tang Dynasty, were particularly good at portraying opulent and plump women in elaborate dresses. Zhou Fang, in his painting *Lady with the Flower in the Hair*, portrayed a beauty with a long bare-shoulder gown and a big-sleeve leno cape lightly covering the breasts, faithfully describing the fine and transparent material, and vividly revealing soft and supple shoulders and arms of the beauty with a realistic painting technique. The Tang aesthetics was that of suppleness and opulence, like peonies in flowers, men and women with short necks and shoulders, and horses with small head, thick neck and a large backside. In Tang paintings, women tried to show their suppleness by pleating their skirts in accordion form, and raised the waist all the way up to under the armpits, so that the waistline was barrel shaped to show a full and round body contour.❶

译笔自测

Men's Skirts

In Han Dynasty (206B.C.-220A.D.), the skirt was not limited to be worn only by women. It was also a common dress for noble men, thus the idiom "Qun Ji Shao Nian" (young man with skirt and wooden shoes) was a synonym of a man coming from a wealthy family. During the Han, Wei and Six Dynasties, the atmosphere of men wearing skirts was extremely popular and fashionable. *Song Shu · Yang Xin Biography* recorded that Yang Xin had liked calligraphy since he was a child, especially good at official script, and was loved by the famous calligrapher Wang Xianzhi. When Yang Xin was 12 years old, his father was the magistrate of Wucheng (now Huzhou City, Zhejiang), and Wang Xianzhi was the prefect of Wuxing County (his office was in Wucheng). In that summer, Yang Xin "wore a new silk skirt and went to bed at night." At the moment, Wang Xianzhi came to Yang Xin's house to visit Yang Xin's father. Before he left, Wang Xianzhi saw that Yang Xin's new dress was very beautiful, and he wrote several "skirt" on a whim. After Yang Xin woke up, he practiced calligraphy according to the "skirt" scrolls written for him by Wang Xianzhi, and thus his calligraphy level was greatly improved.

The Ming Dynasty replaced the Yuan in 1368. Rulers of the new dynasty spared no effort to restore the traditional style of costumes worn by the Han ethnic majority. A set of institutional rules was enforced to this effect. Garments of the Ming Dynasty, elegant in style and beautiful in design, were rated as exemplary of China's traditional costume art. Portraits were once popular in the Ming Dynasty. Thanks to the realistic, portrayal and vivid rendering of the artists, the image and details of the clothing of that time were vividly depicted. Images most frequently found in portraits were those of government officials and scholars, who wore scholar caps or casual square kerchiefs and long robes, sometimes holding a horsetail whisk. In a Ming tomb found in west of Yangzhou, a full set of scholar's dress was unearthed, among which was a scholar's cap with long hanging ends, a gown with dark rimmed round collar and broad sleeves and high boots made with felt. Similar garment style survived through Beijing Opera costumes, so that we can easily tell a Chinese scholar when

---

❶ 译文参考：Hua Mei. *Chinese Clothing*［M］. 北京：五洲传播出版社，2004：30.

we see one. Influenced by the portrait art, other painting styles followed the same realism when reflecting the lives of people at that time.

Ming was the last dynasty in which men wore skirts. In a famous painting, *the Peaceful Pleasures* by Dai Jing, a Song painter, farmers were still wearing short skirts, regardless of whether they were sitting on the buffalo's back or walking in the fields. This type of pleated, knee-length skirts were tied around the waist, and in the skirts pants or shorts were worn. Nowadays, the basic style of the pleated skirts can still be found in the clothing of some clown figures of waiters in Beijing Opera. ❶❷❸

第六节　官服

翻译练习
（一）请将下列汉语句子译成英文

1. In ancient China, how one dressed was not merely a matter of folk customs, but also an integral part of the State rules on ceremony and propriety.

2. In each dynasty there were clearly defined rules and decrees on the material, color, decorative pattern and style of dress, distinguishing the royal, the civil and military officials and the commoners.

3. Female relatives of the officials also wore elaborately decorated dresses. Inlaid brims were lavishly used in their dresses, complemented by pearls, jade ornaments and embroideries on the hemline, the chest and the sleeve edges.

4. Pleats were fixed with silk threads, and even the sole of socks and shoes invisible to others were covered with embroideries.

5. Out of all details in Chinese official uniform, the *buzi* was the most outstanding feature to mark the relationship of garment and power.

6. These ornament patches had birds and beasts of all kinds, both real and mythical. For civil officials, real birds such as cranes, golden pheasants, peacocks, wild geese, silver pheasants, egrets, larks and quails were used, together with mythical birds that look like a cross between an egret and a peacock.

7. In the *buzi* of military officials, there were recognizable animals such as tigers, lions and panthers, as well as beasts apparently coming out of someone's imagination. Different animals were used to signify different ranks.

8. Because of the imperial intention to govern the country with rules and disciplines, several modifications were made to the dressing code of Song Dynasty which became increasingly complicated correspondingly.

9. Although official costumes from Ming Dynasty were forbidden in Qing society, *buzi* was

❶ 译文参考：沈周. 中国红：古代服饰［M］. 合肥：黄山书社，2012：96.
❷ 臧迎春. 中国传统服饰［M］. 北京：五洲传播出版社，2003：113.
❸ Hua Mei. *Chinese Clothing*［M］. 北京：五洲传播出版社，2004：55.

remained to be used with small changes.

10. Institution related to fish-shaped tallies was not established in Song Dynasty but *yudai* was available at that time.

（二）请将下列汉语段落译成英文

The summer hat was a cone-shaped hat worn in summer time. It tended to be large and flat at the beginning of Qing Dynasty. Later the small and tall shape became popular. Its frame was woven by vines, thin bamboo strips and grass with mounted silk and red gauze lining. The hat had gold edge piping. The top of the hat was adorned with a bead, a feather and strips of red. According to the institution of Qing Dynasty, wearing summer hats started in March every year and summer hats were replaced by winter hats in August.❶

译笔自测

Official Costume and Corresponding Status

Official costume also known as *gongfu* was worn by ancient Chinese officials when they dealt with public affairs. *Gongfu* in early times was a garment of single layer with narrow sleeves, which was easy to engage in public affairs, and was different from the memorial ceremony dress and court dress. According to historical records, it had been the main clothing worn by officials until Sui Dynasty (589–618).

Official dress system in Tang Dynasty (618–907) tended to be perfect. The official dress was gown-style with narrow sleeves. Ranks of officials were distinguished by the color, pattern and accessories, which had profound influence for later improvement of the official dress. Official dress of Song Dynasty (960–1279) was gown-style with round collar, loose sleeve and girded waist with a waistband. Following the style of Song Dynasty, official dress had its innovation adding the flower embroidery. The ranks of officials were seen from the category of embroidered flower and its size. The official dress of Ming Dynasty (1368–1644) was gown-style woven by silk and yarn. It had rotating collar, right front piece and three-Chi-wide sleeve (one Chi equals to 33.33cm). Official ranks were distinguished by colors of the dress, sizes of the embroidered flowers and materials of waistband. From the first to the fourth ranks of officials, they had dark red dress; from the fifth to the seventh, they were in cyan; from the eighth to the ninth, they wore green dress. Color regulations on official dress were abolished in Qing Dynasty (1644–1911). Regardless of the ranks, all in blue, with the exception of celebrations, they wore dark red dress. Official dress of Qing Dynasty consisted of unlined garment and round-collar gown with right front piece.

Colors of official costume were taken as one of the standards to distinguish official ranks. Taking Tang Dynasty as an example, the fourth year of the Zhenguan period of Tang in throne (630), four general colors which corresponded to four levels were adopted in the official uniform system.

---

❶ 译文参考：沈周. 中国红：古代服饰［M］. 合肥：黄山书社，2012：56.

Purple is used for the first to the third ranks. Red is for the fourth and the fifth. Green is for the sixth and the seventh. Cyan is for the eighth and the ninth. In late Tang Dynasty these regulations were updated. The color of official costume for third rank and above was purple unchanged. For the fourth was dark red, for the fifth was light red, for the sixth was dark green, for the seventh was light green, for the eighth was dark cyan and for the ninth was light cyan.

Since purple was seen as a noble color, people who wore purple robe were regarded as being in an important position in Tang Dynasty. Here's some detailed introduction about the gown-style dress during that period of time. Red gown had large sleeves, right front piece with edge piping on the cuff and collar. Green robe was for the sixth and the seventh ranks of officials who were in low position, whereas officials with cyan gown ranked the lowest among all the moderate positions. Later on people referred to "cyan gown" as low-rank officials.

Even though there were strict rules for the official costume in Tang Dynasty, it tended to be flexible when these rules were put into practice. There were some lower ranked officials sent on a diplomatic mission. They were authorized to wear a higher ranking costume. It was referred to as "borrowing purple" or "borrowing red".

Official ranks were reflected by the accessories on their official costumes. Taking Tang Dynasty (618–907) as an example, officials based on their ranks wore waistband in different materials such as gold, jade, rhinoceros horn, silver, stone, copper and iron. Their positions were also distinguished by the ornaments on the sashes.

Ancient China boasted lots of accessories. *Peishou* (ribbons hanging jade ornaments) was the most representative one among all accessories. *Pei* referred to jade ornaments such as large jade and groups of jades. *Shou* referred to the ribbons hanging jades and seals. To distinguish hierarchy by *peishou* had been a distinctive feature for the institution of ancient Chinese ornaments. *Peishou* appeared prior to Qin Dynasty. As an institution of ornament, it's passed down with some changes in the color and shape. The institution related to *peishou* wasn't terminated until hat wearing system appeared in Qing Dynasty.

In addition, jade pendant as an essential ornament on formal attire for ancient aristocrats and officials was restricted by many—layered hierarchy. Wearers' identity, shape of jade pendant and the way to wear were clearly defined.❶

第七节　戎装

翻译练习

（一）请将下列汉语句子译成英文

1. History books of that time had recorded the exchange for a paper suit with several sets of fine metal suits, thus validating our estimate of its high quality and value.

---

❶ 译文参考：沈周. 中国红：古代服饰［M］. 合肥：黄山书社，2012：40–52.

2．A protective silk collar is attached for shielding the neck and the ears, decorated with fine embroidery and metal tacks.

3．Incessant wars through the ages led to even greater development in armor suits in the Wei, Jin, Southern and Northern Dynasties.

4．The armor is in two pieces on front and back, covering the chest and the back respectively similar to a vest that runs down to below the belly.

5．*Mingguang* armor is one with round metal plates protecting the chest and the back, worn with a leather belt and wide trousers.

6．The structure of double-layered armor was improved from the previous period with smaller fish-scale chips, and extended to the belly so that leather armor skirt was no longer necessary.

7．There were thirteen types of armor suits designated as official army wear, made with materials from copper to wood, leather and cloth.

8．Iron and leather suits were used in actual wars, whereas decorative armor suits made with silk and cotton, visually pleasing as they were, were used as daily wear or ceremonial suit for generals.

9．As a result of cold forging process, when plates were forged above one third of the original thickness, the end of the plates would leave the wart-like piece, which had the similar size to the head of chopstick, hence its name.

10．By the Ming Dynasty, cotton armor was widely used in the army, which was suited to the battlefields where enormous amounts of firearms were used.

（二）请将下列汉语段落译成英文

Specific armors should be designed for certain types of soldiers. The two or three soldiers within a chariot did not need any great mobility, and so their armor could be heavier and more cumbersome but with the benefit of offering greater protection. The armor could cover all the body and provided the arms with free space to wield weapons such as lances and halberds (a mix of axe and spear). Infantry, meanwhile, had only short tunics and more basic leg protectors which allowed them to move quickly across the battlefield. Cavalry, which began to replace chariots from the 4th century BCE, were traditionally lightly armed with halberds and bow in order to move freely and fire from their primitive saddles while still on the move. Thus, their clothing had to be light and less restrictive.❶

译笔自测

Ancient Chinese Armor of the Qin Dynasty

The Qin Dynasty surfaced after China's Warring States Period. Through many years of war, after conquering rival kingdoms, its first emperor, the absolute monarch Qin Shi Huang unified

---

❶ 译文参考：世界历史百科（World History Enayclopedia）网。

China and established the Qin Dynasty.

During the Qin dynasty, Chinese warriors wore elaborate suits of armor, each one consisting of more than 200 pieces. Much of what historians know about this armor comes from the roughly 7,000 life-sized terracotta warriors found in the mausoleum of Emperor Qin Shi Huang (260 to 210 BCE), which appear to be modeled onto distinct, individual warriors. The Terracotta Army—discovered in 1974 near the city of Xi'an—includes armored infantry, cavalrymen, archers, and chariot drivers. Analysis of the figures reveals much about the ancient Chinese military.

The Qin dynasty dominated the modern-day states of Gansu and Shaanxi from about 221 to 206 BC. The state was the result of several successful conquests during the Warring States Period, which allowed Emperor Qin Shi Huang to consolidate his kingdom. As such, the Qin was known for its powerful warriors. Those above the rank of common soldiers wore special armor made of thin leather or metal plates. Infantry wore suits that covered their shoulders and chest, cavalrymen wore suits that covered their chest, and generals wore armored suits along with ribbons and headdresses.

The armor seemed to have been riveted together and then connected by binding or sewing. The lamellae were small plates (around 2 × 2 inches, or 2 × 2.5 inches) made of leather or metal with a number of metal studs in each plate. In general, larger plates were used to cover the chest and shoulders, and smaller plates were used to cover the arms. For additional protection, some warriors wore extra garments on their thighs in addition to the pants under their coats. There are also some soldiers who would wear knee protectors. Archers may sometimes need to use kneeling posture in need of shooting arrows, so they would wear knee protectors.

The garments on the Terracotta Army were originally lacquered and painted bright colors, including blue and red. Unfortunately, exposure to the elements—air and fire, for example—led to the colors flaking off and getting bleached or discolored. The faded spots remained. Historians are not sure whether Qin soldiers actually wore such bright colors or if the figures of the Terracotta Army were merely painted for decoration.

Qin armor itself was relatively simple in design. Whether a suit covered the chest, shoulders, and arms or only the chest, it was made of small, overlapping scales. To distinguish themselves from lower-ranking soldiers, military leaders wore ribbons around their necks. Some officers wore flat caps, and generals wore headdresses that resembled a pheasant tail.

None of the soldiers in the Terracotta Army carry shields; however, historians believe that shields were used during the Qin dynasty. The soldiers used a variety of weapons, including bows, spears, lances, swords, daggers, battleaxes, etc.. Even among the swords, there was great variety—some were straight like broadswords while others were curved like scimitars. Many of these weapons were made of bronze; others were made of an alloy that included copper and other elements.

Apart from the Qin soldiers' neatly combed and parted hair, their mustaches were exquisite, too. They split their hair in the middle and weave it into a delicate braid, or in a bun on the right side of their head. Sometimes, they wore leather caps, most noticeably on the mounted cavalry. The

Qin soldiers seldom wore helmets. The horsemen used saddles, but no stirrups, and wore, over their leggings, coats that historians believed were shorter than those of Qin foot soldiers.

Generals wore ribbons tied into bows and pinned to their coats in a number of different places. The number and arrangement of bows indicated each general's rank; a small difference in clothing could be the equivalent of the difference between four- and five-star generals.❶

第八节　旗袍

翻译练习

（一）请将下列汉语句子译成英文

1．The basic style of *qipao* is a standing collar, right-side-buttoning front, long sleeves, and straight tailoring from top to bottom only with a wider lower hem but no vents.

2．*Qipao* is a high-necked, close-fitting dress with the skirt slit partway up the side.

3．*Qipao* as known as the mandarin gown during the 1920s and 1930s, then it was modernized by Chinese socialites and upper-class women in Shanghai.

4．*Qipao*, also known as cheongsam in Cantonese, is a one-piece Chinese dress that has its origins in Manchu-ruled China back in the 17th century.

5．In the homeland of the Long, *qipao*, the traditional Chinese dress, is just like a beautiful legend singing praises of the quintessential, profound connotation in Eastern culture of clothing and accessories, and links together all the Chinese people who love China.

6．There were rich colors used in the gown worn by men during the Qing Dynasty, such as bluish white, lilac, lake blue, silver grey and burgundy as well as the popular pale blue.

7．In the 1920s, the cheongsam was modernized and became popular among celebrities and the upper class in Shanghai.

8．The original *qipao*, worn during the Manchu rule, was wide and baggy.

9．The *qipao* was traditionally made of silk and featured intricate embroidery.

10．The modern *qipao* is a one-piece, form-fitting dress that has a high slit on one or both sides.

（二）请将下列汉语段落译成英文

The fashion prevalent in the 1920s clearly reflected that national differences could hardly distinguish between the garments worn by the Manchu and the Han peoples. Casual menswear at that time included four categories: the European style of Western dress; *zhongshan* suit and student uniform; long gown with *magua* or *kanjian*; and Chinese-style jacket and long trousers. At the beginning of the Republic of China womenswear had two categories: Chinese-style jacket and skirt; and *qipao*. But the latter became a popular fashion only in the middle of the 1920s. This also

---

❶ 译文参考：*How the Qing Dynasty Unified Ancient China*, 来源：Thoughtco教育资源网；*Ancient Chinese Armour of the Qing Dynasty*, 来源：Thoughtco教育资源网。

reflected the laws of fashion trends—its up and down cycle. The men's gown changed mainly as follows: decorative accessories greatly simplified; loose waistline and cuffs; moderate length; collar 5.5 cm high in front and 8.0 cm in the back; lining; wider lower hem and higher slit openings to show the inside long trousers made of satin fabric. In the early 1920s the body was much longer, and later became shorter, sometimes just 6 to 12 cm below the knees. The waistline and lower hem became narrower and the position of the slit openings lower as there were no long trousers worn within. The style of collar also changed to become much lower, being 1.5 cm high in front and 2.6 cm high in the back. Later, the changes in style in women's *qipao* were mostly found in the collar: the fashionable "harness" style of 10.5 cm high to partially hide the cheeks in order to present the sophisticated female attraction. ❶

译笔自测

Qipao

Qipao(Qi is Chinese for banner, and *pao*, gown) was originally the common dress worn by the Manchu women of Eight Banner (the Manchu organized all people in a military fashion, namely Eight Banner) around the time when Manchu military forces entered into China proper by route of the Shanhaiguan Pass in the northeast, and *qipao* was also worn by the Mongolian women. The basic style of *qipao* is a standing collar, right-side-buttoning front, long sleeves, and straight tailoring from top to bottom only with a wider lower hem. Usually there was floral embroidery around the collar, front and bottoms of sleeves and lower hem. This kind of *qipao* was at that time a fashion in the north of China, while most of the women of south China still wore Chinese-style jackets and skirts inherited from the Ming Dynasty. After more than three hundred years, Mongolian women still wore gowns fastened with sash for the convenience of horse riding.

After the Revolution of 1911, female students adopted the fashion of wearing jackets in white or green and short skirts in black, but ordinary women whether in the north or the south used to have their simplified *qipao*, especially the cotton-padded *qipao* as a must against the bitter winters.

Looking back to the 1920s, during my elementary and junior-middle school education in Beiping (now Bejing), the school uniform was a white jacket and black skirt, but when I was at home or receiving guests on holidays or visiting relatives with my parents, I had to wear *qipao* which used to have a wider lower hem without a slit in a pagoda-like form, and the sleeves were also in a trumpet style with wider cuffs. If this *qipao* were in white or other solid color, the collar, the front and the lower hem had to be trimmed with small-floral embroidery appliques. At that time there were already cheap machine-made embroidery appliques instead of hand embroidery. If the fabric had decorative pattern, this suit of *qipao* must then be trimmed with piping in dark color. Sometimes, my mother and other middle-aged female relatives were also in fashionable long skirts and jackets, but

---

❶ 译文参考：袁杰英. 中国旗袍［M］. 北京：中国纺织出版社，2000：59-60.

most of the time they wore *qipao*.

Many changes took place in the tailoring methods of *qipao*. In the 1930s when my family moved to Shanghai, from magazines, pictorials, movies, calendars, and advertising cards in cigarettes packs, I discovered a lot of changes in the style of *qipao*. For example, the long sleeve was shortened, even up to the shoulder in a sleeveless style. The dress length was extended even down to the heel, and only by wearing high-heeled shoes could *qipao* not drag on the ground. The most outstanding change in tailoring was the bias of *qipao* began to closely follow the curves of the graceful female figure with a narrower lower hem and right as well as left side-slits up to the knees. This kind of *qipao* was prevalent first in Shanghai's upper-class society, and undoubtedly it had been influenced by the Western-style evening dress with the feature of exposing the curves of the female figure curves.

In 1934 when I was studying at the Art Department of Nanjing Central University, I began to be interested in designing and making dresses on my own. As I remember, that summer I wore a *qipao* I made myself, which was without a standing collar and right-sided-buttoning front, only with neck-baring round collar with piping and short sleeves with slits on the shoulders and vents around the knees. It was a dress that was easy to cut and sew and also let the wearer feel cool. Up to now in the 1990s, it is still the simplest and commonest style prevalent in the summertime both in China and foreign countries. After the 1940s the fashionable length of *qipao* became shorter and shorter, even going above the knees and it was, of course, influenced by the mini skirt.

After World War II, with the increase of visits and dialogue between different countries as well as international cultural exchanges, fashion trends have become globalized. In recent years fashion shows with Italy and France playing the lead and the fashion trends in Europe and America have all explicitly presented the influence of *qipao*. The fashionable style of evening dress is basically the close-fitting gown, and even the casual long skirt has taken on a style of narrow lower hem with slits in the back or on both sides.

For the past 50 years, the standard garment for Chinese women in Hong Kong and Taiwan has been *qipao* which is also the formal dress for special occasions such as the weddings, funerals, and diplomatic activities. In comparison with the style of *qipao* in the 1930s and 1940s, the most popular change has been the height of the slits rising nearly to touch the buttocks. Since the 1980s in Mainland, China with the implementation of the reform and opening up policy to the outside world, *qipao* has been the fashionable dress for those women who host special performances or ceremonies or those receptionists in high-class hotels and restaurants.

The present *qipao* has made its own great developments from its original style in the Qing Dynasty. Taking the traditional style as the foundation China's newly emerging garment industry and the top fashion designers of the new generation will naturally integrate the quintessence of the Chinese art of clothing and accessories to create a more beautiful Chinese *qipao*.

As is said above, national dress is also a kind of cultural packaging which can take the role of communicating with other peoples, just as the language takes the role of presenting people's thoughts and feelings. As one may feel emotional and nostalgic upon hearing his own mother tongue, people can be led to the meaning of Chinese culture by *qipao*'s basic style which brings about some intimacy as if from mother tongue.❶

第九节　中山装

翻译练习

（一）请将下列汉语句子译成英文

1．For 55 years, Hongdu, a time-honored domestic fashion brand, has been providing *zhongshan* suit tailoring service to generations of China's top officials and foreign dignitaries.

2．If you walked down the street in China a couple of decades ago, you would more than likely see both men and women wearing the *zhongshan* suit, the Chinese version of a Western business suit also known as the *mao* suit.

3．After the establishment of the People's Republic of China in 1949, the *zhongshan* suit became widely worn by males and government leaders as a symbol of proletarian unity and an Eastern counterpart to the Western business suit.

4．The *zhongshan* suit, which was also known as the *mao* suit, was first introduced by Sun Zhongshan, the leader of China's 1911 Revolution and was later popularized by Mao following the founding of the People's Republic of China.

5．With the country growing in importance and influence on the international stage, more people were returning to Chinese culture and the *zhongshan* suit.

6．In China, symbolizing solemnity, generosity and steadiness, which were characteristics that leaders need, the *zhongshan* suit was considered a most suitable dress for leaders.

7．The eye-catching dark blue suit, slim-cut with a standing collar, is a simplified and redesigned "*zhongshan* suit".

8．Instead, it is a type of modified Chinese standing-collar outfit. The entire design goes with Chinese style, but some subtleties are tinged with a modern tailoring spirit.

9．The tailors, originally from Ningbo, Zhejiang province, were renowned in the early 20th century for their skills in sewing the *zhongshan* suit and Western-style suits.

（二）请将下列汉语段落译成英文

The *zhongshan* suit became popular with western suit for its conciseness and utility in the 1911 Revolution. The government of Republic of China issued a public order that the *zhongshan* suit was the formal dress. Minor stylistic changes of the *zhongshan* suit were developed, and a revolutionary and patriotic significance was assigned to it. Four pockets, five center-front buttons, three cuff-buttons on each side, all of these own their special meanings from a Chinese ancient classic *Guanzi*,

---

❶ 译文参考：袁杰英. 中国旗袍［M］. 北京：中国纺织出版社，2000：序言.

the Constitution of the Republic of China, and the Three Principles of the People proposed by Mr. Sun Zhongshan.❶

译笔自测

Zhongshan Suit

The *zhongshan* suit is also called the *mao* suit, was designed by Sun Zhongshan on the basis of widely absorbing Western-style suits and integrating the characteristics of Japanese student uniform and Chinese native clothing. It was very popular in the Republic of China (1911–1949). After Sun Zhongshan's death in 1925, the *zhongshan* suit was assigned a revolutionary and patriotic significance.

The style of the *zhongshan* suit was basically formed in the early 1920s. Since the 1920s, although there have been individual changes in the style of *zhongshan* suit, the overall change is not big. The main change is to turn seven buttons to five buttons and remove the belt and slit at the back. *Zhongshan* suit has four pockets, five big central buttons in the front, and three smaller buttons on each sleeve. Sun Zhongshan also gave this Chinese suit very important meaning. The one-piece suit represents the unification of China; The turn-down closed collar, showing the strict and meticulous concept of governing the country. The four pockets were said to represent the Four Virtues cited in the classic *Guanzi*: Propriety, Justice, Honesty and a Sense of Shame. The five center-front buttons were said to represent the five branches of the government—legislation, supervision, examination, administration and jurisdiction—cited in the constitution of the Republic of China and the three cuff-buttons to symbolize Sun Zhongshan's Three Principles of the People: Nationalism, Democracy, and People's Livelihood. After the founding of the People's Republic of China, due to the shortage of materials and manpower, some simplification and improvement have been made to the traditional *zhongshan* suit, which has been popularized nationwide.

Zhongshan suit has many advantages, mainly including balanced and symmetrical shape, elegant and dignified appearance and excellent comfort, etc.. It also highlights men's calmness and fits for a wide range of uses (suitable for both formal and casual occasions). The color of the improved *zhongshan* suit is very rich, in addition to the common black, white and gray, there are camel blue and so on. On different occasions, the choice of color is not the same. The color of the *zhongshan* suit used for formal dress should be solemn, and the color can be bright and lively when it is used for casual dress.

In recent years, the *zhongshan* suit has become more popular. Many successful people like to wear white or beige *zhongshan* suits which look calm and stylish when attending large-scale business activities. The revival of the *zhongshan* suit is attributed to famous movie stars and artists. In every large-scale event, Jackie Chan, the international movie star, always appears in front of the media

---

❶ 译文参考：华鼎中国旅游（Top China Travel）网。

and audience in a *zhongshan* suit, and greatly promotes the Chinese suit culture on the international stage. Director Zhang Yimou also loves the *zhongshan* suit. He often wears a black *zhongshan* suit to appear in major film festivals around the world. Liu Dehua, the singer, is often dressed in black, white and gray *zhongshan* suit on the stage.

There are many traditional costumes in China. Nowadays, people pay more and more attention to cultural life. These costumes are endowed with more significance and new vitality. Traditional culture is thus better preserved and displayed in variety of forms.❶

第三章　传统服饰品汉译英

第一节　头饰

翻译练习

（一）请将下列汉语句子译成英文

1．He was wearing a scholar's square cap, a royal blue forged straight dress, and a pair of boots with black uppers and white soles.

2．*Huadian*, ancient women's fashionable ornaments, prevailed in Tang Dynasty. It mainly had red，yellow and green colors varied in shapes.

3．When spring came, ladies in Chang'an, the former capital of the Tang Dynasty, would compete with each other in flower wearing. The one who wore the most beautiful and diverse flowers would rank the first.

4．The difference among *chaoguan*, *jifuguan* and *changfuguan*, three types of hats, lies in the top adornment for the hats. The top of *chaoguan* mostly has three levels, more specifically, the upper part was a spiny gem, the middle a round bead, and the bottom a metal pedestal. *Jifuguan*'s top, which is simpler, normally has two levels including a round bead and a metal pedestal. *Changfuguan* only has a red wool-weaved round bead at the top.

5．People started to use silk and colored paper to make artificial flowers, considering that the real flowerers withered easily. Artificial flowers were welcomed by women because they were delicate, lifelike and durable.

6．Gold hairpins varied mainly in the head of the hairpin. Often, there were ball-shaped, flower-shaped, phoenix-shaped, fish-shaped, butterfly-shaped and *ruyi*-shaped hairpin heads.

7．Peacock feather as a symbol of high rank was exclusive for high rank officials. Punishment of removing the peacock feather was regarded a felony.

8．*Gugu* coronet is the ceremonial coronet worn by female Mongolia aristocrats in Song and

---

❶ 译文参考：*History of Traditional Chinese Suit –Zhongshan Suit*，来源：听力课堂（tingclass）网；新汉服（New Hanfu）网。

Yuan Dynasties. It shall be taken off when woman wearers go into the tents and carriers because it is in long shape and the framework is made of wire, birch and willow with silk and twigs mounted on the top.

9. Officials of Ming Dynasty wore Loyal Peace Hats (*zhongjingguan*) at home. The framework of this kind of hat is made of wires and covered by black gauze and black velvet. It's in square shape with projecture in the middle. There is a beam pressed by gold thread on the front part of the hat. The shape of the rear looks like two hills.

10. At the end of Eastern Han Dynasty (25–220), because the light scarves were widely used among the rulers and scholars irrespective of restrictions, they became even more popular. Princes and aristocrats used scarves as hairdo known as *fujin* to follow the chic deliberately.

（二）请将下列汉语段落译成英文

Mianfu is a set of garments including the *mianguan*, a crown with a board that leans forward, as if the emperor is bowing to his subjects in full respect and concern. Chains of beads hang at front and back, normally twelve chains each, but also in numbers of nine, seven, five or three, depending on the importance of the occasion and the difference in ranks. The colorful jade beads are threaded with silk, ranging from nine to twelve in number. There are two colorful silk ribbons, each with a jade bead on the end, hanging from both sides of the board above the ears of the wearer, reminding him to listen with discretion. This, like the above mentioned wearing the board that leans forward, has important political significance. ❶

译笔自测

Hair Ornaments

Ji, the oldest hairpin in China, was used to make hairdo and pin on the hat. Both men and women used *ji* in the Shang and Zhou Dynasties. The ancient crown was so small that two hairpins, also named "*hengji*", had to be inserted into the bun from left and right sides to fasten the crown. The emperor, empress and royal family members wore jade *ji*. Scholar-bureaucrat's wife wore ivory *ji*, while the common people used bone *ji*. Therefore, *ji* acted as a tool to distinguish people's social status along with its basic role of decoration and hairdo.

The character ji originally appeared in the Shang and Zhou Dynasties but it was replaced by a synonym—*zan* after the Warring States Period. *Zan* was one of the most common hair accessories for men and women in ancient times. *Zan* was initially used for making hairdo. It became an identification displaying the wearer's wealth and social status in class society. *Zans* made of stone, mussel, bamboo, wood and bone in ancient China were gradually replaced by those made of jade, silver, gold, silver, kingfisher feathers, tortoiseshell, gold inlaid with gems etc. Jade and gold *zans*, which were not affordable by ordinary people, were the most precious. The silver *zan* was worn by middle and lower class women.

---

❶ 译文参考：Hua Mei. *Chinese Clothing* [M]. 北京：五洲传播出版社，2004：57–58.

Chai, another type of head ornament in ancient time, was similar to *zan* and both of them were used to pin hair. *Zan* was single-stranded while *chai* was double-stranded, similar to a branch. A few of them were multi-stranded. There were a number of ways to wear *chai*. They could be pinned in a line, in vertical way, in oblique way and bottom-up way. The number of chai to be worn varied based on different hairstyles. Twelve *chais* could be worn on hair at most with six at each side. *Chai* had various shapes and vivid names, most of which were named after its head, such as swallow-shaped *chai*, floral *chai* and phoenix-shaped *chai*. Such precious materials as gold, silver, copper, jade bone, ivory, coral and colored glaze were worn by the wealthy. What the poor wore was wattles, known as "wattle hairpin".

The design of *buyao* evolved from *zan* and *chai*. Because there were pendants hanging on it, it swayed with women's each step, hence its name. *Buyao* was made of gold, silver and jade. Its material and shape varied with the wearer's social status. *Buyao* was known as "forbidden walking" for the purpose of restraining women's behavior. Women had to walk properly so that *buyao* made rhythmic sound.

Buyao was popular in the court and among aristocrats when it appeared in the Han Dynasty. *Buyao* was so popular in the Wei, Jin and the Southern and Northern Dynasties that it was loved by common women. Therefore, the patterns of *buyao* became diverse. After the Tang Dynasty (618–907), *buyao* was made of gold and jade with the design of bead in phoenix beak, known as "crested hairpin". In addition to gold, it was made of some precious materials such as jade, coral, colored glaze, amber, turquoise and crystal. *Buyao* made a comeback in the Ming Dynasty (1368–1644), known as Shaking-while-walking Hairpins.❶

第二节　肚兜

翻译练习

（一）请将下列汉语句子译成英文

1. Although there is a close correspondence between the traditional costume patterns and the traditional Chinese thoughts, the ancient and mysterious bellybands, along with the changing times,would be fused into the modern costume design through transmission and development for the purpose of promoting the modern fashion.

2. Usually, the head of tiger or *Wudu* are embroidered in kids' bellybands with parents' hope for their thriving growth. Embroidered bellybands with red trimming are more prepared for kids. Generally, Bellybands for boys are often embroidered with such patterns as "plucking cinnamon flowers in the Palace of the Moon(which means becoming a great man)" and "five children's passing the imperial examination", in hope for their excellence in future，while bellybands for girls are quite common for the embroidered patterns of "Finger Citron and Lotus"，or "peony and butterfly", yearning for lifelong happiness.

3. Abundance of lotus patterns and *baoxiang* patterns are embroidered in bellybands, which is

---

❶ 译文参考：戚琳琳. 古代佩饰［M］. 合肥：黄山书社，2013：97–117.

just the embodiment of Chinese mentality towards kindness and the ideal of pursuing eternity.

4．Auspicious culture is an indispensable part of profound Chinese traditional culture, which has penetrated into every aspect of the Chinese social life. The traditional auspicious thought can be found to be brilliantly and effectively conveyed in the traditional bellyband patterns.

5．According to traditional Chinese customs, the picture or statue of *hehe* is more often laid out or hung at the wedding, or hung on the wall of the hall throughout the year. As what bellyband pattern of *hehe* intends to convey, it symbolizes harmony and auspiciousness.

6．Wearing bellybands for the young mainly expresses their yearning for love and happy life. The young ladies tend to send bellybands to their lovers or husbands as the love token, while bellybands designed for kids aim to fend off the evils and pray for happiness. In China, the elderly wear bellybands only to look forward to well-being, good health and longevity.

7．Today, designers have brought out fashioned and sexy looking contemporary versions of the belly-apron. As a part of Chinese style costume, it is being widely accepted. The attractive traditional colors have become more stylized.

8．This bellyband is decorated with embroidered figures, trees and flowers over the yellow satin ground. The composition is well-organized with clear levels and proper density, and the figures are vivid, which reflects the real life.

9．The patch-made bellyband is made, among Chinese people, especially for the hope of prospective future for children, embodying the best wishes from the village neighborhood.

10．Bellybands of the Qing Dynasty were made into lozenge shapes, with a string on top to be hung from the neck. Each side had a string to tie it to the back. The lower half was a triangle pointing downward, reaching to the abdomen covering the navel. Materials used were mostly soft fabrics such as cotton and silk. Red was a favorite color for bellybands.

（二）请将下列汉语段落译成英文

Made of soft silk and cotton, the bellyband is a small single piece hung from the neck with a string to cover the chest and stomach in the Qing Dynasty. Two side strings that go around the waist are used to tie it down to cover the navel. The strings can be made of different materials, such as gold chains for wealthy families, silver and brass for the well-off, and satin ribbons for the ordinary. Generally thought to be a protection to the naval, the bellyband is mainly supposed to bind the breasts so that their curves become less obvious, a measure taken to suppress desires as required by the teachings of feudal China. Restraint from exposure, however, displaying an inner romantic feel, which demonstrates Chinese people's reserved character. The bellyband thus turned out to be even more alluring than it would otherwise, with various materials and methods applied to it, such as pinks and soft pastels, damasks and satins, auspicious flora motifs and tailored into shapes from lozenges to curved triangles, to make it mysteriously charming and romantic.❶

---

❶ 译文参考：薛雁. 时尚百年［M］. 杭州：中国美术学院出版社，2004：44.

译笔自测

The Design Culture of Bellyband Patterns

After thousands of years, many bellyband patterns have become the highly generalized symbol and concentrated reflection of Chinese national spirit. Patterns play the role of identification in dress with the symbolic significance. The vividness of patterns is more intuitive than the structure of garment and the fabric as well. Therefore, it becomes the most important part in demonstrating garments' integral spirituality, and quite easily served to convey information by the wearers.

The traditional bellyband patterns abound in many subject matters. Flowers and birds, animals, gods, fairytales, opera figures and people in daily life are all embroidered in the traditional costume and fused with the aesthetic thoughts, advocating that Human and Nature（including the sky and the land）come from the same origin，stressing the Chinese traditional idea of equality and harmony, which reflects the wishes from the ordinary people looking forward to the happy life, and it also indicates the great effect of the Chinese traditional thoughts on the costume pattern design. The Confucianism and the Taoism in traditional Chinese culture have long dominated and influenced the spiritual life of the Chinese nation by following the common view of fusion in mutual aid, harmony in diversity, and coexistence and mutual prosperity.

Filial piety (xiao)is not only the core of the Confucianism, but also the code of ethics for maintaining the Chinese family harmony for thousands of years, which is adhered to as the traditional virtues. There are many kinds of bellyband patterns promoting the filial piety. Those various patterns present the mental world of people, as well as the significance of adherence to the filial piety.

Besides, the feudal ethics and values under the influence of Confucianism are infiltrated into the bellyband patterns. Confucianism attaches great importance to virtue, that's why many bellyband patterns are related to the idea of sages and upright ministers, heroic maidens and dutiful sons, the three cardinal guides and the five constant virtues and even ethical tales. Self-cultivation is advocated by Confucianism, and good cases in point are the auspicious pattern of three durable plants of winter—pine, bamboo and plum, and that of four nobles of flowers including plum blossom, orchid, bamboo and chrysanthemum. Those patterns are quite popular because of being in conformity with Confucian ethics. Confucianism emphasizes ritual system between a ruler and his ministers, fathers and sons, husbands and wives, and the animals are often taken by Confucian culture as the auspicious signs.

Confucianism values succession of male offspring and continuation of family, advocating the more sons, the more blessings. Therefore, patterns such as *Kylin delivering sons* and *a baby over lotus* are popular in bellyband, praying for auspiciousness. The auspicious patterns of flowers, birds, insects and animals indicate the human-nature harmony for its conformity with Confucian nature-man unity. Of course, the feudal idea of success in imperial examination, official career, and wealth with high rank and substantial emolument in auspicious patterns is the direct result of Confucian influence. Taoism is a combination of sorcery in ancient China, Five Elements and Doctrine of Huang

Lao（Taoism）, and many auspicious patterns are produced along with the pursuit of good luck and happiness, and immortality as well. Immortality is the constant pursuit for Taoism, which attaches great value to the individual life, believing individual cultivation is the preparation for immortality in a fairyland. Immortality, the important message of Taoism, is not only what all previous monarchs pursue, but also what people yearn for. Therefore, patterns related to longevity become the major part of auspicious patterns. "Pursuing good fortune and staying out of disasters" become the main part of Taoism after sorcery is accepted by it. For instance, *Tai Chi* diagram and Eight Trigrams can be often seen in costume patterns as the auspicious pattern of removing ill fortune, which turn out to be the symbol of seeking advantages and avoiding disadvantages. ❶

第四章　丝绸汉译英

翻译练习

（一）请将下列汉语句子译成英文

1.　Patterning techniques had been increasingly enhanced, which enabled production of exquisite textiles, such as the famous Zhangzhou velvet, Zhangzhou satin and cloud brocade. Meanwhile, characteristic local embroideries took shape across China.

2.　Double-layer brocade was the most precious Song brocade, usually applied on dragon chairs and cushions in the palace, with decorations compatible to the imperial court, such as patterns of dragon, cloud, bat and variety of big floral medallions.

3.　As a symbol of the emperor, dragon appeared on imperial silk robes in various forms, such as front viewed dragon, rising dragon soaring over the cloud, descending dragon coming down from the sky, flying dragon among clouds and dragon in roundel.

4.　The floral patterns applied in roundels did not refer to any particular flowers existing in reality, but imagined patterns which transformed, mixed and regrouped different plants and flowers. The floral patterns underwent evolvement from simple to colorful, and from imaginative to more realistic.

5.　Compared to the luxuriance and colorfulness of Tang silk, the most distinctive feature of Song silk lied in its naturalness and grace, which might derive from the political reality of the time: confronting with tough ethnic groups in the north, peace for the gentle and refined Song Dynasty could only be obtained through hermitship in remote villages and indulgence in landscape, flowers and birds.

6.　Among the various Ming silk textiles, embroidered and velvet products were the major woven products that represented the most important technological advances of the Ming Dynasty.

---

❶ 译文参考：訾韦力. 中国传统肚兜服饰文化（汉英对照）［M］. 北京：中国轻工业出版社，2016：130–137.

Minor improvements were also seen in *kesi* and embroidery, based on previous traditions.

7.　The great variety of silk textiles provided ample choices in material support for a highly ritual system which relied heavily on costumes. The silk textiles contributed greatly to the formation of Chinese ritual protocols and codes of etiquette.

8.　On the Eastern Han silk and wool textiles unearthed in the Niya Ruins of Xinjiang, we can see plant and grape vine patterns of the west region, the centaur of Greek mythology, and soldiers of central China, existing together as proof of garment art exchange.

9.　Due to the well-established professions in the silk industry, the Ming Period saw great improvements in techniques. Among the new innovations were the "Chu Kou Gan" (drying during spinning) and "Chu Shui Gan" (drying during reeling) techniques in silkworm raising and silk reeling. In weaving, doffing and twisting techniques, and jacquard-weaving technique were already accomplished.

10.　The first appearance of paintings made of silk in the Spring and Autumn & Warring States Periods is evidenced by two silk paintings *Lady*, *Dragon* and *Phoenix* and *Man Riding the Dragon* unearthed from a tomb of Chu state in Changsha, Hunan province, which are the two earliest complete examples found in China to date.

（二）请将下列汉语段落译成英文

The exceptional role silk played in Chinese culture was above all reflected by its influence on Chinese characters. Pictographic characters of "mulberry", "silkworm" and "silk" originating from mulberry planting, silkworm rearing and silk productions, were already inscribed on the mysterious oracle bones in the Shang Dynasty. Modern Chinese characters with "纟" (meaning silk) as the an essential component, such as经 (*jing*, warp), 纬 (*wei*, weft), 组(*zu*, group), 织(*zhi*, weaving), 线(*xian*, thread), all derive from the sericulture industry. Besides, many common Chinese words are related to sericulture industry as well. For instance, "*jigou*" originally referred to the structure of looms, while now it generally refers to work units such as governmental departments and social groups.❶

译笔自测

Silk and Etiquettes

As a unique integration of practicality and artistry, silk has not only made extraordinary contributions to Chinese literature and art, but is considered as the symbol of dignity and status as well, and has always been closely related with religion and rites since its emergence. On the other hand, grassroots who only afforded clothes made of plain cloth and hemp were therefore referred to as *Buyi* (literally garment made of cloth).

The life process of silkworm, which is transformed from an egg to larva, then to a pupa in the cocoon, and finally to a moth breaking through from the cocoon, also amazed our ancestors. They

---

❶ 译文参考：《中华文明史话》编委会. 丝绸史话［M］. 北京：中国大百科全书出版社，2012：87–88.

connected such transformation with the life circle of human beings, and considered silk a mystic media between men and heaven. As a result, it was believed that the soul must break through a cocoon like a pupa, if it was to ascend to the heaven. Accordingly, bodies of the dead were wrapped up with silk as an artificial cocoon, with which their souls could break through to the heaven.

The ultimate manifestation of silk's application in China's ritual system was the dragon robes—costumes for Chinese emperors, and officials'robes with rank patches. As son of the heaven, emperors had the privilege to wear costumes with dragon patterns and the "twelve emblems". The former implies the authenticity of the emperor as real dragon and thereby son of the heaven, while the twelve emblems, namely the sun, moon, stars, mountain, dragon, *huachong, zongyi*, algae, fire, rice, *fu*(黼) and *fu* (黻), take on the following compliments for emperors: the sun, moon and stars suggest that the mercy of the emperor is as glorious as their rays; mountain stands for his calmness and steadiness in disposition; dragon, which is a legendary creature, symbolizes he is a resourceful and resolute ruler; *huachong*, which is usually in the form of a pheasant, signifies his unusual literary talent; *zongyi* is a sacrificial utensil, usually in pairs, with patterns of tiger and monkey, tiger implying might and monkey implying wisdom; algae stands for nobleness in conduct; fire means open and aboveboard; rice denotes ruler attaches importance to agriculture and he can bring peace and stability to his country; *fu*(黼), which is in the form of an axe, marks efficiency and resolution; while *fu*(黻), which is in the shape of two bows in opposite direction, connotes his pursuit of benignity and break from malignance.

The square rank patches decorated on the front and back of ancient official robes have also become familiar to people today through television. The patterns on the patches were either birds or beasts. Generally speaking, civil officials wore patches with patterns of birds, while military officials wore patterns of beasts, with a particular bird or beast corresponding to a particular official rank. For instance, a patch with pattern of crane implied that its owner was a first rank civil official. In one word, both the dragon robes and the rank patches were applied as tools for distinguishing hierarchy and indicating status, and they clearly reflected the important role silk had played in ancient China, a nation of etiquette.❶

第五章　刺绣汉译英

翻译练习

（一）请将下列汉语句子译成英文

1. If we can compare these Miao embroidery masterworks with the contemporary maestros, we can find to our surprise that "ecstatic self-oblivion" is what in common between the two, because

---

❶ 译文参考：《中华文明史话》编委会. 丝绸史话［M］. 北京：中国大百科全书出版社，2012：91-94.

what they express is thinking and meditation and not material.

2. Thanks to the uniqueness of the milieu of Miao embroidery and brocade as well as the way in which they have been handed down, the designs of some branches retain many features of ancient cultures and primitive arts.

3. Unlike common embroidery with threads pulled through the fabric, braided embroidery involves braiding embroidery threads into plait-likes bands thicker than single threads and arranging them into designs.

4. The universal preference for embroidery and brocade in all the types of Miao dresses may be attributed to the fact that they are superior to the monotonous and two-dimensional batik in the diversity and brilliance of colors, textures and three-dimensional effect achieved by variations in material and technique, and durability.

5. Modern Huaxi type couching is called "new couching". Cloth in different colors is used as the ground fabric, and the colors of embroidery threads are brilliant and diverse, in contrast with other types of couching, which have unvarying designs and colors of ground fabric.

6. The fantastic designs of embroidery and brocade for Miao dresses and adornments reflect mysterious ideas formed in the childhood of mankind and have distinct characteristics of primitive culture.

7. The similar warm colors of the ground fabric and the design evoke an exuberantly festive mood.

8. Every Miao embroiderer is an artist, a maestro without sophisticated professional training but with the same understanding of the aesthetical tastes and understandings. These are their innate instincts.

9. Ethnic Miaos have no written language and when you are gazing at an embroidery work or observing the daily life of the ethnic Miaos in their native land and when you are working hard to understand the implications of the mysterious symbols and patterns on the Miao embroidery, your thoughts have penetrated the boundaries of Miao society.

10. Ethnic Miaos without written language, use their embroidery to express their view of the origin of the universe and the symbols with rich cultural implications survive so many thousand years on the embroidery and pass down to our times and to eternity.

（二）请将下列汉语段落译成英文

Braid Stitch is mainly utilized in Leishan County with the paper cut sample attaching to the embroidery cloth first before the work begins.The color silk lines are usually braided by hand with 8, 9 or 13 lines to the breadth of 3 mm. The braided silk bands are wounded on the paper cut on cloth from outside to inside before stitched to the cloth with the silk lines of the same color.❶

---

❶ 译文参考：曾宪阳. 苗绣：一本关于苗绣收藏与鉴赏的书［M］. 贵阳：贵州人民出版社，2009：45.

译笔自测

Couching

Couching is technically easy and can be done with nothing but a piece of plain weave ground fabric and embroidery threads. The fabric can be rolled up and carried around. For this reason, couching has been adopted by almost all Miao areas and used for the major decoration of costumes by the vast majority of the branches.

Couching must be done on plain weave cloth so that the crossings of its yarn could be used as coordinates for stitching. All old embroideries were done on homespun cotton cloth or fine linen. The latter was widespread in the West Branch because its fabric used to be woven from linen. Some of present embroideries are done on chemical fiber plain weave cloth. Satin is not used because it is too soft and because its texture makes it difficult to count yarn. A single sheet of fabric is used, and it is not mounted on cloth or hard paper. Threads are pulled from the back upwards, and the other side is the obverse. Some couching works look very much the same on the two sides except that one is smoother than the other.

The most common stitch is cross-stitch, i.e. embroidery threads cross at right angles, forming dots which in turn form lines or patches. The stitch is also called cross couching. The threads can be long or short, the usual length being 1-2 mm. Short ones may overlap and form a dot, without any visible cross. There are also satin stitch and long stitch. Satin stitch makes no crosses; each starting or ending point takes up one unit of the yarn grid of the ground fabric, like a node on coordinate paper. Long stitch is somewhat similar to flat embroidery, pulling the thread rather long in every stitch. Couching done entirely with long stitch is so similar to brocade that sometimes it is difficult to tell them apart.

Most couching designs are purely geometric or half abstract, half concrete designs composed of geometric figures. The latter type includes plants, flowers, fruits and occasionally animals, figures and utilitarian articles. Since the images composed of geometric figures are half abstract and half concrete, the same design is called by different names in different places, or even by different persons in the same place. For instance, fretwork or spiral is called bracken flowers, ox whorls or water whorls (i.e. ripples). Animal variations in couching designs mostly represent the butterfly, the frog and birds. The butterfly, in particular, can be formed by diverse variations of geometric lines. The depiction of figures is rare.

The Miao people in Huaxi District in the suburbs of Guiyang City are capable of more couching designs than those in any other place. This type of couching is mostly applied to large areas on aprons, coats or baby carriers. Embroideries of older generations have batik with dim designs as the ground fabric, which has been replaced by white cloth or cloth in other colors in the new generation. Cross-stitch is dominant; designs abound in dots and lines as well as empty places showing the ground fabric. In the local's terminology, there are the following designs: pig trotters, dog teeth, buffalo teeth, bracken flower, buck-wheat flower, purslane, coxcomb, fringed iris, shepherd's purse flower, wheat ear flower, bee, moth, swallow, tit, fish, dragon, frog, lantern, silver hairpin and snowflakes.

In the couching for several types of Miao costumes in Northwest Guizhou, the cross stitches are so short that the crosses are hardly visible and look like dots. Sometimes satin stitch is used instead, also forming dots, each of which resembles a unit 1mm square on coordinate paper. In the entire embroidery, such dots form lines with different thickness, which in turn form shapes. This type of couching, very similar to brocade, has lively colors in sharp contrast and very distinct lines, with a certain measure of substantiality.

The Huangping type of couching in Southeast Guizhou is technically unique. The ground fabric is bluish black or deep purplish black homespun cloth. The design is primarily in red, which is supplemented with blue. The long stitch similar to flat embroidery is used, each thread pulled straight from one border to another of the outline. When the long-stitch couching is done, a border is made with the locking stitch used in the Shidong type and mostly orange threads. Each embroidery has multiple designs formed by minute geometric figures, i.e. the observer can combine the figures in different ways to perceive different yet similar images. For instance, some figures can be combined to form a butterfly, yet they can be combined with others to form different images. Since each geometric unit has a lock-stitched border line, the design appears substantial, with a fine texture.

The couching technique of the Basha type in the lower reaches of the Duliu River is also quite special. It is also unique to Miao couching. The ground fabric is bluish black hand-woven cloth and the long stitch is dominant. Threads can be pulled slantwise, crosswise or lengthwise to form broken or wavy lines. The embroidery threads are primarily in white, mingled with light red, green and yellow, with an effect that is quiet yet not dull. Major designs that can be made include the sunflower, the creel, the dragon, the triangular tooth and the toothed rake. ❶

第六章　新时代中国服饰汉译英

翻译练习

（一）请将下列汉语句子译成英文

1．Jeans also entered into China at the end of the 1970s. And since then, more and more people have started to wear jeans, expanding from fashionable young men to people of all classes and ages.

2．Workers, intellectuals and school students were all proud of wearing army uniforms. They wore green army caps and rubber sole shoes and carried army marching pack bags, army satchel and kettles.

3．The use of national products of hand-made fabrics, the removal of decorative bands and

---

❶ 译文参考：Zhong Tao. *Miao's Embroidery and Brocade*［M］. 贵阳：贵州民族出版社，2010：11–13.

coiled cloth buttons, the introduction of zippers, the lowering of the collar, the shortened length and the soft, light but harmonious colors were all the elements that make up the features of the Chinese *qipao* of the 1940's.

4．Along with the strengthening of consciousness of corporate image design, designers, the theoretical circle of clothes and even users all have realized the importance and promising future of the professional dress.

5．Back in the 60's and 70's, in fact, our clothing was not only monotonous in color, but also extremely poor in variety of modes, looks, fabrics and patterns. The widely spread idiom "wearing it for three years new, wearing it for three years old, and yet for another three years by mending and re-sewing" gave a vivid description of the clothing situation then.

6．Based on the *zhongshan* suit, revised and simplified, the youth uniform was the answer to the search for an even simpler mode in clothing in the 60's. The basic structure of the youth uniform conformed to *zhongshan* suit, with modification being made to have three pockets instead of four, one on top for a handkerchief. The front buttons were hidden inside and invisible, which looked simple and delicate. It was well received among the young people.

7．TV and movies of the first ten years were the primary instructors of fashion. Movies with love stories awakened the youth within people's hearts that had been asleep for many years. Images of the stars and even the smallest moves of their gestures on the screen would have a tremendous impact on millions of viewers, who emulated faithfully.

8．The batman sleeve had base attached to as low as the waist line, looking like the wing of a bat. It was introduced through the media of Hong Kong and Taiwan movies and TV shows. It was a time when fashions were being adopted passively, by copying or emulating. Curiosity and the tendency to follow the public was the main reason the batman sleeve got popular in China.

9．In the early 1980s, drifting in the music of such pop songs as *Sweetheart* and *When Shall You Return* by Teresa Deng, young men began to dress themselves in the trumpet trousers as seen in the movies. The stereotyped image of a young male at the time was one that had long hair, wearing a pair of toad sunglasses，a checked shirt on top and a pair of jeans with tight-fit buttocks and thighs at the bottom.

10．The *qipao* in 1932 was lengthened, and it reached its full length in 1935, with its lower hem close to the floor. Therefore, it was popularly referred to as "the floor-sweeping *qipao*".

（二）请将下列汉语段落译成英文

During the 2001 APEC meetings, Chinese costume attracted lots of attention. This "Tang Style Jacket" is a good example containing both traditional and contemporary features, in traditional form and material but cut using a 3-D approach. It is based on the riding-jacket and combines both Chinese and western advantages in design. The basic mode, the color, pattern, choice of fabrics, coiled cloth fastener and hand stitched fringes are typically Chinese. The material is not limited to silk or satins. Cotton, linen, synthetic material and jeans cloth are incorporated. This is a successfully modified

Chinese costume created by Chinese designers in response to the current prosperity that China is enjoying. Its current popularity signifies a sense of pride and national dignity. It also reflects the need for us, as a nation of great culture, to remain clear as to who we are in this era of globalization, and indicates an interest among the people in a costume that brings a sense of renewed vitality fit for the new era.❶

译笔自测

The World Trend and Chinese Fashion

Till the middle term of the 1980s, there were more and more clothes styles while the popularity circle became shorter and shorter. New styles and materials were kept being introduced to the market. As for upper outer garments, there were all kinds of T-shirts, jackets of mixing color, checkered shirts and cotton pullovers. Suits and ties became social dresses for formal occasions, and accepted by most of the white collars. Under clothes included straight pants, elastic trousers, radish trousers, skirt trousers, 70% trousers, trousers skirts, pleated skirts, eight-piece skirts, western suits skirts, midis and sun skirts. The style changed constantly. By the time when the mini skirt that was born in the western world in 1960s was once again popular in the 1980s, China has kept pace with the world fashion.

At the beginning of the 1990s, the previous dressing order was broken. The sweaters worn inside the jacket in the past entered all kinds of formal occasions as a single dress without going with outer jackets because of their rather loose style. The dressing style of "wearing underwear outside" was not considered as strange any more after two or three years. In the past, if wearing a jacket outside, the sweater or T-shirts inside should be shorter than the jacket. But young people suddenly found that it was hard to find bigger jackets than their loose sweaters, so they covered small jackets outside the long sweaters. The short-sleeve sweaters worn in summer time could also be worn outside the long-sleeve shirts. Very soon, clothes industry began to launch sets of unconventional suits, such as a long jacket and long shirt going with a little waistcoat higher than waist, or the successive style sleeves with outer sleeves shorter than the underclothes sleeves.

During that period of time, sun skirts going with high boots style black satin saddle shoes appeared in Paris garments stage. In the past, sun skirts with tight and small upper part and two thin shoulder straps were worn in beach. But when it appeared on fashion stage, the skirt body had a bigger spread that reached the ankles. Almost at the same time, global fashion experienced first hand-covering clothes, which lengthened sleeves to cover the hands, then the waist-exposing clothes or even navel-exposing clothes, which have short and small upper clothes and a section of skin exposed. This kind of clothes was once popular in China, but the style was not as bold as the neighboring Japan. The navel-exposing clothes popular in Japan even led to new content of body beautification-

---

❶ 译文参考：薛雁. 时尚百年［M］. 杭州：中国美术学院出版社，2004：126.

to beautify the navel. Clothes that expose part of skin inspired by waist-exposing or navel-exposing clothes however became popular in China. A subtle trend was to cover parts that were exposed before such as hands and shanks and expose parts that were covered before such as the waist and navel.

In those years at the turn of century, fashion in China kept close pace with the world trend. Following the international dressing fashion, dressing style tended to be more formal, especially white-collar women who paid particular attention to the charm of being a professional woman. They tried to wear formal and decent dresses. People no longer favored the so-called "original wildness" such as straw hats without edging or tearing thread off the trousers. The trend of exposing certain part of skin was restrained among people of some classes and on some occasions. Though mini-skirts were still popular, many young girls started to favor ankle-long skirts to show female elegance.

Accordingly, some young people held high anti-traditional awareness in the western society. They considered the weird as fashionable on purpose, for example, imitating the hairstyle of *The Last of the Mohicans* to shave head on two sides and leave the middle section hair which was dyed with bright color. Some wore punk dress—another kind of decadent style youth dress in western society after hippies. They glued the hair into animal horn shape using hair gel and embroidered skeleton pattern on black leather jackets; or intentionally tore or burned holes on the clothes. ❶

---

❶ 译文参考：Hua Mei. *Chinese Clothing* [M]. 北京：五洲传播出版社，2004：165–168.

附录2　汉英服饰词汇对译

A形连衣裙　A-line dress

A字形裙　A-line skirt

A字型　A-line

H字型　H-line

O字型　O-line

X字型　X-line

阿拉伯头巾　keffiyeh

阿拉伯长袍　aba; abaya

安全防护服　safety protective coverall

安全帽　hardhat; hard hat; safety hat; safety helmet

按扣　snap[fastener]; popper; dome fastener

暗袋　changing bag; concealed pocket; inside pocket; private pocket

暗缝　blind stitch; closed seam

暗口袋　ticket pocket

暗扣　slim hook

凹顶帽　porkpie [hat]

袄　coat; jacket

八角帽　octagonal cap

巴拿马帽　Panama hat

芭蕾紧身衣　leotard

芭蕾舞鞋　ballet shoes; ballerina shoes; ballet slippers

芭蕾舞长裙　ballerina

白大褂　white garment; consulting gown

百慕大短裤　Bermudas; Bermuda pants; Bermuda shorts

百慕大套装　Bermuda suit

百衲衣　ragged clothes

百褶裙　accordion-pleat skirt; all-round pleated skirt; pleated skirt

办公室制服　official wear; office uniform

半袖　half[-length] sleeve; elbow sleeve

绑腿　legging

棒球服　baseball dress; baseball uniform

棒球帽　baseball cap

包边　overcast

包头巾　turban; kerchief; gotra

薄衫　zephyr

宝塔帽　pagoda hat

保暖衣　thermal cloth

报童帽　apple jack cap; newsboy cap; bebop cap

贝雷帽　beret

背包　back pack; bag pack

背带裤　overalls; suspender pants

背心　vest; singlet; waistcoat; [front] gilet; waist; jerkin

背心式内衣　under vest

比基尼服　bikini dress

比基尼泳装　bikini

毕业服　graduation dress

编织毛衣　sweater; knitted woolen garment

蝙蝠衫　batwing dress

蝙蝠袖　batwing sleeve

便服　casual wear; informal suit; civilian clothes; plain clothes

波浪裙　flared skirt

博士兜帽　Doctor's hood

博士服　Doctor's gown; Doctor's gown and hood

补子　*buzi*; breast plate [of clothes]

布袋装　chemise [dress]; sack dress

草帽　straw hat

草裙　hula skirt; grass skirt

草鞋　straw shoes; straw sandals; rope sandals

插肩　raglan; raglan shoulder; saddle shoulder

茶服　tea jack; tea gown

茶会女礼服　tea gown

衩　placket; vent

缠裹式头巾帽　wrapped turban

缠身衣　clingy

常服　plain clothes; *changfu*

常礼服　cutaway coat; cut-away

常礼帽　derby hat; bowler hat; stiff hat

超短裤　boy shorts; short shorts; mini-pants; mini shorts; monokini

超短连衣裙　micro dress; minidress

超短裙　hot skirt; short short-skirt; skimp skirt; mini-skirt; ultrashort skirt; ultra-mini skirt; microskirt

朝服　court dress; *chaofu*

朝袍　imperial robe

朝圣服　pilgrim apparel

朝靴　courtier boots

朝珠　chu-chu; court beads

衬布　interlining

衬垫　wadding

衬裤　drawers

衬里　backing; liner; lining; under lining

衬裙　slip/slips; patticoat; undergown; under-dress; underskirt;（古）placket

衬衫　（男）shirt;（女）blouse

衬衫裙　shirtwaist dress; shirt-waister

衬衫式外衣　liquette

衬衣式裙装　blouse slip

撑裙　hoop skirt; crinoline; tournure

成衣　ready-to-wear; ready made clothes; ready-mades

冲浪服　surfing suit; wet suit

抽裥裙　gathered dirndl; gathered skirt

传统服装　classic costume; classic clothes; traditional costume

传统婚礼服　classic wedding dress

串珠手袋　beaded hand bag

春装　spring wear; spring wardrobe

刺绣　embroidery

粗布服装　rough wear

粗花呢　tweed

粗呢大衣　duffel coat; duffle coat

褡襻袖　epaulet sleeve

大摆裙　expansion skirt

大氅　cloak; paludament; pella

大襟　Chinese opening; side opening; large front

大衣　overcoat

大众化款式　ford

带盖口袋　flap pockets

单排纽外套　single breasted coat

单鞋　thin shoes

裆　rise；crotch（crutch）

裆裤　crotch pants

道袍；道衣　Taoistic robe ; monk's habit; priest frock

灯笼裤　balloon pants; pantalets; pantalettes; knickerbockers; knickers

灯笼裙　balloon skirt; lantern skirt

灯笼袖　balloon sleeve; lantern sleeve

登山服　alpine jacket; mountaineering suit; mountaineering wear

低领连衣裙　low-cut dress

低腰裤　hipsters

低腰线　dropped waistline

低腰装　hip-hugger

垫肩　shoulder pad; padded shoulder; padding

貂皮大衣　marten coat; mink coat

吊带裤　suspender pants

吊带裙　slip dress

定制服装　custom-made

都市装　urban wear

兜裆　short crotch; short seat

兜肚　abdominal bandage; abdominal belt; stomacher

兜帽　hood

斗牛士装　toreador suit

斗篷　cloak; manteau; mantle; throwover dress; tog

短靴　low boots; demi-boots; ankle boots

短袄　lined short gown

短背心　gilet

短大衣　carcoat; cover coat; semi cope; over sack; topper [coat]

短褂　chaqueta

短夹克　short jacket; spencer; bomber jacket; lumber jack

短裤　bottom piece; golf knickers; short pants; shortie; shorts

短礼服　evening jacket

短马靴　jodhpurs; jodhpur boots; jodhpur shoes; flying jodhpur

短披肩　mantelet

短衫　short-gown; jupon

短项链　choker

短袖　semi-sleeve; short sleeve; quarter sleeve; half sleeve

T恤衫　T-shirt; Tee-shirt

短靴　ankle boots; demi-boots

对襟　front opening; edge to edge front opening

对襟开衫　edge-to-edge cardigan

对襟马褂　*magua*; mandarin coat

多层裙　tiered skirt: layered skirt

多片裙　gore[d] skirt

额带　brow band

额饰　frontayl / frontel

儿童背带裤　crawlers; coogan

儿童短衫　toddlers

儿童连裤背心　under waist

儿童围裙　slip

耳饰　earring

耳坠　ear bob; ear drop

发夹　barette; hairpin

法式羊腿袖　gigot sleeve

翻领　lapel; revers; turndown collar; turnover collar; roll-over collar

翻领衬衫　open-neck shirt

方巾　face cloth; handkerchief

防尘服　dust free coat

防弹夹克　ballproof jacket; bullet-proof jacket; bulletproof jacket; bullet proof vest

防风服　windbreaker

防寒服　winter clothes

防护服　protecting gown; protective clothing; defense apparel

防护面具　protective mask

防水外套　all-weather coat, water-proof coat

防雨大衣　raincoat

飞行员夹克　bomber jacket, flight jacket

分层裙　tiered skirt

分趾凉鞋　slip-flop pumps

风帽　bashlyk; capuche; hood

风帽上衣　hooded jacket

风冠　phoenix crown; phoenix coronet

服饰　apparel & accessories; clothing ornament

服装　apparel; garment; attire; costume; clothes; clothing

福字鞋　Fu Zi shoes

妇女便装　promenade

妇女衬衫　undies

妇女披巾　wimple

妇女胸衣　corset; cuirass; figure-firming fashion; foundation garment; foundation wear

腹带　belt abdominal; binder; cummer-band

高帮鞋　high shoes; high-topped shoes

高顶大礼帽　high hat; top hat

高翻领　turtleneck collar

高级时装　high fashion; high couture/haute couture; couture

高开衩　high slit opening; dizzying slit

高科技服饰　high tech clothes

高领　built-up neck; high collar; high neck; stand-up collar; Mandarin collar

高领衫　turtle neck sweater

高筒袜　high socks

高腰裤　high waist pants; high-rise pants; high waist trousers

高腰款式　high-rise

高腰裙　built-up waistline skirt; high waist skirt

袼褙　gebei lining; pieces of old cloth or rags pasted together to make cloth shoes

格子衬衫　check shirt

格子呢服装　tartan

工装　work clothes; jumpers

工装背带裤　bib top pants

工装短裤　salopette shorts

工装裤　overalls; bib overalls; salopette pants; salopette

工作服　overall; smockfrock; working uniform; work wear

公文包　briefcase

宫廷礼服　court dress

宫廷帽　court hat

瓜皮帽　calotte; Chinese cap; skull cap; skullcap

挂肩式背心　built-up shoulder vest

褂子　short gown; Chinese style unlined upper garment

官服　toga; vestment

官靴　officials' boots

冠　hat; corona; coronet

桂冠　laurel wreath

衮服　royal costume; imperial robe

绲边　binding; bound; piping

裹襟女装　wrap dress

裹襟式大衣　wraparound coat

裹腿　puttee

哈伦裤　harem pants

海滨服饰　beachwear

海魂衫　sailors' striped shirt

海军帽　nanal cap

汗背心　sleeveless undershirt; sleeveless vest; under vest; undershirt; singlet

汗布文化衫　jersey singlet

汗裤　briefs; underpants; drawers

汗衫　singlet; sweat shirt; under-shirt

旱冰靴　roller skate

绗缝　quilting; basting

绗缝充填装　quilted suit

和服　kimono

荷叶边　flounce

后衩　back fork

后裙撑　bustle

蝴蝶结　butterfly bow

蝴蝶领结　bow tie

虎头鞋　tiger head shoes

户外套装　blazer suit

护臂　arm guard; armlet; bracer

护裆　athletic supporter

护目镜　goggles

护身盔甲　corslet

护士服　nurse dress

花样滑冰鞋　figure skate

华达呢　gabardine; gabardeen; gaberdine

华服　Chinese attire; Chinese costume

滑雪服　salopette: skiwear: down-skiwear

滑雪鞋　patten

化妆包　minaudiere

化装舞会服饰　masquerade

黄马褂　yellow jacket

黄袍　Yellow Robe

婚礼服　wedding dress；wedding gown

婚礼帽　wedding hat

婚礼面纱　wedding veil; bridal veil

鸡尾酒会服　cocktail dress

几何纹样　geometric design

加冕礼服　dalmatic

夹袄　lined jacket

夹克　jacket

夹克背心　jacket vest

夹克外衣　surveste

夹衣　lined clothes; lined dress; lined garment

家居服　house wear; home dress; lounge wear;
　at-home wear

袈裟　kasaya

裕袢　Uygur or Tajik robe buttoning down the
　front; qiapan

甲胄　armor

嫁衣　bridal gown；trousseau

尖头鞋　point toe shoes; pointed shoes; sharp
　toe shoes

肩带　shoulder belt; baldric; shoulder strap;
　sash

肩饰　epaulet

茧形外套　cocoon coat

裥　pleat；tuck

裥裙　pinafore pleats

健美服　bodysuit; bodywear

健美裤　tight pants

健美体操服　calisthenic costume

健身服　gymnasium costume

箭裙　shaft skirt

交叉领　crossover collar

角帽　corner cap

节日盛装　Sunday best

解放鞋　liberation shoes

金缕玉衣　jade clothes sewn with gold thread

金银首饰　gold and silver jewelries

紧身芭蕾舞服　tricot

紧身背心　body shirt

紧身大衣　newmarket; sheath overcoat

紧身胴衣　corset bodice

紧身裤　jean pants; tights

紧身连裤袜　body hose

紧身连体衣　bodysuit; tights

紧身胸衣　corset

紧身衣　body clothes; corset

紧身运动衫　fitting sweater

锦袍　brocade robe

警服　police jacket

胫甲　jamb

九分裤　seven-eights pants

救生背心　flotation vest

居家便服　dishabille

居家拖鞋　bedroom slipper

裾　full front

绢　plain silk

军大衣　military coat

军服　battle dress; regi mentals; army uniform

军服肩饰　aiguillette

军服式风雨衣　trenchcoat

军帽　army cap

开衩　vent

开衩裙　slit skirt

开口　placket

铠甲　mail

坎肩　waistcoat; sleeveless jacket; vest

可翻中式领　split mandarin collar

可卸领　detachable collar; separate collar

裤　trousers; pants

裤裆　crotch

裤装　pant; pant garment; slacks suit; trouser suit

宽摆裙　flare skirt; gypsy skirt; peasant skirt

宽大长袍　sagum; gallabiyah; galabia; galabich; boubou; chiton

宽大罩衫　overall

宽肩　extended shoulder

宽裙　full skirt；flared skirt

宽松便服　duster

宽松短裤　trunks

宽松肥大款式　big look

宽松裤　trousers

宽松长袍　sack

宽松直筒无袖连衣裙　shift

盔甲　acton; aketon; armour; armor

阔边帽　broadbrim

阔边遮阳女帽　sun bonnet

阔脚裤　flares; palazzos; palazzo pants

拉链　zipper

拉链夹克　zip front jacket

喇叭裤　bell bottoms; bell-bottom trousers; flared- leg pants; flare trousers

喇叭裙　bell skirt; bell-shaped skirt; flare skirt; trumpet skirt

喇叭袖　bell sleeve

蜡染服装　batik garment

蓝布大褂　blue cotton gown

篮球鞋　basketball shoe

劳动服　corporate clothing; working suits; fatigue uniform; fatigue clothing

劳动裤　dungarees

乐福鞋　moca-loafer

蕾丝　lace

礼服　formal attire; official uniform; ceremonial dress; full dress

礼服背心　dress vest; dress waistcoat; evening vest; formal vest

礼服外套　dress cover

礼服制服　full dress uniform

礼帽　dress headgear; formal hat

立领　stand collar; stand-up collar; banded collar; built-up neck;

立领夹克　closed collar jacket; stand collar jacket

笠帽　bamboo rain hat

连裤装　rompers; one-piece jumpsuit; romper suit

连帽长斗篷　burnoose; burnouse

连身裤　bodysuit; panty set

连身套装　jumpsuit

连衣裤　siren suit; catsuit; jumpsuit

连衣裙　one-piece dress; dress

连衣裙裤　pantdress; pantshift

凉鞋　sandal shoes; sandal[s]; tatbeb

两面穿夹克　reversible clothing

绫罗绸缎　silks and satins

零钱袋　change pocket

领带　neck tie; tie; cravat

领结　tie

领巾　butterfly tie；knot；neck tie；bowtie；muffler; stock; cravat; dickey; dicky

流苏　fringe; tassel; purl

龙袍　Dragon Robe；Imperial Robe

露背裙　backless dress

露脐装　bare-midriff

露腰装　bare-midriff top；midriff jacket；midriff

露指尖的手套　cut-fingered gloves

罗马凉鞋　Roman sandals

罗裙　thin silk skirt

落肩　dropped shoulders

马褂　magua; mandarin coat; mandarin jacket;
　riding jacket

马甲　waistcoat; vest; waist vest; doublet

马裤　jodhpurs

马球衫　polo shirt

马蹄袖　horse-hoof sleeve

马靴　bota; riding boots

满服袖　mandarin sleeve

慢跑运动装　jogging

毛料　wool fabrics

毛皮　fur

毛衫　knitwear

毛线衣　wool knit undercoat；wool jersey;
　woollies

毛衣背心　sleeveless sweater

毛毡帽　wool felt cap

帽　cap; hat

帽舌　burn grace; visor

帽檐　brim

美人鱼式裙子　mermaid dress

门襟　fly; fly-facing; closings; front fly; open
　front; top fly; placket

蒙面纱　veiling

蒙面头盔　burgonet

迷彩服　camouflage wear

迷你裙　mini-skirt/miniskirt; mini-jupe

棉　cotton

棉袄　cotton wadded jacket

冕　crown; diadem

冕状头饰　tiara

面巾　face towel; loop towel

面具　[face] mask; vizard; shade

面纱　veil; izar

民俗服装　folk costume; folk wear

民族风格　ethnic style

民族服装　national clothes; ethnic clothes;

ethnic costume

摩登风貌　modern look; mods look

抹胸　breast plate

木屐　wooden shoes; clogs; sabot; (日本)geta

牧师服　clerical garment

男衬衫　shirt; men's shirt

男大衣　overcoat

男装　menswear; men's clothing; men's wear

男礼服　men's formal attire

男礼服大衣　frock coat

男女通用服装　unisex clothes; unisex garment;
　unisex

男女同款风貌　his and hers look

男三角裤　men's briefs; triangle

男士外穿短夹克　paltock

男式短夹克　slop

男式蝴蝶领结　batwing tie

男式女西服外套　man-tailored jacket

男泳裤　swimmers

男长裤　trousers

南瓜裤　trunk breeches; trunk hose

内衬　interfacing; liner

内衣　under wear; underclothes; undergarment;
　inner wear/innerwear; lingerie

尼姑服　nun habit

尼龙　nylon

牛津布衬衫　oxford shirt

牛仔服　cowboy costume; cowboy wear

牛仔裤　jeans; jeanswear; jean pants;

牛仔帽　cowboy hat

牛仔装　cowboy suit

纽孔　buttonhole

纽扣　button

农民服装　peasant costume

暖袜　foot warmer

女傧相礼服　bridesmaid's dress

女衬衫　blouse

女礼服　gown

女帽　lady's hat; (总称)millinery

女骑装　habit

女裙　skirt; (苏格兰) jupe

女时装店　boutique

女士小提袋　ballantine

女式时装　couture

女式睡袍　lingerie gown

女式小礼服　dinner dress

女式小提袋　reticule

女式长披肩　mantelot

女式罩衫　overblouse

女晚会服　evening gown

女用小包　purse

女游泳衣　swimsuit

女装　women's suit; women's wear; (古)petticoat

派克大衣　parka [coat]

盘花扣; 盘[花]纽[扣]　frog; Chinese frog; frog closure; floral frog closure; frog fastening; fancy mandarin button

盘花纽孔　frog button

盘扣门襟　frog closing

襻（纽襻）　loop; (鞋襻) strap; tap

袍（子）　robe; gown

袍褂　over-robe

跑鞋　canvas rubber shoes; sports shoes; track shoes; trainer; running shoes

泡泡袖　bubble sleeve; puff puffed sleeve

佩带　baldric; baldrik

佩刀剑的腰带　sword belt

佩饰　waist adornments

配白领结的男士夜礼服　white-tie

配套服饰　ensemble

配套服装　matching garment; coordinates

配套时装　unit dress

朋克式样; 朋克款式　punk style; punk fashion

披风　cloak; cape; mant; mantle; mantua;

manteau; (旧法官、教士用)liripipe; liripoop

披风外套　inverness coat

披肩　cape; elbow cloak; shawl; manteau; wraps; pahone; scarf

披巾　scarf; shawl; (肩上的)liripipe, liripoop; gorget; guleron

皮袄　pelt; fur-lined jacket

皮带　belt; leather belt; strap

皮革服装　leather garment; leather wear

皮夹　wallet; (英)notecase; (美)pocket book

琵琶扣　pi-pa button

偏襟　asymmetric closing

平底鞋　flat shoes; flatties

平脚泳裤　boy leg

坡跟　slope heel corkies

坡跟鞋　wedge heel

破旧衣服　duds; schmatte; rag; tatters

齐膝裙　knee-length skirt

骑马服　riding wear; equestrian outfit; hacking

骑士服装　rider wear

棋盘格花纹　even checks

旗袍　qipao; cheongsam; mandarin dress; mandarin robe

旗袍裙　qipao style skirt; slim skirt

旗鞋（中国满族）　thick and sharp shoes; mandarin shoes; bannerman shoes

铅笔裙　pencil skirt

前额发带　headband

潜水服　diving suit; diver's suit; scuba diving; submarine armour

浅口鞋　low topline shoes; low shoes

堑壕夹克　trench jacket

戗驳领　peak lapel; peaked lapel

寝袍　sleeping gown

轻便大衣　topcoat

庆典服　ceremonial garment

囚服　prison garb; prisoner wear

球鞋　gym shoes; sneakers

裘皮大衣　fur coat; fur overcoat

全套制服　fit-out

拳击短裤　boxer pants; boxer shorts; boxer trunks; boxing trunks; boxer

裙（子）　skirt; kirtle

裙摆　sweep; flare

裙撑；裙垫　pannier; panier; bishop; bustle; farthingale

裙裾　train

裙裤　culottes; divided skirt; trousers skirt; split skirt; skirt panties; skirt-like trousers; pantskirt; divided skirt

热裤　hot pants

人造珠宝饰物　costume jewelry

日常服装　informal wear; informal dress; ordinary dress

日常西服　lounge suit

日装或半正式晚装　half dress

绒线帽　woolen hat; gorro

绒线衫　woolen sweater; quernsey

柔道服　judo-gi; judo clothes

肉色（芭蕾舞）紧身衣　flesh tights; fleshings

如意帽　wishing cap

襦　Chinese-type jacket

襦裙　Chinese two-piece dress

软帽　basque beret

赛车夹克　racing jacket

赛服　matching jacket

三件式套装　three piece suit; ensemble suit; three-piece [suit]

三角短裤　briefs

伞兵套装　paratrooper suit

伞兵跳伞服　jump suit

伞裙　umbrella skirt; flared dress

丧服　mourn[ing] dress; sables

僧侣服装　clerical garment

僧侣长袍　clerical gown

沙漏型轮廓　hourglass silhouettes

沙滩装　bathing suit coverups; beach wear

莎丽装（印度）　sari

商务套装　business suit

上班款式　on duty [style]

上衣　coat; upper garment; blouse; top garment; tog; upper outer garment; tops

少女款式　girl's style

少数民族服装　ethnic costume

设计师服装　designer's clothes

射击服　shooting coat

绅士服　gentleman's suit

深海潜水服　deep-sea diving suit

深开衩装　thigh-high slit dress

生态时装　ecology fashion; ecological fashion

圣诞老人套装　Santa Claus suit

圣诞帽　toboggan cap

盛装　dress up; splendid attire

饰边　garnishing; trimming; edging

时髦　chic

时尚　fashion

时装　fashion; stylish clothes; the latest fashion; fashion dress; fashionable dress

实用服装　heavy duty [wear]; practical apparel; utility clothes

饰边　border

饰品　ornament; adornment; accoutrements

饰章　crest

饰针　outh; fibula; tassel; preen

适体性服装　form persuasive garment

室内服装　indoor dress; house coat; room wear; dressing gown

室外服装　outdoor clothes

手工编织服饰　hand knitting

手帕　handkerchief

手术衣　surgical gown

手套　glove/gloves

手提包　gripsack; grip; handbag; pouch

手镯　bracelet; arm ring; bangle; wristlet

首饰　jewelry; head tire

寿鞋　longevity shoes

狩猎服　shooting suit; safari set

绶带　scarf; ribbon; cordon

束带　drawstring

束发带　hairband

束腰　concertina; girdle

束腰紧身衣　girdle

摔跤服　wrestle costume

双刀溜冰鞋　bob

双肩背包　back pack

双面穿衣服　A B wear

双排扣前开襟　double-breasted

水兵服　mariniere

水手服　sailor suit

水兵式衬衫　middy blouse

水手领连衣裙　middy dress; sailor dress

水手帽　sailor hat; gob hat

水袖（中国戏装）　shake sleeve; long sleeve

睡帽　jellybag

睡袍　dorm shirt; gown; nightie

睡衣　pajamas; sleeping wear; night gown; bedgown; night robe

硕士服　master's gown and hood

丝袜　silk stocking

四角帽　square shaped cap

四轮溜冰鞋　skate; skating boots; roller skate

松糕鞋　platform shoes; pantshoes

苏格兰短裙　philabeg; philibeg; filibeg

苏格苏格子呢服装　dress tartan

塑胸衣　bust shaper

蓑衣　palm-bark rain cape; straw rain cape

锁子甲　mail; chain main; chain armour; armor

塔帽　tower

塔裙　tiered skirt

踏脚裤　stirrup pants

太空服　space suit; cosmic suit; penguin suit

太空头盔　space helmet

太阳帽　sun hat

太阳裙　full circle skirt; full circle flare; sun dress

套衫　pullover; pull-on

套领罩衫　jumper

套头衫　slip-on; pull-on overtop

套装　suit; tailleur; ensemble costume

特大号服装　oversized garment

特定场合礼服　occasional dress

特体服装　special measurement clothes

特种工作服　special working uniform

体操服　gym suit; gymnastic suit; leotard; gymnastics kit

体育竞技服　sporter wear

田径服　athlete's suit; track suit

条纹衬衫　stripe shirt

条纹囚服　striped prison garb; stripes

条纹衫　stripes

贴袋　patch pocket; satchel pocket; set-on pocket; sew-on pocket; outline pocket

贴身内衣　intimate apparel; undershirt; body wear

贴身衣　under cover; undershirt; underpinnings

童鞋　children's shoes; children's footwear

童装　children's wear

桶形包　barrel bag

筒裤　straight pants

筒裙　column skirt; straight skirt; tube skirt

头巾　head kerchief; head shawl; head square; headscarf

头巾帽　turban hat

头盔　dicer; helmet; crest; sconce; tilting helmet

头饰　headdress; coiffure; head wear; headgear;

透明薄织物服装　sheer

透明装　transparent clothing; see-through clothes

徒步鞋　hiking shoes

徒步靴　hikers

土[棉]布服装　nankeens

腿甲　cuisse; tassel; leg armour

护腿　tasse; tasset; leggings

臀垫　bustle; hip pad

拖地长度　full length

拖地长裙　draggle-tail; full length skirt

拖鞋　slippers; mule; house shoes; sandals; stuffies; slides

拖鞋式女鞋　mule/mules

娃娃式衬衣　baby-doll shirt

娃娃式女服　baby-doll dress

连裤装　coverall

袜带　stockings elastic braid; stockings band; suspenders; garters

袜裤　pantleg stockings

袜套　ankle warmer; foot cover; sock cover

袜鞋　room socks

袜子　hosiery; leg wear; hose; hosen; leg knit

外科医生手术服　operating gown; surgeon's gown

外套　coat; tog

外套式衬衫　overshirt; coat shirt

外衣　outer garment; outerwear; overall; overclothes; overdress; wraps; dalmatic

弯刀冰鞋　rocker

纨（中国古代）　fine silk clothes

晚礼服　evening dress; cocktail dress; night full dress

晚礼服披肩　evening wrap

宴会[礼]服　dinner dress; party dress

王冠　crown; diadem

网巾　net band

网帽　kall

网球服装　tennis costume

网球裙　tennis dress

围脖　neckpiece

围兜　bib; pinny

围裹裙　wrap skirt; wrapped skirt; wraparound skirt

围裹式服装　wrap; wraparound

围裹式浴袍　bathrobe

围巾　scarf; shawl; muffler; neckerchief; throw; wrappage; wraps

围襟　front girlet

围裙　apron; pinafore; save-all; tablier skirt

文化衫　T-shirt; singlet

文胸　bra

纹章　armorial bearing; blazon; coat armor; emblem

无边帽　cape; elbow cloak; shawl; manteau; wraps; pahom;

无菌衣　germ-free uniform

无省服装　no-dart garment

无性别装　ambisextrous clothing; monosex clothing; monosex fashion

无袖服装　sleeveless garment

无檐小帽　dink/dinky; beanie

五分裤　breeches

舞蹈服　dancing dress; dancewear; fancy dress

舞会服　ball dress; ball gown

舞台服装　stage costume; acting clothes

舞鞋　dancing shoes; dancing slippers; escarpin

西[装]裤　suit trousers; trousers; tailored pants

西服　western-style clothes; European clothes; tailored suit

西服衬衫　dress shirt; coat shirt; European-cut shirt; continental-cut shirt

西服裙　tailored skirt; costume skirt; suit skirt

西服上衣　coat; jacket

西装背心　waistcoat; vest; weskit

膝下裙　above-calf skirt

膝靴　knee boots

嬉皮士服　hippie coat

嬉皮士款式　hippie style; hippy style

洗可穿服装　wash and wear[suit]

喜事服　happy occasion coat

戏剧服装　dramatic costume; theatrical costume; dramatic dress; stage costume

系带　lace; ligament

系带女帽　bonnet

细条纹衬衫　oxford shirt

下摆　bottom; hem; sweep

下装　lower garment; underbody clothing; bottom;

夏威夷式衬衫　Hawaii shirt

现成服装　ready made clothes; ready-to-wear clothes; off-the-peg

现代成衣　modern ready-to-wear

现代服装　modern clothes; nontraditional clothes

线迹　stitch

香奈尔套装　Chanel suit

香囊　sachet; scent bag

箱式外套　box coat

箱式手提包　suitcase bag; box bag

镶滚条　cording

项链　necklace; gorget

项圈　necklet torque

消防服　firefighting garment; fireman uniform

消防靴　fire control boots

小丑装　jester costume; jester dress

小孩围兜　baby pinafore

小脚裤　foot pants

小礼服　dinner coat; dinner jacket

小礼帽　billycock; bowler hat; homburg [hat]

小饰件　charms; trick

小王冠　coronet

孝服　mourning costume

校服　scholastic attire; campus wear; school uniform

楔跟凉鞋　wedge sandals

斜裁服装　bias-cut garment

斜纹　diagonal weave

鞋　shoe/shoes

鞋带　shoe lace; shoe tape; shoestring

鞋底　sole; tread; bottom

鞋跟　[shoe] heel

鞋类（总称）　footwear

新潮款式　chic style; mod style

新潮时装　avant-grade fashion

新娘礼服　bridal dress; bridal gown; wedding gown

新中装　new Chinese style

胸衣　corset; camisa; body; stays; corsage; plastron

胸罩　bra; brassiere

胸罩式上衣　bra top

胸针　breast pin; brooch; preen; brooch; ouch

熊皮帽　bearskin [cap]

度假服　resort wear; leisure clothes

休闲便衣　lounger

休闲套装　country suit; leisure suit

袖子　sleeve

袖衩　sleeve placket; sleeve vent; placket; cuff opening; cuff vent; sleeve slit

袖口　cuff; sleeve opening; sleeve bottom; sleeve head; sleeve end

袖口扣　sleeve button

袖套　sleeve; oversleeve; cuff cover; fore sleeve; sleevelet; arm cover

绣花服装　embroidered dress

靴子　boots

学步鞋 walkers

学士服 bachelor's gown and hood

学士披肩 bachelor's hood

学士袍 bachelor's gown

学位服 degree gown; academic costume

学位装饰披肩 academic hood

雪地靴 snow boots

训练服 exercise suit; workout suit; exercise wear; training wear

训练鞋 training shoes; trainer

压花 embossing

演出服 theatrical costume

燕尾服 swallowtails; swallow-tailed coat; tail coat; tail dress coat; dovetail; claw hammer

燕尾领 swallow-tailed collar

羊毛 wool

羊毛衫 wool knitwear; woolen sweater; sweater; cardigan; golfer

羊绒衫 cashmere knitwear; cashmere sweater

洋装 foreign dress

样衣 sample garment

腰带 waistband; belt; ceinture; girdle; sash; zodiac; abdominal belt

腰带扣 belt buckle

腰线 waistline

曳地礼服 evening gown

一步裙 narrow skirt

一片裙 one piece skirt; one-seam skirt

衣摆 clothing bottom; clothing hem

衣兜 lap; pouch

衣服 clothing; apparel; clothes; garment; attire; dress; costume

衣冠 hat and clothes

衣襟 border; opening

衣扣 dress fastener

衣领 collar

衣身 body

衣饰 costumery; clothing accessories; raiment

医生服 doctor wear

医用服装 hospital garment; medical clothing

婴儿服 babies' clothes; baby dress; babies' wear; infant's wear

婴儿开裆裤 snap panties

婴幼鞋 baby footwear

泳装 swimwear; swimsuit; bathing dress; bathing suit

泳裤 swim pants; swimming pants; [bathing] trunks; swimmers

泳帽 swim cap; swimming cap; bathing cap

有跟鞋 heels

鱼尾裙 fishtail; fishtail skirt

鱼嘴鞋 fish mouth shoes

宇航服 astronaut's garment; space suit

宇航员头盔 space helmet

宇宙服 cosmic wear; gravity suit

羽毛球鞋 badminton shoes

羽绒 down

羽绒服 down garment; down coat; down wear

雨披 rain cape

雨鞋 rain shoes; rubber shoes

雨靴 waterproof boots; rubber boots; rain boots

雨衣 rain wear; showerproof; raincoat; slicker

玉带（中国戏装） jade belt

玉佩 jade pendant

郁金香裙 tulip skirt; bubble skirt

浴衣[袍] bathing costume; bath cape; bath gown; [bath] robe; wrapper

元宝鞋 Yuan Bao shoes

圆领衫 round-neck shirt; T-topper

孕妇服 maternity wear; maternity clothes

运动短裤 exercise shorts

运动服 sportswear; gym suit

运动夹克 bi-swing; blazer; sports jacket

运动手套　action gloves

运动鞋　sports shoes; gym shoes; training shoes; sports pumps

扎染　ikat

扎染服装　tie-dyed clothing

窄脚裤　hose-bottom pants; slim slacks

战斗夹克　battle jacket

战壕外套　trench coat

长发簪　hair sticks

长开衩　deep placket

长女服　maxidress

长袍　role; gown; stole

长衫　*changshan*; cheongsam; long-gown; Chinese robe

长统女袜　stockings

长筒靴　boots; high boots; calf-high boots

长袖罩衫　smock top

罩袍　overgarment; over-robe; cover-up; dust gown

罩裙　overskirt

罩衫　overshirt; blouse; cover-up; dustcoat; frock

遮阳帽　shutter cap; sunshade cap

折叠帽　crusher hat

褶裥　pleat

褶皱裙　draped skirt; sarong skirt

针织服装　knitted garment; knit wear

针织衫　knit shirt; jersey shirt

正式服装　official wear; formal wear

正式礼服　formal gown; ball gown

正装鞋　regular shoes

直筒裙　straight skirt

直筒造型　H-line style

职业装　business wear; job suit; professional garment; career wear

制服　uniform; service dress

中国刺绣　Chinese embroidery

中国宫廷服装　Chinese court costume

中国古代服饰　Chinese ancient costume

中国官服式长衫　mandarin style gown

中国结　Chinese knot

满族发式　Manchu headdress

中国宫廷长袍　Chinese court robe

中国戏装　Chinese theatrical costume

中裤　trouserettes; calf skinners

中山装　Zhongshan suit; Sun-Yat-Sen's uniform

中式服装　Chinese-style clothing; Chinese costume; mandarin dress

中式夹克　Chinese jacket

中式立领　Chinese collar; Mandarin collar

中式连衣裙　Chinese dress

中式廓形　Chinese silhouette

中统靴　calf-length boots; buskin

中性风貌　unisex look

中长裙装　midi dress; longuette

中装　traditional Chinese clothing; Chinese dress; mandarin dress

钟形女帽　cloche

钟形裙　bell skirt; bell-shaped skirt

珠宝饰物　jewelry; head tire

珠翠　pearls and jade

珠绣　beading; beaded embroidery

装饰边　edge

自行车裤　cyclist pants

宗教服装　religious garment

足球鞋　football shoes

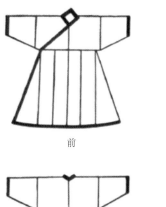

前

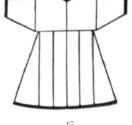

后

深衣前后片示意图（臧迎春提供）

彩图1

彩图2

彩图3

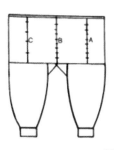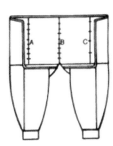

彩图4

彩图5

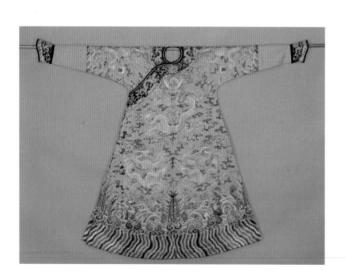

彩图6

彩图7

彩图8

彩图9

彩图10

彩图11

彩图12

彩图13

彩图14

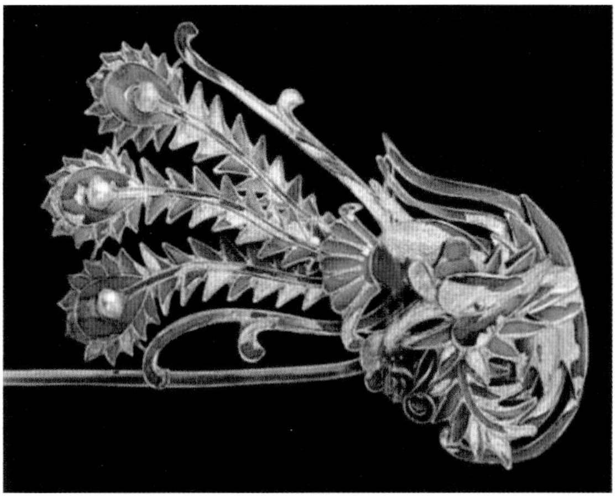

彩图15

彩图16

彩图17

彩图18

彩图19

彩图20

彩图21

彩图22

彩图23

彩图24